PERIODICALLY CORRELATED RANDOM SEQUENCES

The Wiley Bicentennial–Knowledge for Generations

Each generation has its unique needs and aspirations. When Charles Wiley first opened his small printing shop in lower Manhattan in 1807, it was a generation of boundless potential searching for an identity. And we were there, helping to define a new American literary tradition. Over half a century later, in the midst of the Second Industrial Revolution, it was a generation focused on building the future. Once again, we were there, supplying the critical scientific, technical, and engineering knowledge that helped frame the world. Throughout the 20th Century, and into the new millennium, nations began to reach out beyond their own borders and a new international community was born. Wiley was there, expanding its operations around the world to enable a global exchange of ideas, opinions, and know-how.

For 200 years, Wiley has been an integral part of each generation's journey, enabling the flow of information and understanding necessary to meet their needs and fulfill their aspirations. Today, bold new technologies are changing the way we live and learn. Wiley will be there, providing you the must-have knowledge you need to imagine new worlds, new possibilities, and new opportunities.

Generations come and go, but you can always count on Wiley to provide you the knowledge you need, when and where you need it!

William J. Pesce	Peter Booth Wiley
President and Chief Executive Officer	Chairman of the Board

PERIODICALLY CORRELATED RANDOM SEQUENCES

Spectral Theory and Practice

Harry L. Hurd
The University of North Carolina at Chapel Hill

Abolghassem Miamee
Hampton University

WILEY-INTERSCIENCE
A John Wiley & Sons, Inc., Publication

Published by John Wiley & Sons, Inc., Hoboken, New Jersey.
Published simultaneously in Canada.

For general information on our other products and services or for technical support, please contact our Customer Care Department within the United States at (800) 762-2974, outside the United States at (317) 572-3993 or fax (317) 572-4002.

Wiley also publishes its books in a variety of electronic formats. Some content that appears in print may not be available in electronic format. For information about Wiley products, visit our web site at www.wiley.com.

Wiley Bicentennial Logo: Richard J. Pacifico

Library of Congress Cataloging-in-Publication Data:

Hurd, Harry L. (Harry Lee), 1940–
Periodically correlated random sequences : spectral theory and practice / Harry L. Hurd.
p. cm. (Wiley series in probability and statisitcs)
Includes index.
ISBN 978-0-471-34771-2 (cloth)
1. Spectral theory (Mathematics) 2. Sequences (Mathematics) 3. Correlation (Statistics) 4. Stochastic processes. I. Miamee, Abolghassem, 1944– II. Title.
QC20.7.S64H87 2007
515'.24—dc22 2007013742

Printed in the United States of America.

10 9 8 7 6 5 4 3 2 1

To Marcia, Cheryl,
Robert, Olivia, Angela
and to Effie, Goly,
Nazy, and Ali

CONTENTS

PREFACE

Periodically correlated (or cyclostationary) processes are random processes that have a periodic structure, but are still very much random. Roughly speaking, if the model of a physical system contains randomness and periodicity together, then measurements made on the system (over time) will very likely have a structure that is periodically nonstationary, or in the second order case, periodically correlated. For example, meterological systems, communication systems, systems containing rotating shafts, and economic systems all have these properties.

The intent of this work is to introduce the main ideas of *periodically correlated* processes through the simpler periodically correlated sequences. Our approach is to provide (1) motivating and illustrative examples, (2) an account of the second order theory, and (3) some basic theory and methods for practical time series analysis. Our particular view of the second order theory places emphasis on the unitary operator that propagates or shifts the sequence by one period. This view makes clear the well known connection between stationary vector sequences and periodically correlated sequences. But we do not rely completely on this connection and have sometimes chosen methods of proof that are extensible to continuous time or to almost PC processes. As for time series analysis, we suppose that a reader is presented with a sample of a time series and asked to determine if *periodic correlation* is present, and if so, to say something about it, to characterize it. We present the theory, methods, and algorithms that will help the reader answer this question, within the scope of covariance and spectral estimation. The topic of *periodic autoregressive moving average* (or PARMA) became too large for inclusion at this time, especially when we began to consider sequences of less than full rank.

Accordingly, the book is roughly organized into three parts. Chapters 1 and 2 present basic definitions, simple mathematical models, and simulations whose intent is to motivate and give insight. In this we present a number of examples that illustrate that the usual periodogram analysis cannot be expected to reveal the presence of periodic correlation in a time series. We give a historical review of the topic that mainly emphasizes the early development but gives references to application-specific bibliographies. Chapters 3–8 give background and theoretical structure, beginning with a review of Hilbert space including the spectral theorem for unitary operators and correlation and spectral theory for multivariate stationary sequences. We present the (spectral) theory of harmonizable sequences and then the Fourier theory for the covariance of PC sequences. This is naturally followed by representations for PC sequences and here is where the unitary operator plays its part. We then treat the prediction problem for PC sequences and introduce the *rank* of a PC sequence.

The last three chapters (Chapters 9–11) treat issues of time series analysis for PC sequences. We first treat the nonparametric estimation of mean, correlation, and spectrum. Chapter 11 summarizes the methods into a paradigm for nonparametric time series analysis of possibly PC sequences.

MATLAB scripts used in preparing the figures and in conducting the time series analyses, as well as the data used, can be obtained from the website `http://www.unc.edu/~hhurd/pc-sequences`.

The material beginning with Chapter 3 would be useful as a basis for a course of study. It would be helpful for students to have a senior level background in vector spaces, probability, and random processes. The material of Chapter 2 is designed to provide motivation and insight and would probably be helpful to most students except those who may have some familiarity with the topic.

HARRY L. HURD AND ABOLGHASSEM MIAMEE

Chapel Hill, NC and Hampton, VA
January 31, 2007

ACKNOWLEDGMENTS

The authors gratefully acknowledge the support of ONR, USARO, NSA, and the Iranian IPM for work leading to this book. In addition, we acknowledge the encouragement, interest, and helpfulness of Stamatis Cambanis, Harry Chang, Dominique Dehay, Neil Gerr, J. C. Hardin, Christian Houdre, Gopinath Kallianpur, Timo Koski, Douglas Lake, Robert Launer, Jacek Leskow, Andrzej Makagon, P. R. Masani, Antonio Napolitano, M. Pourahmadi, M. M. Rao, H. Salehi, and A. M. Yaglom.

HLH and AGM

GLOSSARY

X_t	A univariate process (or sequence).
$\mathbf{X}_t$	A vector (or multivariate) sequence.
$\mathbf{X}_n$	The T-variate sequence formed from blocking.
m_t, $m(t)$	The mean of X_t; that is, $m(t) = E\{X_t\}$.
$R_X(s,t)$	The covariance of X_t evaluated at (s,t).
$\mathbf{F}$	The matrix spectral distribution function of the T-variate vector stationary sequence arising from the blocking (lifting) of a univariate PC-T sequence.
$\mathbf{f}$	The matrix spectral density of the T-variate vector stationary sequence arising from the blocking (lifting) of a univariate PC-T sequence.
$\mathcal{F}$	The matrix spectral distribution function of the T-variate vector stationary sequence $\{Z_t^j, j = 0, 1, \ldots T-1, t \in \mathbb{Z}\}$ resulting from Gladyshev's transformation.
rank (X)	The rank of the PC-T sequence X_t.
rank (A)	The rank of the matrix A.
$\mathcal{H}_X$	Hilbert space generated by the sequence X_t.

Periodically Correlated Random Sequences:Spectral Theory and Practice. By H.L. Hurd and A.G. Miamee

$\mathcal{L}$	Generic set with a linear structure.
$\mathcal{M}$	Generic subspace of a Hilbert space.
NND	Nonnegative definite.

CHAPTER 1

INTRODUCTION

Periodically correlated (PC) random processes are random processes in which there exists a periodic rhythm in the structure that is generally more complicated than periodicity in the mean function. We will begin with an illustration of some meteorological data.

The top trace of Figure 1.1 shows a 40 day record of hourly solar radiation levels taken at meteorological station DELTA on Ellsmere Island, N.W.T., Canada.

A daily (24 hour period) rhythm may be observed in this data in two ways: in the periodic average (or mean) and in the variation about the periodic mean. Since solar radiation can be expected to have a 24 hour period, let us compute the average of the 40 measurements for each of the 24 hours. Precisely, if the time series is denoted by $X_t, t = 1, 2, ..., NT$, where $NT = 960$, then the *sample periodic mean* (with period $T = 24$) is computed by

$$\hat{m}_N(t) = \frac{1}{N} \sum_{p=0}^{N-1} X_{t+pT}, \quad t = 1, 2, ..., T, \tag{1.1}$$

Periodically Correlated Random Sequences:Spectral Theory and Practice. By H.L. Hurd and A.G. Miamee

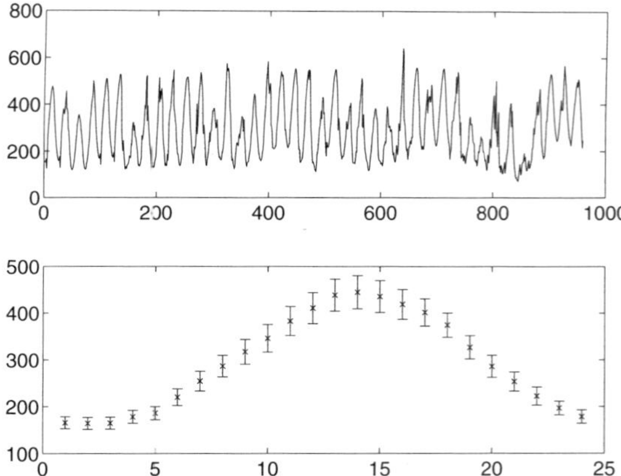

Figure 1.1 (Top) Solar radiation from station DELTA of the Taconite Inlet Project [211]. (Bottom) $\hat{m}_N(t)$ with 95% confidence intervals determined by the Student's t. $T = 24, N = 40$.

and plotted in the bottom trace of Figure 1.1. For t not in the base interval, $\hat{m}_N(t)$ is defined periodically. It is visually clear that the sample periodic mean is not constant (but properly periodic) and a simple hypothesis test for difference in mean, say, between hour 1 and hour 13, indicates a difference with much significance.

We postpone the details of testing for a proper fluctuation in the mean (i.e., for rejection of the hypothesis that the true mean $m(t)$ is constant) to Chapter 9.

The top trace of Figure 1.2 is the deviation $Y_t = X_t - \hat{m}_N(t)$ of X_t from the sample periodic mean $\hat{m}_N(t)$. The bottom trace presents the sample periodic variance,

$$S_N^2(t) = \frac{1}{N-1} \sum_{p=0}^{N-1} Y_{t+pT}^2, \quad t = 1, 2, ..., T, \tag{1.2}$$

and it too appears to have a significant (with the details again postponed) variation through the period. So it is not just the mean that appears to have a periodic rhythm, the variance does too, suggesting that the entire probability law may have a periodic rhythm. We will state this more precisely following some discussion of notation.

First, a stochastic (or random) process $X(t, \omega)$ is taken to be a function $X : \mathbb{I} \times \Omega \to \mathbb{C}$, where $\mathbb{C}$ is the set of complex numbers, $\mathbb{I}$ is called the *index set*, and Ω is a space, on which a sigma-algebra $\mathcal{F}$ of subsets and a probability measure P are defined. An $\mathcal{F}$-measurable function is called a *random variable*, and for a stochastic process, the function $X(t, \cdot)$ is assumed to be a random variable for each $t \in \mathbb{I}$. Although the focus of this book is random sequences

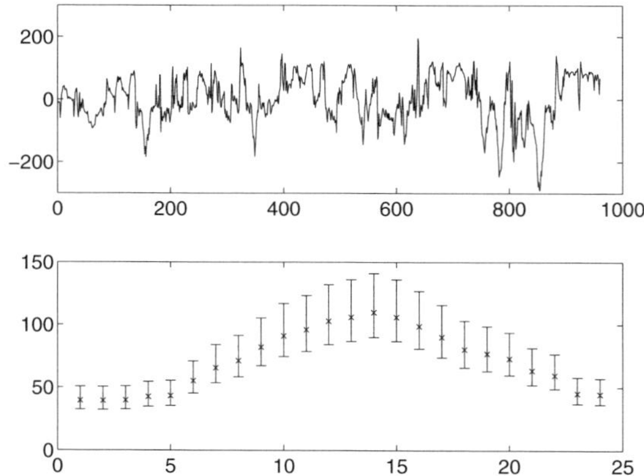

Figure 1.2 (Top) Deviation around sample periodic mean. (Bottom) $S_N(t)$ with 95% confidence limits determined by the chi-squared distribution with $N-1=39$ degrees of freedom.

($\mathbb{I}=\mathbb{Z}$) having a periodic rhythm, extensions of the ideas to fields ($\mathbb{I}=\mathbb{Z}^2$), to processes ($\mathbb{I}=\mathbb{R}$), to multivariate sequences, and to *almost periodic* sequences are briefly described in the supplements to this chapter.

We will most often denote the element of the random sequence by X_t so that the dependence on ω is suppressed and the index is the subscript symbol t, conveying time. The essential structure needed to characterize a stochastic process is its probability law, meaning the collection of finite dimensional distributions, defined as the probabilities

$$P_{t_1,t_2,\ldots,t_n}(A_1,A_2,\ldots,A_n)=P[X_{t_1}\in A_1,X_{t_2}\in A_2,...,X_{t_n}\in A_n] \quad (1.3)$$

for arbitrary n, collection of times $t_1,t_2,...,t_n$ in $\mathbb{Z}$, and Borel sets $A_1,A_2,...,A_n$ of $\mathbb{C}$.

Definition 1.1 (Strict Stationarity) *A stochastic process $X_t(\omega)$ is called* (strictly) stationary *if its probability law is invariant with respect to time shifts, or more precisely, if for arbitrary n, collection of times $t_1,t_2,...,t_n$ in $\mathbb{Z}$, and Borel sets $A_1,A_2,...,A_n$ of $\mathbb{C}$ we have*

$$P_{t_1+1,t_2+1,\ldots,t_n+1}(A_1,A_2,\ldots,A_n)=P_{t_1,t_2,\ldots,t_n}(A_1,A_2,\ldots,A_n). \quad (1.4)$$

Now we can formalize the structure suggested by Figures 1.1 and 1.2.

Definition 1.2 (Periodically Stationarity) *A stochastic sequence $X_t(\omega)$ is called* (strictly) periodically stationary *with period T if, for every n, any collection of times $t_1,t_2,...,t_n$ in $\mathbb{Z}$, and Borel sets $A_1,A_2,...,A_n$ of $\mathbb{C}$,*

$$P_{t_1+T,t_2+T,\ldots,t_n+T}(A_1,A_2,\ldots,A_n)=P_{t_1,t_2,\ldots,t_n}(A_1,A_2,\ldots,A_n), \quad (1.5)$$

and there are no smaller values of $T > 0$ for which (1.5) holds.

Synonyms for *periodically stationary* include *periodically nonstationary*, *cyclostationary* (think of cyclically stationary), *processes with periodic structure*, and a few others. For a little more on this nomenclature, see the historical notes (Section 1.2) at the end of this chapter.

If (1.5) holds for $T = 1$, then the process (or sequence) is stationary and it is clear that if X_t is periodically stationary with period T, then it is also for period $kT, k \in \mathbb{Z}$. And so we say that a sequence is properly periodically stationary if the least T for which (1.5) holds exceeds 1. Most often we will be considering second order random sequences, so that

$$E\{|X_t|^2\} = \int_\Omega |X_t(\omega)|^2 P(d\omega) < \infty, \quad \text{for all } t \in \mathbb{Z}.$$

We will sometimes just write that $X_t \in L^2$. The mean exists for second order sequences

$$m(t) := \int_\Omega X_t(\omega) P(d\omega), \quad \text{for all } t \in \mathbb{Z}$$

and we define the *covariance* of the pair (X_s, X_t) to be

$$R(s,t) := \text{Cov}\,(X_s, X_t) = E\{[X_s - m_s]\overline{[X_t - m_t]}\}.$$

If there is no ambiguity, we will write $m(t)$ and $R(s,t)$ for the mean and covariance of X_t. Sometimes, in order to conserve space, we will write variables as subscripts rather than in parentheses, such as m_t for $m(t)$ and $R_{s,t}$ for $R(s,t)$.

Since, for a zero mean sequence X_t, the covariance

$$\text{Cov}\,(X_s, X_t) = E\{X_s \overline{X_t}\}$$

is clearly the L^2 inner product, our conclusions about zero mean second order random sequences can be interpreted for sequences of vectors in a Hilbert space. For some topics (e.g., those involving shift operators) it will be more natural to think of X_t in this manner.

The notion of stationarity for second order sequences is expressed in terms of the first two moments.

Definition 1.3 (Weak Stationarity) *A second order random process $X_t \in L^2(\Omega, \mathcal{F}, P)$ with $t \in \mathbb{Z}$ is called (weakly) stationary if for every $s, t \in \mathbb{Z}$*

$$m(t) \equiv m \quad \textit{and} \quad R(s,t) \equiv R(s-t).$$

If X_t is of second order, periodic stationarity induces a rhythmic structure in the mean and covariance.

Definition 1.4 (Periodically Correlated) *A second order process $X_t \in L^2(\Omega, \mathcal{F}, P)$ is called periodically correlated with period T (PC-T) if for every $s, t \in \mathbb{Z}$*

$$m(t) = m(t+T) \tag{1.6}$$

and

$$R(s,t) = R(s+T, t+T) \tag{1.7}$$

and there are no smaller values of $T > 0$ for which (1.6) and (1.7) hold.

It is clear that if the period is T , then (1.6) and (1.7) also hold when T is replaced by kT, for any integer k. If X_t is PC-1 then it is stationary (weakly) because then $R(s,t)$ is a function only of $s-t$. Clearly a stationary sequence is PC with every period.

We will write an indexed collection $\{X_t^j, j = 1, 2, \ldots, q\}$ of random sequences as the vector sequence $\mathbf{X}_t = [X_t^1, X_t^2, \ldots, X_t^q]'$.

Definition 1.5 (Multivariate Stationarity) *A second order q-variate random sequence $\mathbf{X}_t$ with $t \in \mathbb{Z}$ is called (weakly) stationary if*

$$E\{X_t^j\} \equiv m^j \tag{1.8}$$

and

$$R^{jk}(s,t) = \mathrm{Cov}\,(X_s^j, X_t^k) = R^{jk}(s-t) \tag{1.9}$$

for all $s, t \in \mathbb{Z}$ and $j, k \in \{1, 2, \ldots, q\}$. If this is the case, we denote

$$\mathbf{m} = [m^1, m^2, \ldots, m^q]' \quad \textit{and} \quad \mathbf{R}(\tau) = [R^{jk}(\tau)]_{j,k=1}^q$$

.

Multivariate (or vector) sequences obtained from the *blocking* of univariate (or scalar) sequences will be indexed by n and thus denoted as $\mathbf{X}_n$. That is, the univariate sequence X_t is related by T-blocking to the T-variate sequence $\mathbf{X}_n$ by

$$[\mathbf{X}_n]^j = X_{j+nT}, \ n \in \mathbb{Z}, \ \ j = 0, 1, \ldots, T-1. \tag{1.10}$$

The following proposition is a simple matter of following the indices.

Proposition 1.1 (Gladyshev) *A second order random sequence $\{X_t : t \in \mathbb{Z}\}$ is PC with period T if and only if the T is the smallest integer for which the T-variate blocked sequence $\mathbf{X}_n$ (1.10) is stationary.*

Proof. Considering the covariance $\mathrm{Cov}([\mathbf{X}_n]^j, [\mathbf{X}_m]^k) = \mathrm{Cov}(X_{j+nT}, X_{k+mT})$, then stationarity of $\mathbf{X}_n$ implies

$$\mathrm{Cov}\,([\mathbf{X}_n]^j, [\mathbf{X}_m]^k) = R^{jk}(n-m) = \mathrm{Cov}\,(X_{j+nT}, X_{k+mT}),$$

which implies (1.7) holds for X_t, and conversely. The same argument applies to the mean. ∎

Periodically correlated sequences are generally *nonstationary* but yet they are nonstationary in a very simple way that, when the period T is known, makes them equivalent to vector valued stationary processes.

The term *periodically correlated* was introduced by E. G. Gladyshev [77], but the same property was introduced by W. R. Bennett [12] who called them *cyclostationary*.

Since PC sequences are so closely related to stationary vector sequences, which are rather well understood, then one can legitimately ask: why go to the effort to study the structure of these processes? There are several answers. First, the value of T, required to transform a PC sequence to a vector stationary sequence, sometimes is not known prior to the analysis of an observed time series. Thus studying the time and spectral structure of the process using its natural time organization can provide clues to help us develop tests for PC structure and estimators for the period T. Second, the issues concerning innovation rank are more easily understood for PC sequences than for multivariate sequences because the natural time order eliminates some ambiguity. Third, the methods developed here for sequences naturally carry over to continuous time and to the almost periodic case; and in those cases it is not generally possible to block the process into a stationary sequence of finite dimensional vectors.

We will often assume that $E\{X_t\} \equiv 0$ as it is the covariance (or quadratic) structure that is of most interest. However, we shall carefully discuss the issue of the additive periodic terms of a PC sequence, and how they can be conceptually viewed, and how they can be treated in the analysis of time series.

There are several ways in which two sequences can be considered equal. For example, two random processes X_t and Y_t can be called equal if for each $\omega \in \Omega$ their respective sample paths $X_t(\omega)$ and $Y_t(\omega)$ are the same. However, throughout this book, unless otherwise specified, we take two processes X_t and Y_t to be equal if

$$E \mid X_t - Y_t \mid^2 = 0, \quad \text{for every } t \in \mathbb{I}.$$

1.1 SUMMARY

This summary provides a little more detail about the contents with enough precision to make our direction clear, but not with the same care we will give subsequently. And it also provides further discussion of notation.

Chapter 1: Introduction. Gives an introductory empirical example to motivate the definitions, and then this summary followed by a historical development of the study of these processes. In this we do not attempt a complete bibliography but concentrate on the beginnings of the topic and give additional references that contain more complete bibliographies.

Chapter 2: Examples, Models, and Simulations. Presents simple models for constructing PC sequences, usually by combining randomness (usually through stationary sequences) with periodicity. Some important examples are sums and products of periodic sequences and stationary sequences, time scale modulation of stationary sequences, pulse amplitude modulation, periodic autoregressions, periodic moving averages, and periodically perturbed dynamical systems.

For most of these examples, results of simulations are presented to show the extent to which some sort of periodic rhythm is visually perceptible in the time series. These also illustrate that the usual periodogram typically does not reveal the presence of the periodic structure in PC sequences, and the periodogram of the squares sometimes can reveal the periodic structure, but not always.

Chapter 3: Review of Hilbert Spaces. Presents the basic facts about Hilbert space that will be needed. After definitions of vector space, inner product, and Hilbert space, general properties of (linear) operators are discussed. Of particular interest are projection operators, which have an important use in prediction, and unitary operators, which have a fundamental role in stationary and PC sequences. Finally, we review the spectral theory for unitary operators, including spectral measures, integrals, and the representation

$$U = \int_0^{2\pi} e^{i\lambda} E(d\lambda). \tag{1.11}$$

This spectral representation plays a critical role in the spectral theory for stationary and PC sequences.

Chapter 4: Stationary Random Sequences. Emphasizes the role of the unitary operator and its spectral representation as we believe this helps to give a clear view of PC sequences. The core result is that if $X_t^j, j = 1, 2, \ldots, q$ are *jointly* (weakly) stationary and $\mathcal{H} = \overline{sp}\{X_t^j : j = 1, 2, ..., q,\ t \in \mathbb{Z}\}$, the stationary covariance structure allows one to prove quite easily that there exists a *unitary* operator $U : \mathcal{H} \mapsto \mathcal{H}$ for which

$$X_{t+1}^j = UX_t^j \tag{1.12}$$

for every $j = 1, 2, \ldots, q$ and $t \in \mathbb{Z}$. Iterating (1.12) gives $X_t^j = U^t X_0^j$ for all t, and by applying the spectral representation (1.11) we obtain the spectral representation of the sequence

$$X_t^j = \int_0^{2\pi} e^{i\lambda t}\, \xi^j(d\lambda) \tag{1.13}$$

where ξ_j is orthogonally scattered.

We then discuss the main topics connected with prediction, regularity and singularity, the Wold decomposition, innovations, the predictor expressed by innovations, the connection between spectral theory and prediction, and finally, finite past prediction. We also discuss the issue of rank in connection with innovations and spectral theory.

Chapter 5: Harmonizable Sequences. Presents the main facts about *harmonizable* random sequences with emphasis on what is important to PC sequences. As a generalization of the spectral representation for stationary sequences (and also for continuous time), M. Loève [138], who also wrote about (strongly) harmonizable processes in the first edition of Probability Theory [139], defined a sequence to be *harmonizable* if it has a spectral representation

$$X_t = \int_0^{2\pi} e^{i\lambda t}\, \xi(d\lambda), \tag{1.14}$$

where $\xi(\cdot)$ is an $L^2(\Omega, \mathcal{F}, P)$ valued measure but no longer has orthogonally scattered (or uncorrelated) increments as it does in the stationary case. In order to convey the precise meaning of (1.14), we discuss vector valued measures and integration with respect to such measures. Then we discuss *weakly* and *strongly* harmonizable sequences, their connection to projections of stationary sequences, and spectral representation

$$R(s,t) = \int_0^{2\pi} \int_0^{2\pi} e^{i\lambda_1 s - i\lambda_2 t} F(d\lambda_1, d\lambda_2) \tag{1.15}$$

of the covariance, where the sense of integration depends on whether X_t is weakly or strongly harmonizable.

Finally, we show how time invariant linear filtering affects the spectral representation of a harmonizable sequence (and of its covariance).

Chapter 6: Fourier Theory of the Covariance. This is a topic introduced and mainly completed by Gladyshev [77]. The bijection between PC-T sequences and T-variate stationary vector sequences makes it no surprise that the Fourier theory for the covariance for PC sequences is very much related to the Fourier theory for the covariance of stationary vector sequences.

The PC structure in the covariance (1.7) implies easily that

$$R(t+\tau, t) = \sum_{k=0}^{T-1} B_k(\tau) e^{i2\pi kt/T}, \tag{1.16}$$

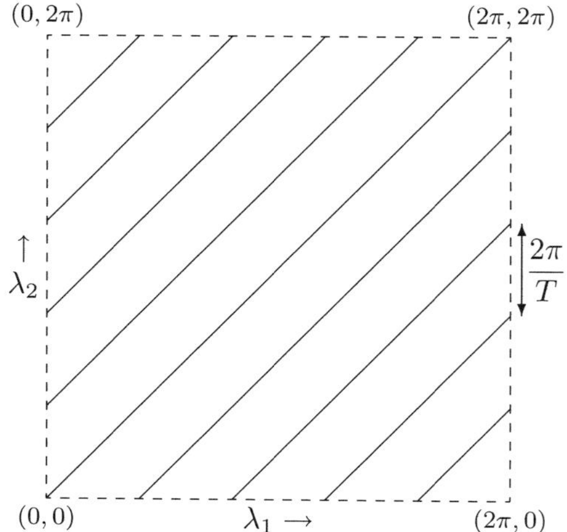

Figure 1.3 Support S_T of the spectral measure F for a periodically correlated sequence.

where $B_k(\tau) = T^{-1} \sum_{t=0}^{T-1} e^{-i2\pi kt/T} R(t+\tau, t)$. Using the connection to stationary vector sequences, Gladyshev argued that the coefficient functions $\{B_k(\tau) : k = 0, 1, \ldots, T-1\}$ are Fourier transforms

$$B_k(\tau) = \int_0^{2\pi} e^{i\lambda\tau} dF_k(\lambda). \tag{1.17}$$

We show it by use of a characterization of Fourier transforms based on a theorem of Riesz.

The plausibility that $R(s,t)$ given by (1.16) can be put into the form (1.15), which would make the covariance strongly harmonizable, turns out to be a fact, so every PC sequence is strongly harmonizable. The defining rhythm (1.7) associated with a PC sequence constrains the support set of the spectral measure F appearing in (1.15) to the $2T-1$ diagonal lines

$$S_T = \{(\lambda_1, \lambda_2) \in [0, 2\pi)^2 : \lambda_2 = \lambda_1 - 2\pi k/T, k = -(T-1), \ldots, T-1\}, \tag{1.18}$$

as illustrated in Figure 1.3. The support lines of F may be identified with the sequence $\{F_k(\cdot) : k = 0, \ldots, T-1\}$ of complex measures whose Fourier transforms are $B_k(\tau)$.

We discuss the Lebesgue decomposition of F and the issue of point masses in the random spectral measure $\xi(\cdot)$, some of which are produced by the mean $m(t)$. The effects of time invariant and periodic filtering, sampling, and random time shifting of PC sequences are examined. We also give the mapping between the spectral measure F and the matrix valued spectral measure $\mathbf{F}$ of the (blocked) vector stationary sequence $\mathbf{X}_n$.

Chapter 7: Representations of PC Sequences. Addresses various representations of PC sequences, with an emphasis on the connection to the unitary operator of a PC sequence. The basic covariance structure (1.7) implies that on the Hilbert space $\mathcal{H} = \overline{sp}\{X_t : t \in \mathbb{Z}\}$ there exists a unitary operator $U : \mathcal{H} \mapsto \mathcal{H}$ for which

$$X_{t+T} = UX_t \tag{1.19}$$

for every $t \in \mathbb{Z}$. Thus U is a shift operator for X *but only for shifts of length* T. The most basic consequence of (1.19) is that we can find (derived from U) another unitary operator $V : \mathcal{H} \mapsto \mathcal{H}$ and a periodic function P_t taking values in $\mathcal{H}$ for which

$$X_t = V^t P_t, \text{ for every } t \in \mathbb{Z}. \tag{1.20}$$

Using the spectral theorem for the unitary V leads to a spectral representation $X_t = \int_0^{2\pi} e^{i\lambda t} \xi_1(d\lambda, t)$, where $\xi_1(\cdot, t)$ is orthogonally scattered for all t whereas harmonizability implies that PC sequences *also* have a spectral decomposition (1.14) with respect to a *time invariant* random spectral measure ξ that is not orthogonally scattered. With the aid of (1.20) we explicitly construct the time independent random measure ξ. By expanding P_t in a Fourier series we obtain the Gladyshev representation $X_t = \sum_{k=0}^{T-1} Z_t^k e^{i2\pi kt/T}$ as a Fourier series having jointly stationary coefficients $\{Z_t^k : k = 0, 1, \ldots, T-1\}$. We show (see [160]) how to explicitly construct a dilated sequence Y_t such that X_t can be recovered by projection, $X_t = PY_t$.

Chapter 8: Prediction of PC Sequences. Treats the prediction problem for PC sequences, again with the help of the unitary operator U. We discuss regularity and singularity, the Wold decomposition and innovations, where we find that, at any t, the dimension d_t of the innovation space is either 0 or 1, and $d_t = d_{t+T}$. It follows that a regular PC-T sequence has an infinite moving average representation with respect to the orthonormal sequence $\{\xi_t : t \in D^+\}$,

$$X_t = \sum_{j \geq 0 : t-j \in D^+} a_t^j \xi_{t-j},$$

where the ℓ^2 sequence of coefficients $A_t = \{a_t^j : j \geq 0\}$ is periodic $A_t = A_{t+T}$ and $D^+ = \{t : d_t > 0\}$ is the set of times where nontrivial innovation occurs. The number $r = \sum_{j=1}^{T} d_{t+j}$ is a constant (independent of t) and is defined to be the *rank* of a PC-T sequence. A PC-T sequence is of *full rank* whenever $r = T$. For a simply constructed PC sequence of less than full rank, let $\{\xi_t, t \in \mathbb{Z}\}$ be an orthonormal sequence; the sequence $\{\ldots \xi_{-1}, \xi_{-1}, \xi_0, \xi_0, \xi_1, \xi_1, \cdots\}$ is PC-2 but of rank 1. We discuss the prediction problem for infinite and finite sets of predictors and give some illustrative results for periodic autoregressions of order 1, which, although simple, may also be of less than full rank. We then discuss prediction based on a finite past, periodic partial autocorrelations, and the periodic Durbin–Levinson algorithm. We also give the innovation algorithm for nonnegative definite (and hence possibly of deficient rank) covariances, along with a Cholesky decomposition for NND matrices.

Chapter 9: Estimation of Mean and Covariance. Addresses the problems of estimation of the time-varying mean $m_t = E\{X_t\} = m_{t+T}$ and covariance $R_{t+\tau,t} = E\{[X_{t+\tau} - m_{t+\tau}][X_t - m_t]\} = R_{t+T+\tau,t+T}$, and their Fourier coefficients $\widetilde{m}_k = T^{-1}\sum_{t=0}^{T-1} m_t e^{-i2\pi kt/T}$ and $B_k(\tau) = T^{-1}\sum_{t=0}^{T-1} R_{t+\tau,t}e^{-i2\pi kt/T}$. Here, X_t is taken to be a real valued PC-T sequence. The corresponding estimators, which may be motivated by the lifting to the stationary vector sequence $\mathbf{X}_n$, are given by the following: for $\widehat{m}_{t,N}$ see (1.1),

$$\widehat{\widetilde{m}}_{k,N} = \frac{1}{NT}\sum_{j=0}^{NT-1} X_j e^{-i2\pi kj/T} = \frac{1}{T}\sum_{t=0}^{T-1} \widehat{m}_{t,N}e^{-i2\pi kt/T},$$

$$\widehat{R}_N(t+\tau,t) = \frac{1}{N}\sum_{k=0}^{N-1}[X_{t+kT+\tau} - \widehat{m}_{t+\tau,N}][X_{t+kT} - \widehat{m}_{t,N}],$$

$$\widehat{B}_{k,NT}(\tau) = \frac{1}{NT}\sum_{t\in I_{NT,\tau}} [X_{t+\tau} - \widehat{m}_{t+\tau,N}][X_t - \widehat{m}_{t,N}]e^{-i2\pi kt/T}.$$

For $\widehat{m}_{t,N}$ and $\widehat{\widetilde{m}}_{k,N}$, we give conditions for mean square consistency in spectral terms, express the limits spectrally, and discuss the connection to the mean ergodic theorem. We show how to use the random time shift to give almost sure consistency in terms of B_0 and F_0 from known stationary results. The random time shift is used again to obtain asymptotic normality via mixing for linear PC sequences. The practical estimation programs `permest.m` and `permcoeff.m` are presented and demonstrated. To test for a proper periodic mean (null is $m(t) \equiv m$), the former produces confidence intervals based on Student's t and an ANOVA test; the latter uses a variance contrast method applied to the periodogram to produce p-values for $\widetilde{m}_k = 0$.

For $\widehat{R}_N(t+\tau,t)$ we give conditions for consistency in probability for a linear PC sequence using the lifted correlations and the fact that a linear PC sequence, when lifted, is a linear T-variate stationary sequence so known results may be applied. For X_t with bounded fourth moments, various conditions on the second moments of $Z^{\dagger}_{t,\tau} = [X_{t+\tau} - m_{t+\tau}][X_t - m_t] - R(t+\tau,t)$ ensure mean square consistency; and other conditions give almost sure consistency. Asymptotic normality is obtained using either of two approaches: (1) a condition on the covariance of $Z^{\dagger}_{t+jT,\tau}$ along with ϕ-mixing, and (2) normality of X_t and a summability condition on the covariance, namely, $\sum_{t=0}^{T-1}\sum_{\tau=-\infty}^{\infty}|R(t+\tau,t)|^2 < \infty$. Similar results are obtained for consistency of $\widehat{B}_{k,NT}(\tau)$.

The practical estimation programs `persigest.m` and `Bcoeff.m` are presented and demonstrated. Program `peracf.m` computes $\widehat{R}_{N_{t,\tau}}(t+\tau,t)$ and $\widehat{\rho}_{N_{t,\tau}}(t+\tau,t)$. Assuming normal X_t, confidence limits for the latter are computed (and plotted) by use of the Fisher transformation. Also computed are tests for (1) equality of correlations $\rho(t+\tau,t) \equiv \rho(\tau)$, where $\rho(\tau)$ is some unknown constant; and (2) for $\rho(t+\tau,t) \equiv 0$ for some specific τ ($\equiv$ means for all t in a period). Program `Bcoeff.m` computes (and plots) $\widehat{B}_{k,NT}(\tau)$ for $k = 0, 1, \ldots, \lfloor (T-1)/2 \rfloor$ (real X_t) via the sample Fourier transform applied to $Y_{t,\tau} = [X_{t+\tau} - \widehat{m}_{t+\tau,N}][X_t - \widehat{m}_{t,N}]$. This permits the computing of p-values for the test $\widehat{B}_{k,NT}(\tau) = 0$, based on the variance contrast method of Section 9.2.2. Also, program `persigest.m` computes $\widehat{\sigma}_N(t)$ along with confidence intervals based on normal (χ^2 distribution with $N-1$ degrees of freedom) and the Bartlett test for heterogeneous variances.

Chapter 10: Spectral Estimation. Addresses the problems of estimation of the possibly complex density functions $f_k(\lambda)$ when the $F_k(\cdot)$ in (1.17) are absolutely continuous with respect to Lebesgue measure. The principal idea for the estimation of $f_k(\lambda)$ is based on smoothing the two-dimensional periodogram

$$f(N, \lambda_1, \lambda_2) = \frac{1}{2\pi N} \tilde{X}_N(\lambda_1) \overline{\tilde{X}_N(\lambda_2)} \tag{1.21}$$

along lines of support of F in $[0, 2\pi) \times [0, 2\pi)$, where

$$\tilde{X}_N(\lambda) = \sum_{t=0}^{N-1} [X_t - m_t] e^{-i\lambda t}$$

is the sample Fourier transform of $[X_t - m_t], t = 0, 1, \ldots, N-1$. Note the usual estimators for the spectral density in the stationary case are formed by smoothing $f(N, \lambda_1, \lambda_2)$ along the main diagonal $\lambda_1 = \lambda_2$. We begin by showing that $f_{k,N}(\lambda) = f_N(\lambda, \lambda - 2\pi k/T)$ is the Fourier transform of $\widehat{B}_{k,N}(\tau)$ and if $\sum_{\tau=-\infty}^{\infty} \sum_{t=0}^{T-1} |R(t+\tau,t)| < \infty$, then $f_{k,N}(\lambda)$ is an asymptotically unbiased estimator for $f_k(\lambda)$. By assuming X_t is Gaussian, we obtain $\lim_{N\to\infty} \text{Var}\,[f_{k,N}(\lambda)] = f_0(\lambda) f_0(\lambda - 2\pi k/T)$ if $\lambda \neq \pi n/T$ and the limit is $f_0(\lambda) f_0(\lambda - 2\pi k/T) + |f_{n-k}(\pi n/T)|^2$ if $\lambda = \pi n/T$, thus showing, as in the stationary case, that the estimator is not consistent. However, as in the stationary case, consistency can be achieved by smoothing $f_{k,N}(\lambda)$ by the Fourier transform $W(\lambda)$ of a summable weight sequence $w(j)$, $\widehat{f}_{k,N}(\lambda) = \frac{1}{\mu_N} \int_0^{2\pi} W((\sigma - \lambda)/\mu_N) f_{k,N}(\sigma) d\sigma$, and where μ_N is a positive sequence with $\mu_N \to 0$ and $N\mu_N \to \infty$ as $N \to \infty$. We give conditions under which estimators formed in this manner are consistent and asymptotically normal. If X_t is a Gaussian PC-T sequence for which $\sum_{\tau=-\infty}^{\infty} \left[\sum_{t=0}^{T-1} |R(t+\tau,t)|^2 \right]^{1/2} < \infty$, then there exists a $K > 0$ for which $N\mu_N \text{Cov}\,[\hat{g}_j(\lambda_1), \hat{g}_k(\lambda_2)] \leq K$ for any $j, k \in [0, 1, ..., T-1]$ and $\lambda_1, \lambda_2 \in [0, 2\pi)$. If X_t is periodically stationary with fourth moments and uniformly ϕ-mixing with $\sum_{n=-\infty}^{\infty} (\bar{\phi}_n)^{1/2} < \infty$ and $k(j)$ is any sequence with $\sum_{j=-\infty}^{\infty} k(j)|j|^{1/2} < \infty$, then

$$\lim_{N\to\infty} \text{Cov}\,[\hat{f}_j(\lambda_1), \hat{f}_k(\lambda_2)] = 0$$

if $\mu_N \to 0$, $N\mu_N^3 \to \infty$ as $N \to \infty$. In addition, we discuss the empirical spectral analysis of harmonizable sequences, which leads naturally to the notion of spectral coherence, which may be defined theoretically as

$$\begin{aligned}\gamma(\lambda_1, \lambda_2) &= \lim_{N\to\infty} \frac{\text{Cov}\,[\widetilde{X}_N(\lambda_1), \widetilde{X}_N(\lambda_2)}{\text{Var}^{\,1/2}[\widetilde{X}_N(\lambda_1)]\text{Var}^{\,1/2}[\widetilde{X}_N(\lambda_2)]} \\ &= \lim_{N\to\infty} \text{Corr}\,[\widetilde{X}_N(\lambda_1), \widetilde{X}_N(\lambda_2)],\end{aligned}$$

and whose squared magnitude may be estimated by

$$|\gamma(\lambda_p, \lambda_q, M)|^2 = \frac{|\sum_{m=1}^{M} \widetilde{X}_{p-M/2+m}\overline{\widetilde{X}}_{q-M/2+m}|^2}{\sum_{m=1}^{M} |\widetilde{X}_{p-M/2+m}|^2 \sum_{m=1}^{M} |\widetilde{X}_{q-M/2+m}|^2}$$

where $\widetilde{X}_p$ is the sample Fourier transform of X_t of length N. For PC-T sequences, setting $\lambda_1 = \lambda$ and $\lambda_2 = \lambda_1 - 2\pi k/T$, we obtain

$$\gamma(\lambda, \lambda - 2\pi k/T) = \frac{f_k(\lambda)}{f_0^{1/2}(\lambda) f_0^{1/2}(\lambda - 2\pi k_0/T)}.$$

Spectral coherence is useful because (1) it may be estimated in a simple manner, (2) its distribution is known under a null hypothesis condition, and (3) in the case of PC sequences, it gives a way to judge the largeness of $\widehat{f}_{k,N}(\lambda)$. If a harmonizable sequence has a jump (or atom) in its random spectral measure at λ_a and at λ_b, then the theoretical spectral coherence at (λ_a, λ_b) will be unity. Hence it becomes important to sense the presence of discrete spectral components in a time series and to remove them. Such methods are also discussed in this chapter. Finally, we present programs `fkest.m` and `scoh.m` that implement the estimator $\widehat{f}_{k,N}(\lambda)$ and the empirical spectral coherence $|\gamma(\lambda_p, \lambda_q, M)|^2$ given above.

Chapter 11: A Paradigm for Nonparametric Analysis of PC Time Series. Suppose one is given a sample of a time series and asked the question: Does this series exhibit the PC property or not? If so, what can we say about it? This chapter summarizes and organizes the methods discussed in previous chapters into a procedural outline, or paradigm, for answering these questions only within the scope of nonparametric time series analysis. That is, we consider only the tools of mean, correlation, and spectral measurements. Obviously, our ability to answer these questions, especially regarding characterization, will substantially improve by the inclusion of PARMA time series analysis, a topic to be addressed in future writings.

1.2 HISTORICAL NOTES

The notion of PC processes seems to have begun with W. R. Bennett [12] who observed their presence in a communication theoretic context and called them *cyclostationary*. L. I. Gudzenko [84] initiated the subject of nonparametric spectral analysis for PC processes. V. A. Markelov [149] addressed some level crossing problems for Gaussian PC processes. A short time later E. G. Gladyshev [77] published the first analysis of spectral properties and representations based on the connection between PC sequences and stationary vector sequences. He gave necessary and sufficient conditions, in the spirit of A. Khintchine [129], for a doubly indexed sequence $R(s,t)$ to be the correlation of a PC sequence and argued that all PC sequences are strongly harmonizable and showed that their spectral support consists of a family of lines parallel to the main diagonal and having spacing of $2\pi/T$. He also gave two representations for the processes and conditions for the processes to be purely nondeterministic. In 1963 Gladyshev [78] treated continuous time PC processes and introduced the almost periodically correlated processes.

In a series of papers L. J. Herbst [90–96] explored sequences and processes whose variances may be periodic or almost periodic with respect to time; this work was done without the benefit of the PC structure. W. M. Brelsford [25] obtained, for PC sequences, a spectral-like representation of mixed summation and integral form. He also presented methods for estimation of the periodic coefficients in periodic autoregression models.

In various investigations of asymptotic stationarity, J. Kampé de Fériet [126], J. Kampé de Fériet and F. N. Frenkiel [127], and Parzen [178, 180] mentioned processes that are PC in nature but their work concentrated on the estimation of the asymptotic correlation and spectral density functions (i.e., on estimation of $B_0(\tau)$ and $f_0(\lambda)$; see the preceding summary of Chapter 6 for this notation).

To give some of the early connections to applications, Markelov [149] states that the noise output of a parametric amplifier has the PC property if the noise input is stationary; along similar lines, Parzen [179] suggested that a Poisson process with time periodic parameter would be a model for electron emissions from the cathode of a temperature limited diode whose filament was heated by an alternating current. A. S. Monin [165] suggests using PC processes as models for meteorological time series and R. H. Jones and Brelsford [122] do this for a time series of temperatures.

Several additional books touch on various aspects of PC processes. First, A. Papoulis [176] discusses various properties of PC processes in an early edition of his book on probability and stochastic processes; he calls them periodically stationary. L. E. Franks [58] discusses cyclostationary processes in a book on communication theory.

For the continuous time case, H. L. Hurd [101] showed the nature of the spectral support (an extension of Gladyshev's theorem) for strongly harmonizable PC processes, identified the connection between PC processes and those that can be made stationary by an independent uniformly distributed time shift, and obtained consistency results for estimation of the coefficient functions $B_k(\tau)$ and the densities $f_k(\lambda)$. H. Ogura [174] presented some of the spectral theory based on harmonizable processes. W. A. Gardner, in his dissertation [63] developed various representations of continuous time PC processes and used them in the solution of estimation problems. Much of this appears in the paper by Gardner and Franks [64]. After the initial work by Gladyshev, the topic was seriously examined in the former Soviet Union by Y. A. Dragan and his colleagues; much of their work seems to be summarized in three books [50–52], all in Russian. A. M. Yaglom [227] gives many references and a nice exposition of many of the basic relationships; we recommend this as a reference for those who wish to work in the topic.

Much work on cyclostationary processes followed, mainly motivated by communications problems and lead by Gardner, resulting in two books [66,67]. In these books Gardner principally takes a viewpoint much like Wiener's generalized harmonic analysis, where an observed sequence is considered to be a nonrandom sequence. The usual notion of probability is replaced with a limit of occupation time above a threshold, or fraction of time [67,166]. The approach produced understanding and solutions to many problems and is therefore interesting and useful. A discussion of the two views (random and nonrandom) may be found in [69]. Some later efforts to clarify the Wold isomorphism [67,226] applied to random and nonrandom cyclostationary sequences are given in [116,117].

In 1992 Gardner initiated a large meeting, whose subject was *Cyclostationarity in Communications and Signal Processing* [73], which brought together engineers, statisticians, and mathematicians. Many new problems and collaborations came from this meeting.

Subsequent work took several directions, in addition to continued work on communications problems. In statistics, work ensued on structural theory of PC and almost PC processes [23,44,46,65,97,98,102–104,106,110–113,116–118,143,145,157,160] on spectral and covariance estimation [5–7,30,39–41, 43,75,105,108,109,135,212,228], on testing, [7,37,72,107] and on PARMA time series [10,137,140,175,205,212,213,217–219].

Interesting work on PC processes, a little under-represented here, exists in some other fields of study. As pointed out earlier by Monin [165] and Jones and Brelsford [122], there is a natural application in meteorology due to the obvious daily or annual forcing. Applications of both parametric and nonparametric methods are indicated and have been used rather extensively. We find the connection between PC processes and Bloch's theorem, pointed out by K. Kim, G. North, and J. Huang [130, 131] to be of great interest. An extensive body of related work on periodic control [15–17] has a direct relation to PC sequences. For connected work in economics, see the book by P. H. Franses [60] and the references therein.

A recent survey on cyclostationarity by Gardner, A. Napolitano, and L. Paura [74] contains a very complete bibliography.

During this period of development of PC processes, the theory of harmonizable processes also matured. H. Cramér [36] attributes the word *harmonizable* to M. Loève [138], who also wrote about (strongly) harmonizable processes in [139]. Yu A. Rozanov [200] made an important early contribution, before the more recent developments [1, 31, 38, 82, 99, 100, 144, 155, 169, 189].

PROBLEMS AND SUPPLEMENTS

1.1 PC fields indexed on $\mathbb{Z}^2$. A collection of second order random variables $X_{s,t}$ indexed on $\mathbb{Z}^2$ is called a (strongly) PC field with period (S,T) if its mean and covariance functions satisfy

$$m(s,t) = m(s+kS, t+lT), \tag{1.22}$$

$$R(s,t,s',t') = R(s+kS, t+lT, s'+kS, t'+lT) \tag{1.23}$$

for all integers s, t, s', t' and k, l in $\mathbb{Z}$.

The second order random field $X_{s,t}$ is called *weakly PC* with *period* (S,T) if

$$R(s,t,s',t') = R(s+S, t+T, s'+S, t'+T) \tag{1.24}$$

for every s, t, s', t'. Here we require $S \geq 0$ and $T \geq 0$ but do not permit $S = T = 0$ because this would put no constraint on the covariance structure of the field.

A weakly PC random field is essentially a countable collection of PC sequences arranged along parallel lines of slope T/S in $\mathbb{Z}^2$. If X is strongly PC, then it is also weakly PC.

1.2 Multivariate PC sequences. The multivariate sequence $\mathbf{X}_t = [X_t^1, X_t^2, \ldots, X_t^N]'$ is PC with period T if

$$m^j(t) = E\{X_t^j\} = m^j(t+T) \tag{1.25}$$

and

$$R^{jk}(s,t) = E\{[X_s^j - m^j(s)]\overline{[X_t^k - m^k(t)]}\} = R^{jk}(s+T, t+T) \tag{1.26}$$

for every $j, k = 1, 2, ..., N$ and $s, t \in \mathbb{Z}$.

1.3 Almost PC sequences. A complex valued nonrandom sequence f_t is called *almost periodic* in the Bohr sense if for every $\varepsilon > 0$ the set

$$E(\varepsilon) = \{u : \sup_t |f_{t+u} - f_t| < \varepsilon\} \tag{1.27}$$

has bounded gaps, meaning there is a real number A for which every interval of length A intersects $E(\varepsilon)$. A L^2 sequence is *almost PC* if for every τ the sequence $R(t + \tau, t)$ is Bohr AP with respect to t.

1.4 Continuous time processes and fields. For $\mathbb{I} = \mathbb{R}$, the defining equations (1.6) and (1.7) as well as (1.22) and (1.23) read exactly the same. Continuous time APC processes are L^2 processes for which $R(t+\tau, t)$ is Bohr AP with respect to t for each τ. A function $f : \mathbb{R} \mapsto \mathbb{C}$ is Bohr almost periodic if it is continuous and for every $\varepsilon > 0$ the set $E(\varepsilon)$ defined by (1.27) has bounded gaps.

CHAPTER 2

EXAMPLES, MODELS, AND SIMULATIONS

With the objective of building intuition, we will now consider some mathematical models that produce PC sequences and present some simulated sample paths based on these models. As expected, if the PC structure is *strong enough*, it can be perceived by viewing the time series. The simulated sample paths are also used to demonstrate that the *classical periodogram*[1] is unable to determine the presence of the PC structure in a time series. Its modification for PC sequences is given in Chapter 10.

[1] Although there are several useful methods, the periodogram is probably the most widely used method for determinining the presence of periodicities in a time series. A review of its use in this context will be given in Chapter 10.

2.1 EXAMPLES AND MODELS

Although the more interesting and most general examples of PC processes come from combining periodicity with stationary random processes, we begin with *random periodic sequences*.

2.1.1 Random Periodic Sequences

■ **EXAMPLE 2.1**

If $X_t \in L^2(\Omega, \mathcal{F}, P)$ is a random periodic sequence for which

$$X_t = X_{t+T}, \text{ for every } t \in \mathbb{Z}, \tag{2.1}$$

then X_t is PC with period T. To see this, recall that (2.1) means that the random variables X_t and X_{t+T} are the same modulo L^2; that is,

$$\| X_t - X_{t+T} \|_{L^2} = 0 \tag{2.2}$$

for every t. Then it becomes clear that for every s, t

$$m(t) = \int_\Omega X_t(\omega)P(d\omega) = \int_\Omega X_{t+T}(\omega)P(d\omega) = m(t+T) \tag{2.3}$$

and

$$R(s,t) = \int_\Omega X_s(\omega)\overline{X_t(\omega)}P(d\omega) - m(s)\overline{m(t)} = R(s+T, t+T) \tag{2.4}$$

so X_t is PC-T. A special case of a periodic sequence is given by

$$X_t = X \cdot f_t,$$

where X is a second order random variable and f_t is a scalar periodic sequence $f_t = f_{t+T}$. This special case gives $E\{X_t\} = f_t E\{X\}$ and $\text{Var}[X_t] = \text{Var}[X] f_t^2$. By taking X to be a constant random variable, say, $X = 1$, we see that a scalar periodic sequence is PC.

Note that a periodic sequence has a periodicity in the covariance that is stronger than the condition (1.7) given above. That is,

Proposition 2.1 *A sequence X_t is periodic if and only if $m(t)$ is periodic and $R(s,t)$ is doubly periodic, in symbols,*

$$\begin{aligned} m(t) &= m(t+T), \\ R(s,t) &= R(s+mT, t+nT) \end{aligned} \tag{2.5}$$

for every $s, t, m, n \in \mathbb{Z}$.

Periodic sequences are also present, along with unitary operators, in general representations of PC sequences. This is the subject of Chapter 7.

2.1.2 Sums of Periodic and Stationary Sequences

■ EXAMPLE 2.2

Suppose X_t and Y_t are uncorrelated random sequences. If X_t is T-periodic and Y_t is wide sense stationary with mean m_Y and covariance $R_Y(u)$, then

$$Z_t = X_t + Y_t$$

is PC with period T. By (2.3) it is easy to see

$$m_Z(t) = m_X(t) + m_Y = m_X(t+T) + m_Y = m_Z(t+T)$$

and

$$\begin{aligned} R_Z(s,t) &= R_X(s,t) + R_Y(s-t) \\ &= R_X(s+T, t+T) + R_Y(s+T-t-T) = R_Z(s+T, t+T). \end{aligned}$$

So the conditions (1.6) and (1.7) are satisfied for Z_t. Here, and in some of the following examples, we write the time arguments using parentheses to help legibility. Sums of periodic and stationary sequences are among the simplest of PC sequences, although it is important to understand them.

Perhaps a little more interesting are the cases where additive periodic components are not present. Such are the following cases.

2.1.3 Products of Scalar Periodic and Stationary Sequences

■ EXAMPLE 2.3

If X_t is wide sense stationary with $E\{X_t\} \equiv 0$ and f_t is a scalar periodic sequence $f_t = f_{t+T}$, then

$$Y_t = f_t \cdot X_t \tag{2.6}$$

is PC with period T.

The required computations yield

$$m_Y(t) = f_t E\{X_t\} \equiv 0, \quad \text{for every } t \in \mathbb{Z}$$

and

$$\begin{aligned} R_Y(s+T, t+T) &= f_{s+T}\overline{f_{t+T}} \cdot R_X(s+T-t-T) \\ &= f_s \overline{f_t} \cdot R_X(s-t) \\ &= R_Y(s,t). \end{aligned} \tag{2.7}$$

In engineering, it is sometimes said that Y_t is produced by *amplitude modulation* of a stationary process by a scalar periodic function. Let us observe that the variance of Y_t is *properly*[2] periodic with period T whenever $|f_t|$ is *properly* periodic since

$$R_Y(t,t) = |f_t|^2 r_X(0) = |f_{t+T}|^2 r_X(0) = R_Y(t+T, t+T). \tag{2.8}$$

The sample paths can be just as complicated as those of stationary sequences, but the amplitude is periodically scaled. Of course, information about X_t is lost if $f_t = 0$ for some $t = 0, 1, 2, \ldots, T-1$.

2.1.4 Time Scale Modulation of Stationary Sequences

EXAMPLE 2.4

If X_t is wide sense stationary with $EX_t \equiv 0$ and f_t is a scalar periodic sequence $f_t = f_{t+T}$ taking values in $\mathbb{Z}$, then

$$Y_t = X_{t+f_t} \tag{2.9}$$

is PC with period T. For every s, t in $\mathbb{Z}$,

$$m_Y(t) = E\{X_{t+f_t}\} \equiv 0$$

and

$$\begin{aligned} R_Y(s+T, t+T) &= E\{X_{s+T+f_{s+T}}\overline{X_{t+T+f_{t+T}}}\} \\ &= R_X(s+T+f_{s+T}-t-T-f_{t+T}) \\ &= R_X(s+f_s-t-f_t) \\ &= R_Y(s,t), \end{aligned} \tag{2.10}$$

thus showing Y_t is PC-T. In engineering, time scale modulation is related to phase or frequency modulation. In contrast to the amplitude modulation case, the variance of Y_t is never properly periodic; it is always constant:

$$R_Y(t,t) = R_X(t+f_t-t-f_t) = R_X(0). \tag{2.11}$$

Hence there exist simply constructed PC processes whose variance function is constant in time.

The next example seems to be of both amplitude scale and time scale type.

[2] A sequence g_t will be called *properly periodic with period T* if $g_t = g_{t+T}$ for every t but g is not constant.

■ **EXAMPLE 2.5**

If X_t is stationary with $EX_t \equiv 0$ and f_t is a scalar periodic sequence $f_t = f_{t+T}$ defined by

$$f_t = \begin{cases} 1 & \text{if} \quad 0 \le t < [T/2] \\ -1 & \text{if} \quad [T/2] \le t < T \end{cases}, \tag{2.12}$$

where $[x]$ denotes greatest integer less than x, then

$$Y_t = f_t \cdot X_t$$

is PC with period T (from the amplitude modulation example). Although Y_t is formed by multiplication of a stationary sequence by a periodic function, we still get $R_Y(t,t) = |f_t|^2 R_X(0) = R_X(0)$ so in this way Y_t acts like time scale modulation.

2.1.5 Pulse Amplitude Modulation

This example is closely related to the amplitude modulation example.

■ **EXAMPLE 2.6**

If X_n is a zero mean stationary sequence and f_r is a nonrandom real valued function defined on $\{0, 1, \ldots, T-1\}$, then for $t = nT + r, 0 \le r < T$, the sequence

$$Y_t = X_n f_r \tag{2.13}$$

is PC with period T. To see this, for arbitrary integers s, t write $s = mT + q$ and $t = nT + r$, where $0 \le q, r < T$. Then

$$\begin{aligned} E\{Y_s Y_t\} &= E\{X_m X_n\} f_q f_r \\ &= r_X(m-n) f_q f_r \\ &= E\{Y_{s+T} Y_{t+T}\}. \end{aligned} \tag{2.14}$$

Bennett [12] studied sequences with this form, where the sequence $\{f_r : 0 \le r < T\}$ is called a *pulse* and the random sequence $\{X_n : n \in \mathbb{Z}\}$ contains information that is carried by the amplitude of the pulse over consecutive intervals of length T; hence these sequences are called *pulse amplitude modulation.* Bennett's work, which was in the data communications context, began the study of *cyclostationary* signals. From (2.14) it follows easily that

$$R_Y(t,t) = r_X(0) f^2_{t \bmod T} = R_Y(t+T, t+T),$$

so if f_t^2 is not constant these sequences will have a properly periodic variance.

2.1.6 A More General Example

This example is a generalization of the simple amplitude modulation discussed earlier and requires the following definition.

Definition 2.1 *The second order sequences* $\{X_t^j : j = 1, 2, \ldots, m,\ t \in \mathbb{Z}\}$ *are called* jointly stationary *if*

$$m_j(t) = E\{X_t^j\} = m_j(0)$$

is constant with respect to t *and*

$$R_{jk}(s,t) = E\{X_s^j \overline{X_t^k}\}$$

depends only on $s - t$ *for every* $j, k = 1, 2, \ldots, m$ *and* s, t *in* $\mathbb{Z}$.

Jointly stationary sequences are discussed further in Chapter 4, where they are called a *multivariate stationary sequence.*

■ EXAMPLE 2.7

Suppose the sequences in the collection $\{X_t^j : j = 1, 2, ..., N,\ t \in \mathbb{Z}\}$ are jointly stationary and the scalar functions $f_t^j : j = 1, 2, ..., N$ are all periodic with period T. Then the sequence

$$Y_t = \sum_{j=1}^{N} f_t^j X_t^j \tag{2.15}$$

is PC with period T.

To see this, let $m_j = E\{X_t^j\}$ so that

$$m_Y(t) = \sum_{j=1}^{N} f_t^j m_j = m_Y(t+T)$$

and

$$\begin{aligned} R_Y(s+T, t+T) &= \operatorname{Cov}\left(\sum_{j=1}^{N} f_{s+T}^j X_{s+T}^j, \sum_{k=1}^{N} f_{t+T}^k X_{t+T}^k \right) \\ &= \sum_{j=1}^{N} \sum_{k=1}^{N} f_{s+T}^j \overline{f_{t+T}^k} R_{jk}(s+T-t-T) \\ &= \sum_{j=1}^{N} \sum_{k=1}^{N} f_s^j \overline{f_t^k} R_{jk}(s-t) = R_Y(s,t). \end{aligned} \tag{2.16}$$

In Chapter 7 it will be shown that every PC sequence can be given in this form with $N \le T$.

2.1.7 Periodic Autoregressive Models

A very important class of PC sequences are those given by the periodic parametric models. Here we will begin to examine the periodic autoregressive (PAR) models of order 1 and in the next subsection we will introduce periodic moving average (PMA) processes. In this book, we will only treat simple periodic autoregressive models and periodic moving average models, leaving the treatment of periodic autoregressive moving average (PARMA) models for future work.

Definition 2.2 *A zero mean second order sequence X_t is called a* periodic autoregression *of order 1 (PAR(1)) if*

$$X_t = \phi(t)X_{t-1} + \sigma(t)\xi_t, \tag{2.17}$$

where $\phi(t) = \phi(t+T)$ and $\sigma(t) = \sigma(t+T)$ are real and $\{\xi_t : t \in \mathbb{Z}\}$ is an orthonormal sequence.

Defining $\{\sigma(t)\xi_t\}$ as the shock sequence, here we set $\sigma(t) \equiv \sigma$ and denote the resulting model as the PAR(1) with constant variance shocks, or PAR(1)-CVS. Assuming X_t is causal with respect to the shocks in the sense $E\{X_t\overline{\xi_s}\} = 0$ whenever $s > t$, we can obtain the main result of this section:

■ **EXAMPLE 2.8**

A PAR(1)-CVS sequence X_t is PC-T if and only if $|A| < 1$, where

$$A = \prod_{t=0}^{T-1} \phi(t). \tag{2.18}$$

The proof is contained in the following discussion.

Let us first recall that when $\phi(t)$ is constant, $\phi(t) \equiv \phi$, the sequence X_t is a homogeneous autoregression of order 1 (AR(1)) and (2.17) becomes $X_t = \phi X_{t-1} + \sigma\xi_t$. Then input shocks are always of constant variance and the sequence X_t is stationary with autocorrelation

$$E\{X_s\overline{X_t}\} = \phi^{|s-t|}\sigma^2$$

if and only if $|\phi| < 1$.

Taking $s > t$ to be specific, let us now calculate $E\{X_s X_t\}$ for a PAR(1)-CVS sequence. By recursion of (2.17),

$$\begin{aligned}
E\{X_s\overline{X_t}\} &= E\{[\phi(s)X_{s-1} + \xi_s]\overline{X_t}\} \\
&= E\{\phi(s)X_{s-1}\overline{X_t}\} + E\{\xi_s\overline{X_t}\} \\
&= E\{\phi(s)X_{s-1}\overline{X_t}\} + 0 \quad \text{for } s > t \\
&\vdots \\
&= \phi(s)\phi(s-1)\dots\phi(t+1)E\{X_t\overline{X_t}\}
\end{aligned} \tag{2.19}$$

and we see from the periodicity of $\phi(t)$ that we will obtain

$$E\{X_s\overline{X_t}\} = E\{X_{s+T}\overline{X_{t+T}}\}$$

provided

$$R_X(t,t) \equiv E\{X_t\overline{X_t}\} = E\{X_{t+T}\overline{X_{t+T}}\} < \infty$$

for every t. So X_t will be PC-T provided the variance function $R_X(t,t)$ is periodic. Periodicity and boundedness of the variance are always necessary for a PC sequence because $R(t,t) = R(t+T, t+T) < \infty$ for all t. The following lemma completes our current discussion of $|A|$. In Chapter 8 we explore more thoroughly the connection between causality, boundedness of $\| X_t \|$, and $|A| < 1$.

Lemma 2.1 *A necessary and sufficient condition for the variance $\sigma_X^2(t)$ of a PAR(1)-CVS to be periodic, $\sigma_X^2(t) = \sigma_X^2(t+T)$, is that $|A| < 1$, where A is given by (2.18).*

Proof. First, for arbitrary t let us denote

$$\begin{aligned}
A_0(t) &= 1 \\
A_1(t) &= \phi(t) \\
A_2(t) &= \phi(t)\phi(t-1) \\
&\vdots \\
A_{T-1}(t) &= \phi(t)\phi(t-1)\cdots\phi(t-T+2)
\end{aligned} \tag{2.20}$$

and we can see that $A_T(t)$ is the product of $\phi(t)$ over exactly one cycle so that

$$A_T(t) = \phi(t)\phi(t-1)\cdots\phi(t-T+1) = A. \tag{2.21}$$

Using (2.17) we obtain

$$\sigma_X^2(t) = A_1^2(t)\sigma_X^2(t-1) + 1, \tag{2.22}$$

which may be continued recursively to obtain

$$\begin{aligned}\sigma_X^2(t) &= 1 + A_1^2(t) + \cdots + A_{T-1}^2(t) + A^2 + A^2 A_1^2(t) + \cdots \\ &= \lim_{N\to\infty} \left(\sum_{p=0}^{T-1} A_p^2(t)\right) \sum_{j=0}^{N} A^{2j},\end{aligned} \tag{2.23}$$

and this converges if and only if $|A| < 1$. Hence, if $|A| < 1$, $\sigma_X^2(t)$ is bounded. And again from the (convergent) representation (2.23), $\sigma_X^2(t) = \sigma_X^2(t+T)$ for all t because $A_p(t) = A_p(t+T)$ for all t and $p = 0, 1, \ldots, T-1$.

Conversely, if $\sigma_X^2(t) = \| X_t \|^2$ is bounded, then for all t

$$\sum_{p=0}^{T-1} A_p^2(t) \geq 1,$$

we must conclude that

$$\sum_{j=0}^{\infty} A^{2j} < \infty$$

and hence $|A| < 1$. ∎

Given $|A| < 1$ we may solve for $\sigma_X^2(t)$ by doing the aforementioned recursion for only T times.

$$\sigma_X^2(t) = 1 + A_1^2(t) + \cdots + A_{T-1}^2(t) + A^2 \| X_{t-T} \|^2 \tag{2.24}$$

and so if $\sigma_X^2(t) = \sigma_X^2(t-T)$ then easily

$$\sigma_X^2(t) = \frac{1 + A_1^2(t) + \cdots + A_{T-1}^2(t)}{1 - A^2}. \tag{2.25}$$

2.1.8 Periodic Moving Average Models

A second order random sequence X_t is a *periodic moving average of order q* (PMA(q) and with period T if it satisfies

$$X_t = \sum_{j=0}^{q} \theta_j(t)\xi_{t-j} \tag{2.26}$$

where $\theta_j(t) = \theta_j(t+T)$ for every j, t and where the ξ_t are taken to be of zero mean and orthonormal. Then X_t has mean zero for all t and its covariance is easily computed to be

$$\begin{aligned} R(s,t) = E\{X_s\overline{X_t}\} &= \sum_{j=0}^{q}\sum_{k=0}^{q}\theta_j(s)\overline{\theta}_k(t)E\{\xi_{s-j}\overline{\xi_{t-k}}\} \\ &= \sum_{j\in I_{s,t}} \theta_j(s)\overline{\theta}_{t-s+j}(t) \\ &= \sum_{j\in I_{s+T,t+T}} \theta_j(s+T)\overline{\theta}_{t+T-s-T+j}(t+T), \end{aligned} \tag{2.27}$$

where $I_{s,t} = \{j : 0 \le j \le q \text{ and } j = s-t+k \text{ for some } 0 \le k \le q\}$. Since $I_{s,t} = I_{s+T,t+T}$, we conclude that X_t is PC-T and its variance is

$$R(t,t) = \sum_{j=0}^{q} |\theta_j|^2(t). \tag{2.28}$$

We may also conclude from (2.27) that $R(s,t) = 0$ if $|s-t| > q$. Usually ξ_t and $\theta_j(t)$ are taken to be real.

2.1.9 Periodically Perturbed Dynamical Systems

The PAR and PMA models we have just briefly discussed are simple examples of periodically time varying systems that produce PC sequences when driven by random shocks. This leads quite naturally to the question: Do *nonlinear* dynamical systems, when periodically perturbed, give rise to orbits that are theoretically or empirically consistent with *periodic correlation* or *cyclostationarity*? An obvious physical example that motivates this question is the perturbation of meteorological processes by the daily variation in solar radiation. Below we will see that periodically perturbing a simple family of maps (the logistic maps) yields orbits that exhibit *periodic correlation.* To make the notion of a periodically perturbed map more precise, suppose we have a family $g_\alpha : \mathbb{R} \to \mathbb{R}$ of maps, and a finite collection of parameters $\{\alpha_0, \alpha_1, \dots, \alpha_{T-1}\}$. The orbit of x under the *periodically perturbed* map (or family of maps) is the sequence given by

$$\begin{aligned} x_1 &= g_{\alpha_0}(x) \\ x_2 &= g_{\alpha_1} \circ g_{\alpha_0}(x) \\ &\vdots \\ x_T &= g_{\alpha_{T-1}} \circ \cdots g_{\alpha_1} \circ g_{\alpha_0}(x) \\ &\vdots \\ x_n &= g_{\alpha_r} \circ g_{\alpha_{r-1}} \cdots g_{\alpha_0} \circ g_{\alpha_{T-1}} \cdots g_{\alpha_0}(x), \end{aligned} \tag{2.29}$$

where $r = (n-1) \bmod T$. Simulated orbits for the logistic family $g_\alpha(x) = \alpha x(1-x)$ are presented in the next section.

2.2 SIMULATIONS

The main goal of this section is to provide some simulated sample paths of real valued series generated from the models described in the previous section. We first present plots of the simulated series to illustrate what can be visually perceived about the presence of PC structure in a time series. Finally, for some of the series we will show the periodograms in order to illustrate that they are ineffective in revealing the presence of PC structure unless there are additive periodic components.

2.2.1 Sums of Periodic and Stationary Sequences

Figure 2.1 presents some time series formed from the sum of a very simple periodic sequence and a very simple stationary sequence,

$$Y_t = A\cos(\pi t/16) + \xi_t,$$

where ξ_t is white noise with $E\{\xi_t\} = 0$, $\sigma_\xi = 1$. In Figure 2.1(a), where $A = 0.5/512$, the additive periodic sequence is not clearly perceived from the raw time series. However, it is easily perceived in Figure 2.1(b), where $A = 2.5/512$.

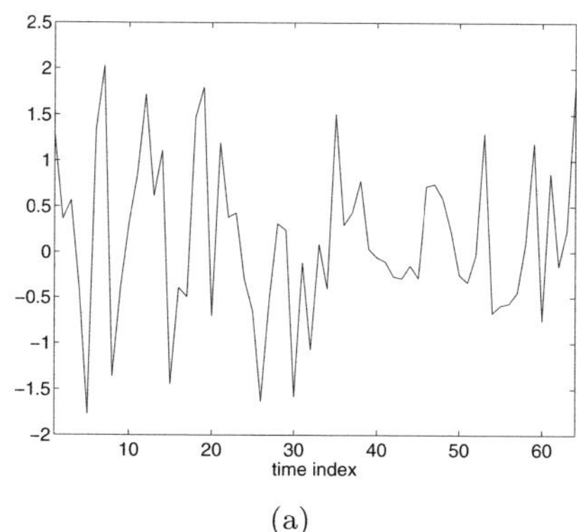

(a)

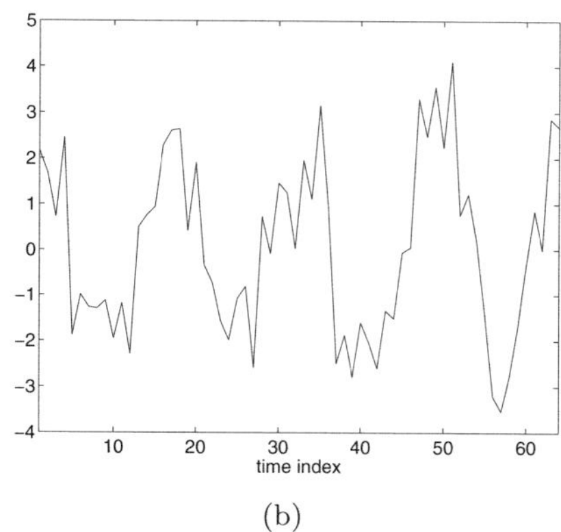

(b)

Figure 2.1 Time series $Y_t = A\cos(\pi t/16) + \xi_t, \sigma_\xi = 1$. (a) $A = 0.5/512$. (b) $A = 2.5/512$.

Figure 2.2 shows that even the weak periodic sequence can be visually perceived by comparing the squares of the sample Fourier transform at frequency index $j = 32$ to those near it, a procedure we will formalize later. In this figure, the Fourier transform was taken of a sample of 512 points so that the period 16 gives $j = 512/16 = 32$ periods.[3] The squares of the sample Fourier transform are called the *periodogram*. More on the periodogram will be given in Chapters 9 and 10.

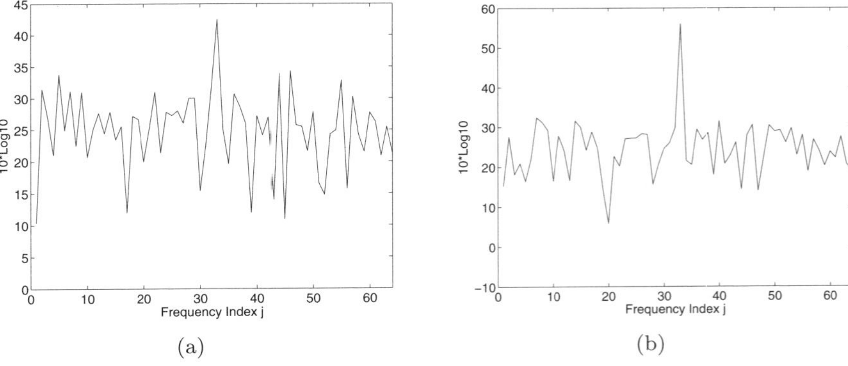

Figure 2.2 Periodograms of $Y_t = A\cos(\pi t/16) + \xi_t$, $\sigma_\xi = 1$ based on one FFT of length 512. (a) $A = 0.5/512$. (b) $A = 2.5/512$.

2.2.2 Products of Scalar Periodic and Stationary Sequences

Figure 2.3 presents some time series formed from a simple periodic amplitude modulation of white noise,

$$Y_t = [1 + m\cos(2\pi t/64)]\,\xi_t, \tag{2.30}$$

where ξ_t is mean zero white noise with unit variance, $\sigma_\xi^2 = 1$. In Figure 2.3(a), where $m = 0.25$, the multiplicative periodic structure is not clearly perceived from the raw time series. However, it is easily perceived in Figure 2.3(b), where $m = 1$. But even in the case of strong modulation ($m = 1$), Figure 2.4 shows that the periodogram of a 512 point sample does not reveal the presence of the period 64 modulation. (It would be expected at $j = 512/64 = 8$.)

However, Figure 2.5 shows that the period $j = 8$ is clearly perceived in the periodogram of the squared series Y_t^2. Consideration of the time series of squares immediately reveals the reason; the series of squares has a periodic mean,

$$E\{Y_t^2\} = [1 + 2m\cos(2\pi t/64) + m^2\cos^2(2\pi t/64)]\sigma_\xi^2.$$

[3]The peak is actually at $j = 33$ because the MATLAB computing environment begins indexing at 1 rather than 0.

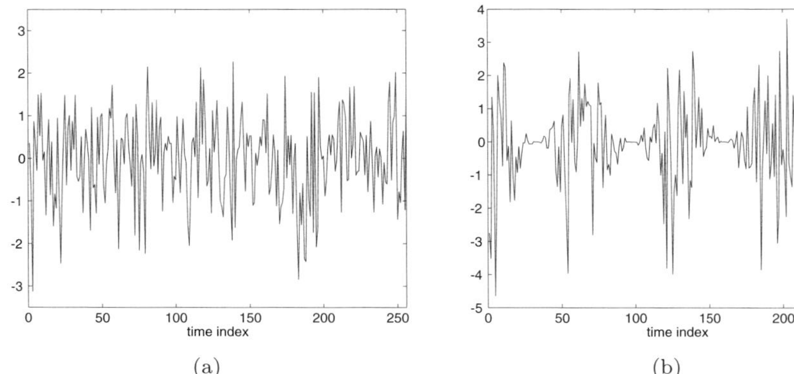

Figure 2.3 Time series $Y_t = [1 + m\cos(2\pi t/64)]\ \xi_t$. (a) $m = 0.25$ (weak modulation). (b) $m = 1$ (strong modulation).

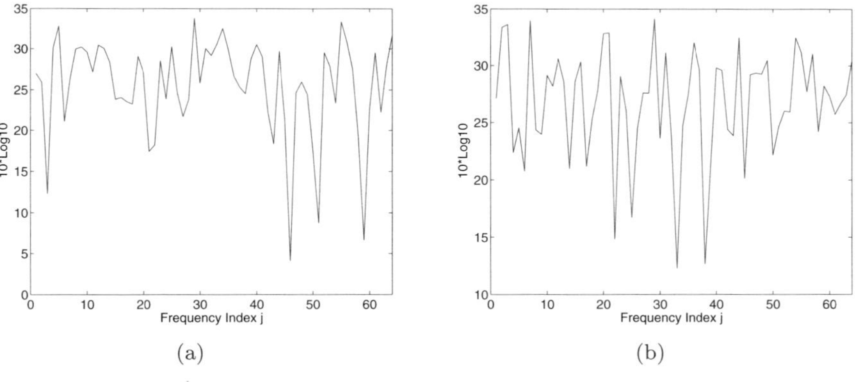

Figure 2.4 Periodogram of $Y_t = [1 + m\cos(2\pi t/64)]\ \xi_t$, based on one FFT of length 512. (a) $m = 0.25$ (weak modulation). (b) $m = 1$ (strong modulation).

Can this simple modification, forming the periodogram of the squared series Y_t^2, always permit us to perceive the presence of any arbitrary PC structure? The answer is no; its success depends on Y_t^2 having a properly periodic mean, that is, $E\{Y_t^2\} = R_Y(t,t)$ must be a proper periodic function. This proper periodicity typically occurs for amplitude modulation models but never occurs for time scale modulation models.

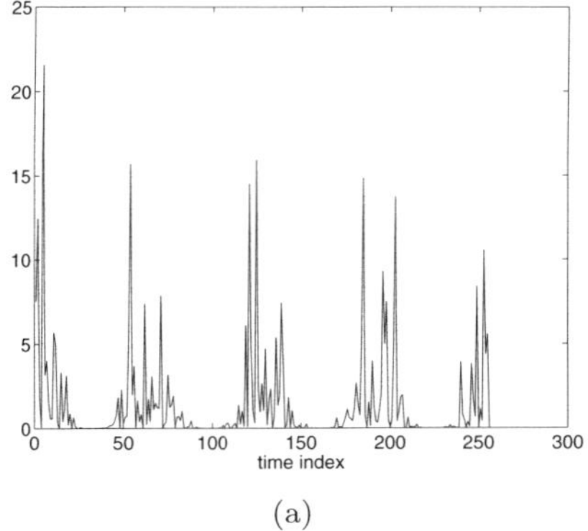

(a)

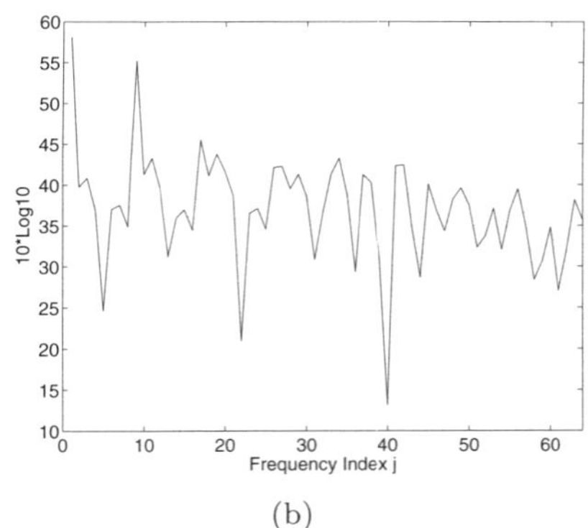

(b)

Figure 2.5 (a) Time series of squares Y_t^2 for $Y_t = [1 + m\cos(2\pi t/64)]\ \xi_t$, with $m = 1$ (strong modulation). (b) Periodogram of Y_t^2, based on one FFT of length 512, with strong peak at index $j = 8$.

2.2.3 Time Scale Modulation of Stationary Sequences

Figure 2.6(a) shows the time series of a stationary first order autoregression

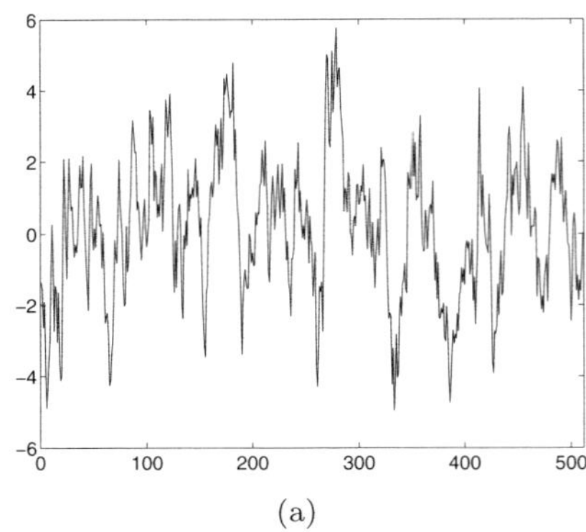

(a)

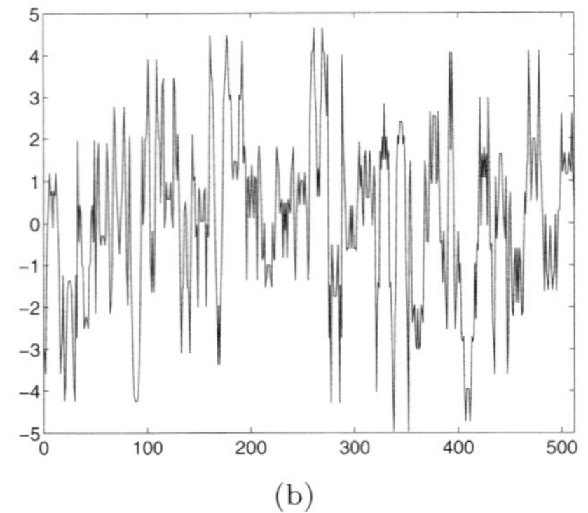

(b)

Figure 2.6 Time series. (a) $X_t = 0.9X_{t-1} + \xi_t$ (AR(1)). (b) $Y_t = X_{t+f_t}$ (time scale modulation).

$X_t = 0.9X_{t-1} + \xi_t$ with covariance $R_X(k) = 0.9^k\ \sigma_X^2$, and Figure 2.6 is the series $Y_t = X_{t+f_t}$, where $f_t = f_{t+32}$ is a permutation of the integers $1, 2, \ldots, 32$. Since from (2.11) we have $R_Y(t,t) = R_X(0)$, then it is no surprise

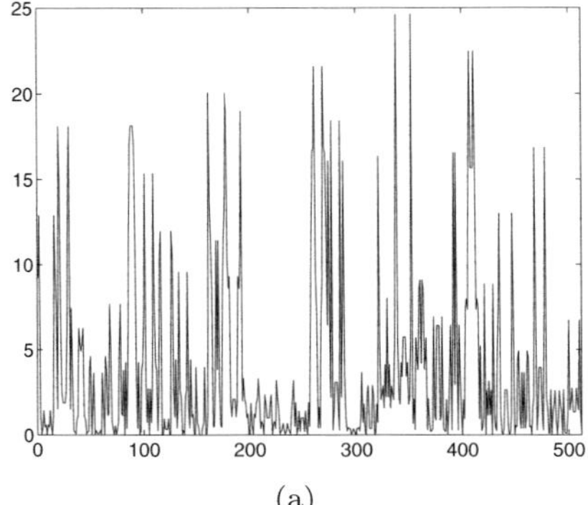

(a)

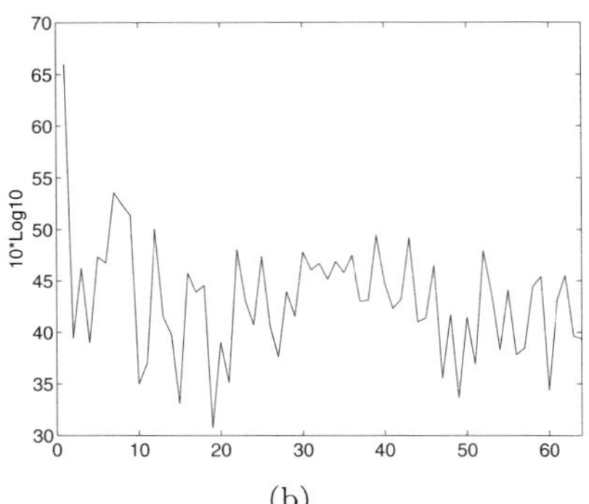

(b)

Figure 2.7 Y_t^2 for $Y_t = X_{t+f_t}$ (time scale modulation). (a) time series. (b) periodogram based on one FFT of length 512.

that we cannot perceive any periodicity in the mean of the squares Y_t^2 shown in Figure 2.7(a), nor in the periodogram in Figure 2.7(b).

As another example, Figure 2.8 shows that neither the time series of the squares of $Y_t = f_t \cdot X_t$, nor the periodogram, show any evident sign of periodicity when we take X_t to be a stationary first order autoregression $X_t = 0.8X_{t-1} + \xi_t$ with correlation $R_X(k) = 0.8^k\ \sigma_X^2$, and we take f_t to be the ± 1 sequence of (2.12). Although Y_t is most certainly PC with period $T = 32$, $E\{Y_t^2\} = |f_t|^2 R_X(0) = R_X(0)$ is constant with respect to t (and thus not properly periodic).

The last two examples show that in the case of time scale modulation we cannot expect to perceive the PC structure visually from the time series nor from the periodogram of the squares, even though both cases represent severe modulation of the original sequences. The contents of Chapters 6, 9 and 10 will give a better understanding of these and we will give methods that do permit the perception of their PC structure.

2.2.4 Pulse Amplitude Modulation

Figure 2.9(a) is a time series resulting from the pulse amplitude modulation $Y_t = X_n f(r)$, $t = nT + r, 0 \le r < T$ of a triangular $f(r)$ of duration 32 by an uncorrelated noise sequence X_n having $R_X(0) = 1$. The periodogram of Y_t (based on fast Fourier transform (FFT) length 1024) shows no clear component at $j = 1024/32 = 32$. The large spectral "hole" at 64 periods (in 1024 point FFT) is caused by the exact symmetry of $f_t = f_{T-t+1}$. The

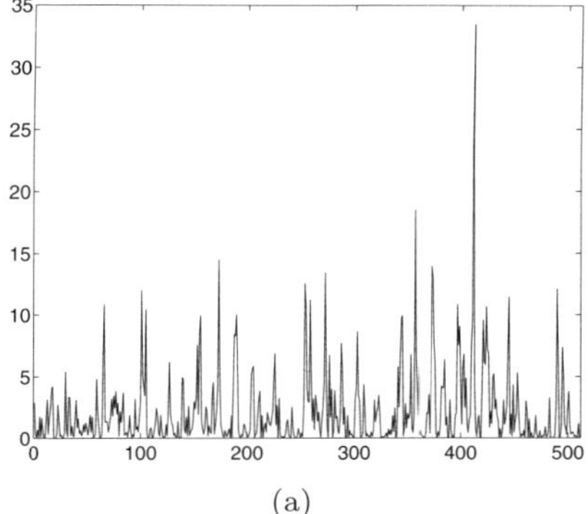

(a)

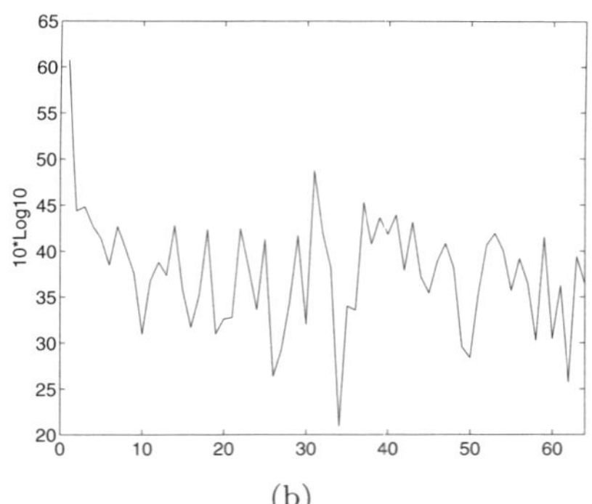

(b)

Figure 2.8 Y_t^2 for $Y_t = f_t \cdot X_t$ with $f_t = f_{t+32}$ and $f_t = \pm 1$ according to Eq. (2.12). (a) Time series. (b) Periodogram based on one FFT of length 512. A period 32 component would be seen at frequency index $512/32 = 16$ or harmonics thereof.

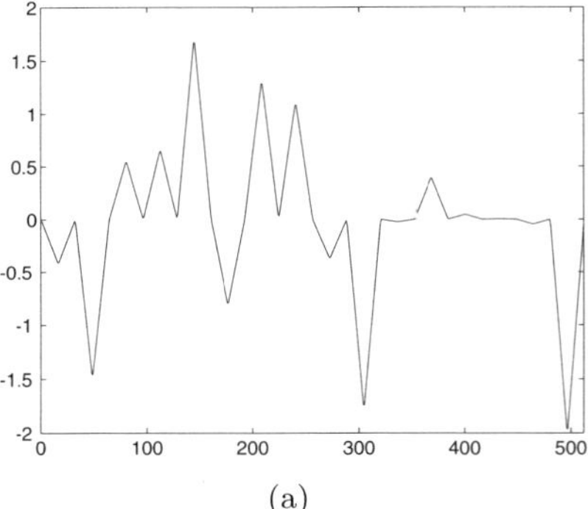

(a)

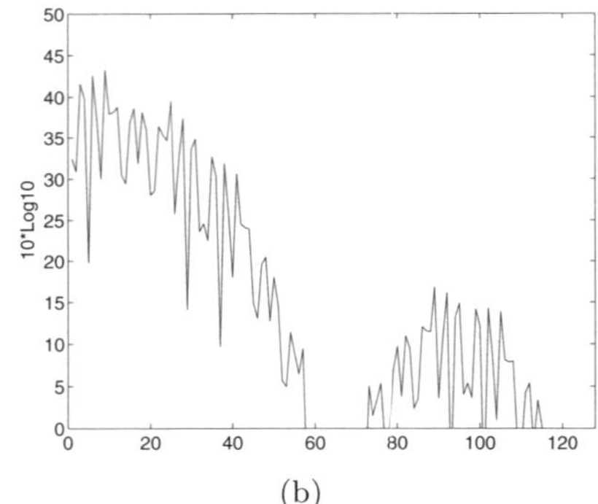

(b)

Figure 2.9 PAM sequence $Y_t = X_n f(r),\ t = nT + r, 0 \le r < T$ for triangular $f(t)$ and X_n uncorrelated. (a) Time series. (b) Periodogram based on one FFT of length 1024.

periodogram of Y_t^2 shown in Figure 2.10(b) clearly shows the component at $j = 32$.

The previous examples have illustrated that PC sequences can arise from stationary sequences whose amplitudes or time scales are given a periodic rhythm. This example, aside from the historical importance, shows that PC

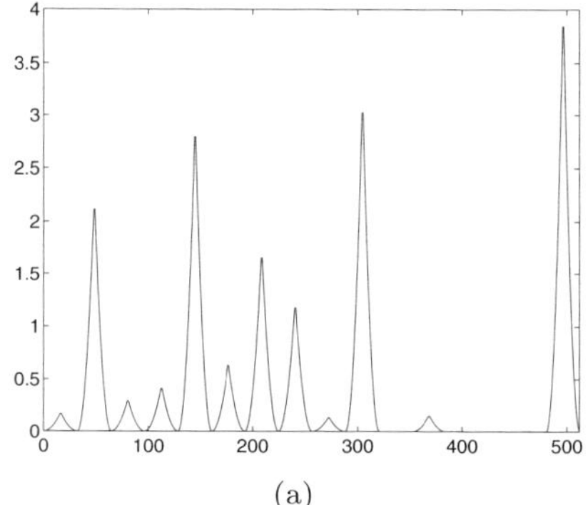

(a)

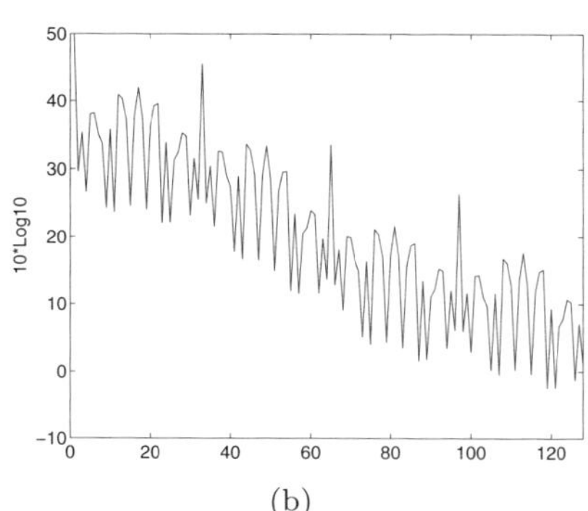

(b)

Figure 2.10 Y_t^2 for PAM sequence. (a) Time series. (b) Periodogram based on one FFT of length 1024.

sequences can arise when random events (such as a pulse with a random amplitude) occur on a periodic schedule.

2.2.5 Periodically Perturbed Logistic Maps

In the experimental results that follow, we shall take $g_\alpha(x)$ to be the family of logistic maps

$$g_\alpha(x) = \alpha x(1 - x) \tag{2.31}$$

for $2 \le \alpha \le 4$. These maps carry $[0, 1]$ into $[0, 1]$ and have been extensively studied (e.g., Devaney [45]). For $\alpha = 4$, the logistic map is chaotic (has sensitive dependence on initial conditions) but converges to fixed points or periodic points for many other values of α; for example, $g_{2.6}$ has a single attracting fixed point at $x = 8/13$.

To go immediately to our objective, we chose $T = 20$ and set

$$\alpha_0 = \alpha_1 = \cdots \alpha_9 = 4$$

to give chaotic behavior for the first half of the period and

$$\alpha_{10} = \alpha_{11} = \cdots \alpha_{19} = 2.6$$

to give stable behavior for the second half of the period. Figure 2.11 presents 200 consecutive values of x_n and the estimated periodic mean $\hat{m}_n : 0 \le n \le 19$ based on $N = 100$ periods.

We can easily perceive some sort of periodic character in the raw orbit and that the estimated periodic mean is significantly nonzero (based on normal

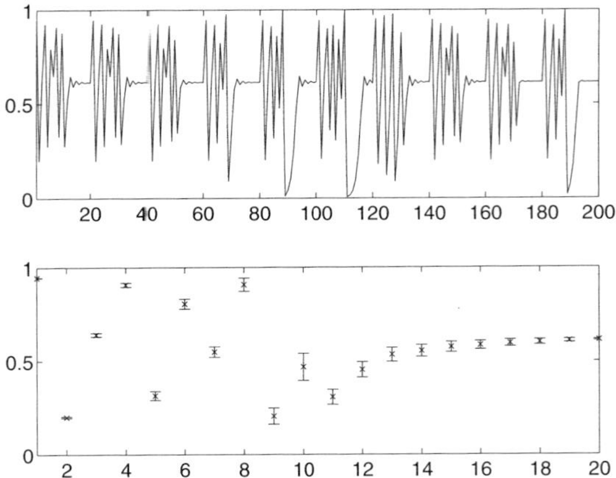

Figure 2.11 Periodically perturbed logistic map. (Top) Two hundred points from a 2000 point simulation. (Bottom) Estimated periodic mean $\hat{m}_n : 0 \leq n \leq 19$ with 95% confidence intervals based on $N = 2000/20 = 100$ periods.

distribution of errors in the estimated periodic mean). The orbit seems to behave rather chaotically (qualitatively) in the first half of each cycle, and then appears to be rapidly converging in the second half. However, the confidence intervals of $\hat{m}_n$ for $0 \leq n \leq 4$ suggest that the orbit is rather stable in this interval from one period to the next, even though $\alpha = 4$ in this region. The reason for the apparent stability is that x_{20k+20} (the first point in a new cycle) always determined by g_4 operating on the last point, x_{20k+19}, in the previous cycle, which is always very close to the fixed point for $g_{2.6}$ at $x = 8/13$. The sensitive dependence of g_4 causes the confidence intervals of $\hat{m}_n$ to continue to increase until $n = 10$ when the map switches to $g_{2.6}$ whose stability causes the confidence intervals to decrease until the map switches again to g_4 at the start of the next cycle.

The demeaned orbit, $\tilde{x}_n = x_n - \hat{m}_n$ shown in the top trace of Figure 2.12, appears to be comprised of some irregular pulses occurring on a periodic schedule, a qualitative clue for periodic correlation or cyclostationarity. The estimated periodic $\hat{\sigma}_n$ in the bottom trace clearly shows the sample variance is not constant through the period, as did the confidence intervals we discussed earlier.

And, as we now might expect, the usual periodogram shown in the top trace of Figure 2.13 shows no evidence of periodicity at period $T = 20$ (which would produce $1000/20 = 50$ periods in 1000 points and would present at index 51 on the periodogram). However, in the periodogram of the squares, the peak

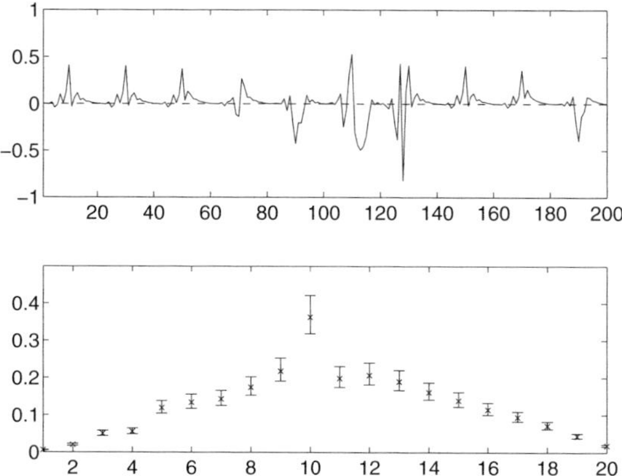

Figure 2.12 (Top) Two hundred points of the demeaned orbit $\tilde{x}_n$. (Bottom) Estimated $\sigma_X(t)$ with 95% confidence intervals.

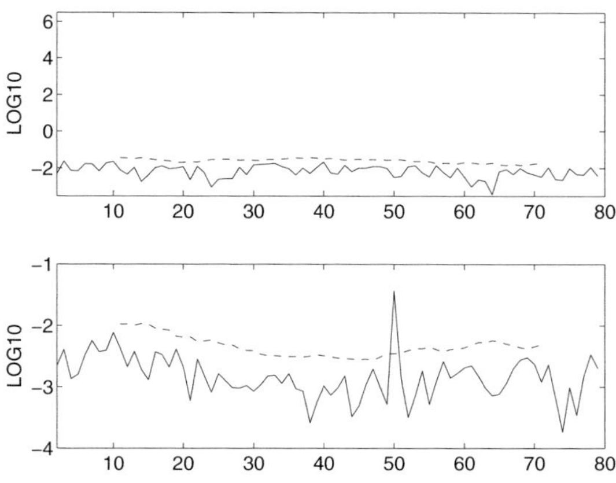

Figure 2.13 Periodograms based on one FFT of length 1000. (Top) x_n. (Bottom) $\tilde{x}_n^2$.

at index 51 is very significant (p-value $< 10^{-6}$). The dashed lines are the $\alpha = 0.01$ threshold for test of variance contrast based on a half-neighborhood of size $m = 8$.

2.2.6 Periodic Autoregressive Models

Some simulated PAR(1) time series illustrate our ability to perceive evidence of periodic correlation, either visually in the series or by detecting periodicity in the sample periodic variance. Figures 2.14 and 2.15 present 256 samples of the simulated series, the theoretical (see (2.25)) and estimated values of $\sigma_X(t)$, and periodograms of X_t and of X_t^2 for the model $X_t = \phi(t)X_{t-1} + \xi_t$ with $\phi(t) = [\phi_0 + \phi_1 \cos(2\pi t/T)]$ for $T = 32$ and Cov $(\xi_s, \xi_t) = \delta_{s-t}$. Under this model, $\phi(t)$ changes smoothly through the period.

Summarizing the content of Figure 2.14 ($(\phi_0, \phi_1) = (0.6, 0.4)$, $A = 1.019e-009$), periodicity is not visually perceptible in the raw time series. Strong evidence of nonconstant variance is seen in $\hat{\sigma}_X(t)$. No evidence of periodicity is seen in the usual periodogram but extremely significant (p-value$= 0$) evidence of periodicity is seen in the periodogram of X_t^2.

In Figure 2.15 ($(\phi_0, \phi_1) = (-0.6, 0.4)$, $A = 1.019e - 009$), periodic character is visually perceptible by the large excursions, which are synchronized to the period $T = 32$. These excursions appear to be bursts of random amplitude occurring late in the period, which is consistent with the fluctuation in the variance seen in $\hat{\sigma}_X(t)$. There is no significant evidence of periodicity in the periodogram of X_t but there is extremely significant (p-value$= 0$) evidence of periodicity in the periodogram of X_t^2; higher harmonic terms are now significant.

The extremely small value of $|A| \sim 10^{-9}$ in both of these simulations indicates that the memory of past events had, in effect, decayed to zero in the duration of one period. In order to see a contrasting case we use a switching model to produce a situation with large $|A|$ and also with $\sigma_X(t) \equiv 1$. In Figure 2.16 we set

$$\phi(t) = \begin{cases} \phi_a, & 0 \le t < [T/2] \\ \phi_b, & [T/2] \le t < T, \end{cases} \tag{2.32}$$

where $\phi_a = 0.95$ and $\phi_b = -0.95$ and $T = 16$. This yields $A = 0.4403$ which is substantially larger than the largest value of $|A|$ in the group with sinusoidally varying $\phi(t)$.

Note the values of $\hat{\sigma}_X(t)$ and their confidence intervals strongly support the hypothesis of a constant variance, which is indeed the true situation. But still there is strong visual evidence of something periodic in the time series. Neither the usual periodogram nor the periodogram of the squares reveals any hint of periodicity with period $T = 32$. So here is a case in which the detection of periodic correlation must wait for the more general methods we shall discuss in later chapters.

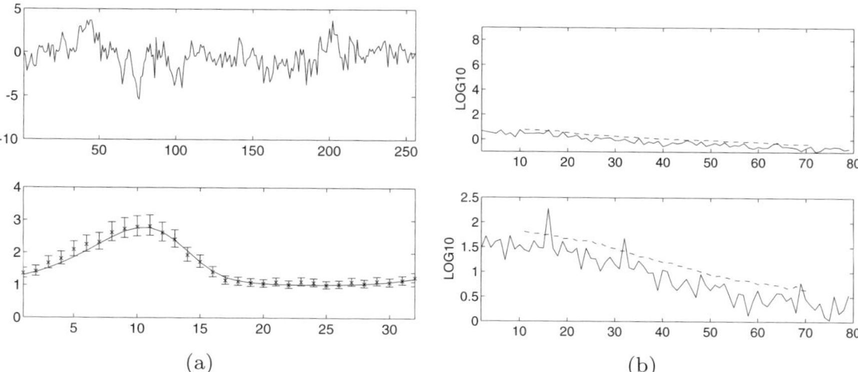

Figure 2.14 Simulated $X_t = \phi(t)X_{t-1} + \xi_t$ where ξ_t is $N(0,1)$ white noise and $\phi(t) = [\phi_0 + \phi_1 \cos(2\pi t/T)], T = 32,\ \phi_0 = 0.6, \phi_1 = 0.4$. (a) (Top) A sample of 256 points from a 5120 point simulation. (Bottom) Theoretical and estimated $\sigma_X(t)$ with 95% confidence intervals based on $N = 5120/32 = 160$ periods. (b) Average of 10 periodograms using FFTs of length 512 for X_t (Top) and X_t^2 (Bottom). Dashed lines are the $\alpha = 0.01$ threshold for test of variance contrast based on a half neighborhood of size $m = 8$.

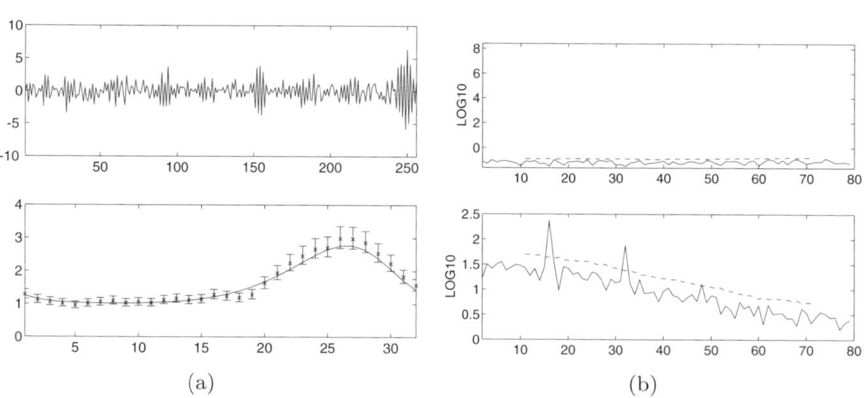

Figure 2.15 Same model as Figure 2.14 except $\phi_0 = -0.6$, $\phi_1 = 0.4$. (a) Time series and $\sigma_X(t)$ using same parameters as Figure 2.14(a). (b) Average of 10 periodograms as in Figure 2.14(b).

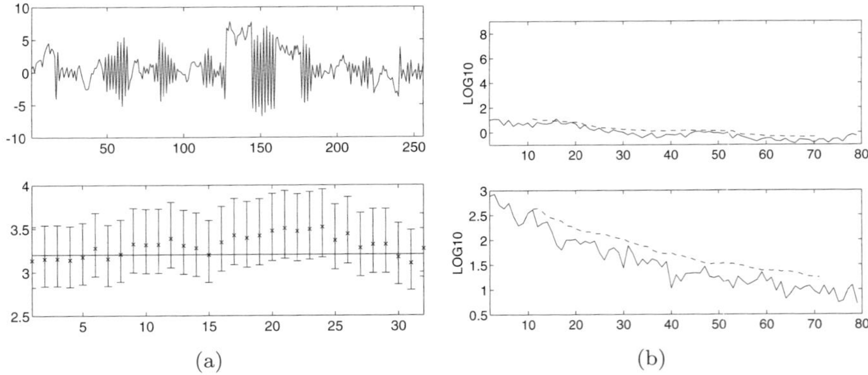

Figure 2.16 Simulated X_t with $\phi(t) = 0.95$ for $0 \le t < T/2$ and $\phi(t) = -0.95$ for $T/2 \le t < T$, with $T = 32$ and Cov $(\xi_s, \xi_t) = \delta_{s-t}$. (a) Time series and $\sigma_X(t)$ using same parameters as Figure 2.14(a). (b) Average of 10 periodograms as in Figure 2.14(b).

2.2.7 Periodic Moving Average Models

Again some simulated series illustrate our ability to visually perceive evidence of periodicity in the time series, in sample periodic variance or in the periodograms of X_t and X_t^2 . Figure 2.17 presents the now familiar views of a simulated X_t for the slowly varying model

$$X_t = \xi_t + \theta_1(t)\xi_{t-1}$$

with

$$\theta_1(t) = 1 + 0.25\cos(2\pi t/T),$$

$T = 32$, and Cov $(\xi_s, \xi_t) = \delta_{s-t}$.

We see no clear visual evidence of the periodicity in the time series trace of Figure 2.17(a) although it is reasonably clear in $\widehat{\sigma}_N(t)$. Again, the usual periodogram is unable to detect any evidence of periodicity at $\alpha = 0.01$, but the periodicity in X_t^2 is clear; the p-value at 16 periods (in 512 point FFT) is 5.4×10^{-4}.

Note from the form of (2.28) that it is easy to construct PMA models that have constant variance. For example, the model $\theta_0(t) \equiv 1$, $\theta_1(t) = \cos(2\pi t/T)$, and $\theta_2(t) = \sin(2\pi t/T)$ would give

$$R(t,t) = \sum_{j=0}^{2} \theta_j^2(t) = 1 + \cos^2(2\pi t/T) + \sin^2(2\pi t/T) \equiv 2.$$

Figure 2.18 presents a $N = 600$ point simulation of such a series with no clear evidence to reject $\sigma(t) \equiv \sigma$ from $\widehat{\sigma}_N(t)$ nor from the periodogram of squares.

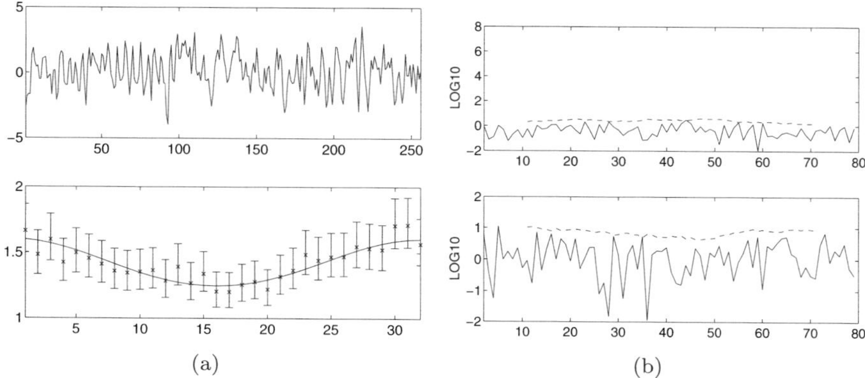

Figure 2.17 Simulated $X_t = \xi_t + \theta_1(t)\xi_{t-1}$ with $\theta_1(t) = 1 + 0.25\cos(2\pi t/T)$ for $T = 32$ and $Cov(\xi_s, \xi_t) = \delta_{s-t}$. (a) Time series and $\sigma_X(t)$ using same parameters as Figure 2.14(a). (b) One periodogram using same parameters as in Figure 2.14(b).

Another constant variance model,

$$(\theta_0(t), \theta_1(t)) = \begin{cases} (\theta_0, \theta_1) & \text{if } 0 \le t < [T/2] \\ (\theta_0, -\theta_1) & \text{if } [T/2] \le t < T \end{cases}, \tag{2.33}$$

permits the switching between two MA models in a way similar to the switching AR model which we can understand from the analysis of stationary models. By choosing the parameters $(\theta_0, \theta_1) = (1, 1)$, the switching model (2.33) gives a two-point averaging filter for the first half of the period and a two-point differencing filter for the second half. We leave experimentation with this model to the reader.

A Closing Question. Does periodicity in the variance imply the presence of periodic correlation? Based on all the simple models, the presence of a periodic variance is a good indicator that PC structure is present. But strictly, the answer is no, as seen from the following example for which we acknowledge Andrzej Makagon. Suppose, for $t \in \mathbb{Z}$, X_t is a stationary sequence, f_t is a periodic scalar sequence with $|f_t|$ properly periodic, and g_t is a nonrandom scalar sequence of values taking ± 1 with

$$\lim_{N\to\infty} \frac{1}{2N+1} \sum_{t=-N}^{N} g_t = 0.$$

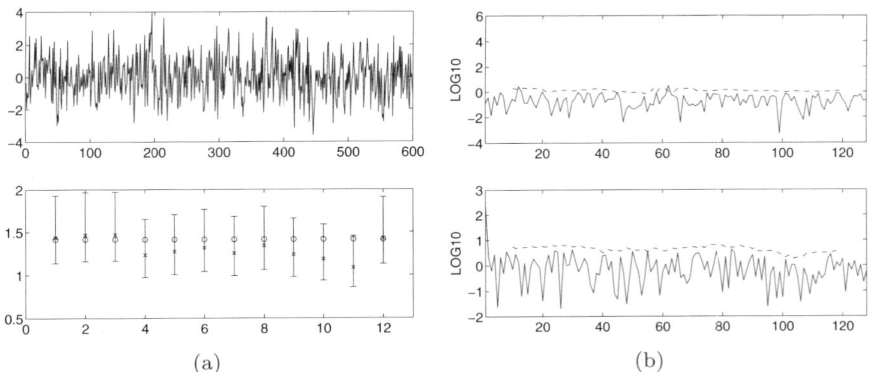

Figure 2.18 Simulated PMA $X_t = \xi_t + \cos(2\pi t/T)\xi_{t-1} + \sin(2\pi t/T)\xi_{t-2}$, where ξ_t is white noise and $T = 12$. The resulting variance is constant $R_X(t,t) \equiv 2$. (a) (Top) Simulated time series. (Bottom) Theoretical and estimated $\sigma_X(t)$ with 95% confidence intervals based on $N = 600/12 = 50$ periods. (b) Periodograms of X_t (Top) and X_t^2 (Bottom) based on one FFT of length 600. Dashed lines are the $\alpha = 0.01$ threshold for test of variance contrast based on a half neighborhood of size $m = 8$.

Then $Y_t = f_t g_t X_t$ has standard deviation $\sigma_Y(t) = |f_t|^2 \sigma_X$, but generally

$$\begin{aligned} E\{Y_s Y_t\} &= f_s f_t g_s g_t r_X(s-t) \\ &\neq f_{s+T} f_{t+T} g_{s+T} g_{t+T} r_X(s+T-s-T) \end{aligned}$$

for all s, t because $g_t \neq g_{t+T}$ for all t. It would seem that we need to look pretty hard to find a sequence with periodic variance that is not PC. And so for practical purposes we use it as an important clue that periodic correlation is present.

PROBLEMS AND SUPPLEMENTS

2.1 Prove Proposition 2.1.

2.2 What is the spectral representation of a periodic $L^2(\Omega, \mathcal{F}, P)$-valued sequence? What is the spectral representation of a sequence that is periodic and also stationary?

2.3 Use the program `parbatch.m` to further explore the effects of ϕ_0 and ϕ_1 in the PAR(1)-CVS simulation of $X_t = [\phi_0 + \phi_1 \cos(2\pi t/T)]X_{t-1} + \xi_t$.

2.4 Construct versions of the elementary examples for the cases of continuous time PC processes, for PC fields, and for almost PC processes.

2.5 Continuous time wide sense stationary processes that are not continuous in quadratic mean are often considered pathological because of the usual assumption of continuity of the covariance $R(u)$ at $u = 0$. This exercise illustrates that this is not the case for continuous time PC processes. Show that if X_t is stationary and continuous in quadratic mean, and $Y_t = f_t X_t$, where f_t has a simple discontinuity at t_0, then Y_t is not continuous in quadratic mean at t_0. Hence there are simple PC processes that are not continuous in quadratic mean.

CHAPTER 3

REVIEW OF HILBERT SPACES

The purpose of this chapter is to study those aspects of Hilbert space theory that are needed for understanding the rest of this book. It is directed toward the spectral theorem for unitary operators. We give some proofs, the remainder of which can be found in the indicated pages of Akheizer and Glazman [2].

3.1 VECTOR SPACES

Definition 3.1 *A vector space over the field* $\mathbb{C}$ *of complex numbers consists of a set* $\mathcal{X}$ *in which two operations called addition and scalar multiplication are defined so that for each pair* $\mathbf{x}$ *and* $\mathbf{y}$ *in* $\mathcal{X}$ *and each complex number* α *there are unique elements* $\mathbf{x}+\mathbf{y}$ *and* $\alpha\mathbf{x}$ *in* $\mathcal{X}$ *such that the following conditions hold for any vectors* $\mathbf{x}, \mathbf{y}, \mathbf{z} \in \mathcal{X}$ *and any complex numbers* α *and* $\beta \in \mathbb{C}$*:*

(a) $\mathbf{x} + \mathbf{y} = \mathbf{y} + \mathbf{x}$;

(b) $(\mathbf{x} + \mathbf{y}) + \mathbf{z} = \mathbf{x} + (\mathbf{y} + \mathbf{z})$;

Periodically Correlated Random Sequences:Spectral Theory and Practice. By H.L. Hurd and A.G. Miamee

(c) *there is an element* $\mathbf{0} \in \mathcal{X}$*, called zero, such that* $\mathbf{x} + \mathbf{0} = \mathbf{x}$*;*

(d) $\alpha(\mathbf{x} + \mathbf{y}) = \alpha\mathbf{x} + \alpha\mathbf{y}$*;*

(e) $(\alpha + \beta)\mathbf{x} = \alpha\mathbf{x} + \beta\mathbf{x}$*;*

(f) $\alpha(\beta\mathbf{x}) = (\alpha\beta)\mathbf{x}$ *;*

(g) $0\mathbf{x} = \mathbf{0}$*;*

(h) $1\mathbf{x} = \mathbf{x}$.

The elements of a vector space $\mathcal{X}$ are called vectors. Here are examples of some important vector spaces.

Examples.

1. It is easy to check that the Euclidean space $\mathbb{C}^k = \{\mathbf{x} = [x^1, x^2, \ldots, x^k] : x^i \in \mathbb{C}, i = 1, 2, \ldots, k\}$ with component-wise addition and scalar multiplication is a vector space and its zero vector is $\mathbf{0} = [0, 0, ..., 0]$.

2. The space of square summable sequences

$$\ell^2 = \left\{ \mathbf{x} = (x_j)_{j=1}^{\infty} : x_j \in \mathbb{C} \ \text{ and } \ \| \mathbf{x} \|^2 = \sum_{j=1}^{\infty} |x_j|^2 < \infty \right\}$$

with component-wise addition and scalar multiplication is a vector space. Its closedness under scalar multiplication and addition follows from

$$\| \alpha\mathbf{x} \|^2 = |\alpha|^2 \ \| \mathbf{x} \|^2,$$

which is easy to check, and

$$\| \mathbf{x} + \mathbf{y} \|^2 \leq 2(\| \mathbf{x} \|^2 + \| \mathbf{y} \|^2)$$

which is immediate from Problem 3 at the end of this chapter. The rest of requirement for a vector space can be easily checked. In particular here, the $\mathbf{0}$ vector is the sequence with all zero entries.

3. The space of square integrable functions $\mathbf{X}$ on a probability space $(\Omega, \mathcal{F}, P)$, that is, the space of all random variables $\mathbf{X}$ on Ω satisfying the condition

$$E|\mathbf{X}|^2 = \int_{\Omega} |\mathbf{X}(\omega)|^2 P(d\omega) < \infty$$

with the usual point-wise addition and scalar multiplication, is a vector space. As in the last example, checking the vector space properties becomes straightforward once we notice that for any two square integrable functions $\mathbf{X}$ and $\mathbf{Y}$ and any complex number α

$$E|\alpha \mathbf{X}|^2 = |\alpha|^2 E|\mathbf{X}|^2 < \infty,$$

which is easy to check and

$$E|\mathbf{X} + \mathbf{Y}|^2 \leq 2E|\mathbf{X}|^2 + 2E|\mathbf{Y}|^2,$$

which is an immediate consequence of problem 3 at the end of this chapter.

Definition 3.2 *Let $\mathcal{X}$ be a vector space and L be a subset of $\mathcal{X}$. L is called a linear manifold if $\alpha\mathbf{x} + \beta\mathbf{y} \in L$ for any $\alpha, \beta \in \mathbb{C}$ and any $\mathbf{x}, \mathbf{y} \in L$.*

Definition 3.3 *Given a subset E of a vector space $\mathcal{X}$, the set of all finite linear combinations from E is called the span of E and is denoted by*

$$\text{sp}(E) = \left\{ \sum_{j=1}^{n} \alpha_j \mathbf{x}_j : n \in \mathbb{N}, \alpha_j \in \mathbb{C}, \mathbf{x}_j \in E \right\}.$$

One can easily verify that $sp(E)$ turns out to be a vector space, which is called the vector space (or linear manifold) generated by E.

Remark. If $\mathcal{L}_\gamma$ are all the linear manifolds containing E, then $\cap_\gamma \mathcal{L}_\gamma$ is a linear manifold containing E and it is easy to check that it is the smallest among all such linear manifolds.

Definition 3.4 *A set E is called linearly independent if for every finite subset $\{\mathbf{x}_1, \mathbf{x}_2, \cdots, \mathbf{x}_n\}$ of E, the condition $\alpha_1\mathbf{x}_1 + \alpha_2\mathbf{x}_2 + \cdots \alpha_n\mathbf{x}_n = \mathbf{0}$ implies $\alpha_1 = \alpha_2 = \cdots = \alpha_n = 0$.*

Definition 3.5 *A subset E of a vector space $\mathcal{X}$ is called a basis for $\mathcal{X}$ if it is linearly independent and* $\text{sp}(E) = \mathcal{X}$.

3.2 INNER PRODUCT SPACES

Here the notion of inner product space is introduced, its properties are given and some specific inner product spaces are discussed.

Definition 3.6 *A map $\langle \cdot, \cdot \rangle : \mathcal{X} \times \mathcal{X} \to \mathbb{C}$, where $\mathcal{X}$ is a vector space, is called an inner product on $\mathcal{X}$ if for all complex numbers α, β and all vectors $\mathbf{x}, \mathbf{y}, \mathbf{z} \in \mathcal{X}$ we have*

(a) $\langle \mathbf{x}, \mathbf{y} \rangle = \overline{\langle \mathbf{y}, \mathbf{x} \rangle}$;

(b) $\langle \alpha \mathbf{x} + \beta \mathbf{y}, \mathbf{z} \rangle = \alpha \langle \mathbf{x}, \mathbf{z} \rangle + \beta \langle \mathbf{y}, \mathbf{z} \rangle$;

(c) $\langle \mathbf{x}, \mathbf{x} \rangle \geq 0$;

(d) $\langle \mathbf{x}, \mathbf{x} \rangle = 0$ *if and only if* $\mathbf{x} = \mathbf{0}$.

If $\mathcal{X}$ is a vector space and $\langle \cdot, \cdot \rangle$ is an inner product on $\mathcal{X}$, then the pair $(\mathcal{X}, \langle \cdot, \cdot \rangle)$ is called an inner product space. For a vector $\mathbf{x} \in \mathcal{X}$ we define its norm to be

$$\| \mathbf{x} \| = \sqrt{\langle \mathbf{x}, \mathbf{x} \rangle}.$$

Two vectors $\mathbf{x}$ and $\mathbf{y}$ in an inner product space $\mathcal{X}$ are said to be orthogonal, written $\mathbf{x} \perp \mathbf{y}$, if $\langle \mathbf{x}, \mathbf{y} \rangle = 0$.

Some properties of inner product and norm are given in the following proposition (See [2, Chapter I].)

Proposition 3.1 *For any vectors* $\mathbf{x}, \mathbf{y}$ *in an inner product space* $\mathcal{X}$ *and any complex number* α *we have*

(a) Cauchy–Schwarz inequality: $|\langle \mathbf{x}, \mathbf{y} \rangle| \leq \| \mathbf{x} \| \, \| \mathbf{y} \|$;

(b) triangle inequality: $\| \mathbf{x} + \mathbf{y} \| \leq \| \mathbf{x} \| + \| \mathbf{y} \|$;

(c) $\| \alpha \mathbf{x} \| = |\alpha| \, \| \mathbf{x} \|$;

(d) $\| \mathbf{x} \| = 0$ *if and only if* $\mathbf{x} = \mathbf{0}$;

(e) parallelogram law: $\| \mathbf{x} + \mathbf{y} \|^2 + \| \mathbf{x} - \mathbf{y} \|^2 = 2 \| \mathbf{x} \|^2 + 2 \| \mathbf{y} \|^2$.

From the triangle inequality for norms it follows that $d(\mathbf{x}, \mathbf{y}) = \| \mathbf{x} - \mathbf{y} \|$ is a metric. The topology of an inner product space is the one induced by this metric.

Definition 3.7 *A sequence* $\mathbf{x}_n$ *of vectors in an inner product space* $\mathcal{X}$ *is said*

(a) *to converge to* $\mathbf{x} \in X$ *if* $\| \mathbf{x}_n - \mathbf{x} \| \to 0$, *as* $n \to \infty$;

(b) *to be Cauchy if for any* $\epsilon > 0$ *there exists* $N > 0$ *such that*

$$\| \mathbf{x}_n - \mathbf{x}_m \| < \epsilon, \text{ whenever } n, m > N.$$

Proposition 3.2 (Continuity of Inner Product and Norm) *If* $\{\mathbf{x}_n\}$ *and* $\{\mathbf{y}_n\}$ *are two sequences in an inner product space converging to* $\mathbf{x}$ *and* $\mathbf{y}$, *respectively, then*

(a) $\| \mathbf{x}_n \| \to \| \mathbf{x} \|$;

(b) $\langle \mathbf{x}_n, \mathbf{y}_n \rangle \to \langle \mathbf{x}, \mathbf{y} \rangle$.

Proof. From the triangle inequality it follows that

$$\| \mathbf{x} \| = \| \mathbf{x} - \mathbf{y} + \mathbf{y} \| \leq \| \mathbf{x} - \mathbf{y} \| + \| \mathbf{y} \|$$

and similarly $\| \mathbf{y} \| \leq \| \mathbf{x} - \mathbf{y} \| + \| \mathbf{x} \|$. Combining these two inequality one arrives at

$$| \| \mathbf{x} \| - \| \mathbf{y} \| | \leq \| \mathbf{x} - \mathbf{y} \|,$$

from which (a) is immediate. Using the Cauchy–Schwarz inequality we can write

$$\begin{aligned} |\langle \mathbf{x}_n, \mathbf{y}_n \rangle - \langle \mathbf{x}, \mathbf{y} \rangle| &= |\langle \mathbf{x}_n, \mathbf{y}_n - \mathbf{y} \rangle + \langle \mathbf{x}_n - \mathbf{x}, \mathbf{y} \rangle| \\ &\leq |\langle \mathbf{x}_n, \mathbf{y}_n - \mathbf{y} \rangle| + |\langle \mathbf{x}_n - \mathbf{x}, \mathbf{y} \rangle| \\ &\leq \| \mathbf{x}_n \| \| \mathbf{y}_n - \mathbf{y} \| + \| \mathbf{x}_n - \mathbf{x} \| \| \mathbf{y} \|, \end{aligned}$$

from which (b) follows. ∎

3.3 HILBERT SPACES

Definition 3.8 *A Hilbert space $\mathcal{H}$ is an inner product space that is complete with respect to the norm induced by the inner product; that is, a Hilbert space is an inner product space in which every Cauchy sequence $(\mathbf{x}_n)$ converges in norm to some vector $\mathbf{x} \in \mathcal{H}$.*

Here are some examples of Hilbert spaces.

Examples.

1. The Euclidean space $\mathbb{C}^k = \{\mathbf{x} = [x^1, x^2, \ldots, x^k] : x^i \in \mathbb{C}, i = 1, 2, \ldots, k\}$ equipped with

$$\langle \mathbf{x}, \mathbf{y} \rangle = \sum_{i=1}^{k} x^i \overline{y^i}$$

becomes an inner product space. Here the norm of a vector $\mathbf{x} = [x^1, x^2, \ldots, x^k]$ in $\mathbb{C}^k$ is given by $\| \mathbf{x} \| = \sqrt{\sum_{j=1}^{k} |x^j|^2}$. This inner product space turns out to be a Hilbert space. To verify this, which means verifying its completeness, let $\mathbf{x}_n = [x_n^1, x_n^2, \ldots, x_n^k]$ be a Cauchy sequence in $\mathbb{C}^k$. Given any positive ϵ there exists positive integer N such that

$$\epsilon^2 > \| \mathbf{x}_n - \mathbf{x}_m \|^2 = \sum_{j=1}^{k} |x_n^j - x_m^j|^2, \text{whenever } m, n > N.$$

Hence for each integer $j = 1, 2, \ldots, k$ we have

$$|x_n^j - x_m^j| < \epsilon, \text{whenever } m, n > N.$$

Now by completeness of complex numbers $\mathbb{C}$, for each positive j there is a complex number x^j to which x_n^j converges. Taking $\mathbf{x} = [x^1, x^2, \ldots, x^k]$ we get

$$\| \mathbf{x}_n - \mathbf{x} \| \to 0, \text{ as } n \to \infty,$$

which completes the proof.

2. The space $\ell^2 = \left\{ \mathbf{x} = (x_j)_{j=1}^{\infty} : x_j \in \mathbb{C}, \sum_j |x_j|^2 \langle \infty \right\}$ equipped with

$$\langle \mathbf{x}, \mathbf{y} \rangle = \sum_{j=1}^{\infty} x_j \overline{y_j}$$

becomes an inner product space with norm $\| \mathbf{x} \| = \sqrt{\sum_{j=1}^{\infty} |x_j|^2}$. Using an argument similar to the one used in the example of Euclidean space $\mathbb{C}^k$ above, one can verify that the space ℓ^2 is a Hilbert space.

This turns out to be a prototype Hilbert space in the sense that every separable Hilbert space is essentially the same as (isomorphic to) this one. However, the next example, which is the cornerstone of the theory developed in this book, is not of this type (because it is not separable) and thus deserves our attention.

3. Consider the space $L^2(\Omega, \mathcal{F}, P)$ of square integrable functions (random variables with finite second moments) on a probability space $(\Omega, \mathcal{F}, P)$ and define the inner product to be

$$\langle \mathbf{X}, \mathbf{Y} \rangle = E\mathbf{X}\overline{\mathbf{Y}} = \int_{\Omega} \mathbf{X}(\omega)\overline{\mathbf{Y}(\omega)} dP(\omega). \tag{3.1}$$

It is easy to check that $\langle \mathbf{X}, \mathbf{Y} \rangle$ satisfies requirements (a)-(c) in Definition 3.6. However, requirement (d) does not hold. That is, $\langle \mathbf{X}, \mathbf{X} \rangle = 0$ in general does not imply that $\mathbf{X}$ is identically zero. It only implies that $\mathbf{X}$ is zero almost surely. This problem can be settled by saying two random variables are equivalent if they are equal almost surely and by reinterpreting $L^2(\Omega, \mathcal{F}, P)$ as the collection of all these equivalence classes. Each class is uniquely represented by any specific random variable in that class and we shall use notation $\mathbf{X}, \mathbf{Y}$, and so on for elements of $L^2(\Omega, \mathcal{F}, P)$ and call them random variables, although it is sometimes necessary to remember that $\mathbf{X}$ stands for the class of all random variables equivalent to $\mathbf{X}$. We similarly reinterpret the definition for the inner product $\langle \mathbf{X}, \mathbf{Y} \rangle$. To show that this modified space $L^2(\Omega, \mathcal{F}, P)$ is a Hilbert space, it is necessary to establish its completeness, a task beyond the scope of this presentation, but readily found in measure theory texts. See, for example, [202, Theorem 3.11].

This Hilbert space is of particular importance to us because most of the stochastic processes in this book are just two-sided sequences of vectors in such a Hilbert space.

Definition 3.9 *If E is a subset of a Hilbert space $\mathcal{H}$, then the closure of* $\text{sp}(E)$*, denoted by* $\overline{\text{sp}}(E)$*, is called the span closure of E.*

Definition 3.10 *Any closed linear manifold $\mathcal{M}$ of a Hilbert space $\mathcal{H}$ will be called a subspace. The inner product $\langle \cdot, \cdot \rangle$ of $\mathcal{H}$ induces an inner product on $\mathcal{M}$. The subspace $\mathcal{M}$ together with this inherited inner product is itself complete (this follows from completeness of $\mathcal{H}$) and hence is a Hilbert space.*

3.4 OPERATORS

Definition 3.11 *Let $\mathcal{H}$ and $\mathcal{K}$ be two Hilbert spaces. A mapping (or transformation) T from $\mathcal{H}$ into $\mathcal{K}$ is said to be additive if $T(\mathbf{x}+\mathbf{y}) = T\mathbf{x} + T\mathbf{y}$ for all $\mathbf{x}, \mathbf{y} \in \mathcal{H}$ and homogeneous if $T(\alpha\mathbf{x}) = \alpha T\mathbf{x}$ for all complex numbers α and all vectors $\mathbf{x} \in \mathcal{H}$. A transformation T that is both additive and homogeneous is called linear.*

It is easy to check that a transformation $T : \mathcal{H} \to \mathcal{K}$ is linear if and only if $T(\alpha\mathbf{x} + \mathbf{y}) = \alpha T\mathbf{x} + T\mathbf{y}$ for all $\mathbf{x}, \mathbf{y} \in \mathcal{H}$ and all complex numbers α. A linear transformation from a Hilbert space $\mathcal{H}$ into the complex numbers $\mathbb{C}$ is called a *linear functional.*

Definition 3.12 *A linear transformation $T : \mathcal{H} \to \mathcal{K}$ is called bounded if there exists a positive number M such that*

$$\| T\mathbf{x} \| \leq M \| \mathbf{x} \|, \text{ for all } \mathbf{x} \in \mathcal{H}.$$

The smallest such M is called norm of T and is denoted by $\| T \|$. A bounded linear transformation from $\mathcal{H}$ into itself is called an operator on $\mathcal{H}$.

Proposition 3.3 *For any linear transformation $T : \mathcal{H} \to \mathcal{K}$ the following statements are equivalent:*

(a) *T is bounded;*

(b) *T is uniformly continuous on $\mathcal{H}$;*

(c) *T is everywhere continuous on $\mathcal{H}$;*

(d) *T is continuous at* $\mathbf{0}$.

Proof. Let T be a bounded linear transformation and ϵ be a positive number. Take $\delta = \epsilon / \parallel T \parallel$. Now whenever $\parallel \mathbf{y} - \mathbf{x} \parallel \leq \delta$, we can write

$$\parallel T\mathbf{y} - T\mathbf{x} \parallel = \parallel T(\mathbf{x} - \mathbf{y}) \parallel \leq \parallel T \parallel \parallel \mathbf{y} - \mathbf{x} \parallel \leq \parallel T \parallel \frac{\epsilon}{\parallel T \parallel} = \epsilon,$$

which means T is uniformly bounded. This shows (a)$\Rightarrow$(b). Implications (b)$\Rightarrow$(c) and (c)$\Rightarrow$(d) are obvious. To show (d) $\Rightarrow$(a) suppose T is continuous at $\mathbf{0}$. Since $T\mathbf{0} = \mathbf{0}$, there exists some $\delta > 0$ such that $\parallel T\mathbf{x} \parallel$ $= \parallel T\mathbf{x} - T\mathbf{0} \parallel < 1$, whenever $\parallel \mathbf{x} \parallel \leq \delta$. Now for any $\mathbf{x} \in \mathcal{H}$ we can write

$$\parallel T\mathbf{x} \parallel = \frac{\parallel \mathbf{x} \parallel}{\delta} T\left(\frac{\delta \mathbf{x}}{\parallel \mathbf{x} \parallel}\right) \leq \frac{\parallel \mathbf{x} \parallel}{\delta}.$$

So taking $M = 1/\delta$ we get

$$\parallel T\mathbf{x} \parallel \leq M \parallel \mathbf{x} \parallel, \text{ for all } \mathbf{x} \in \mathcal{H},$$

which means T is bounded. This completes the proof of (d)$\Rightarrow$(a). ∎

Definition 3.13 *A bounded linear transformation $T : \mathcal{H} \to \mathcal{K}$ is called an isomorphism if it is onto and $\langle T\mathbf{x}, T\mathbf{y} \rangle = \langle \mathbf{x}, \mathbf{y} \rangle$ for all $\mathbf{x}, \mathbf{y} \in \mathcal{H}$. Two Hilbert spaces $\mathcal{H}$ and $\mathcal{K}$ are called isomorphic if there exists an isomorphism from $\mathcal{H}$ onto $\mathcal{K}$.*

For the proof of the following two important theorems one can refer to [202, Theorem 4.12] or [87, Section 22].

Theorem 3.1 (Riesz Theorem) *If φ is a bounded functional on $\mathcal{H}$, there exists a unique vector $\mathbf{y} \in \mathcal{H}$ such that*

$$\varphi(\mathbf{x}) = \langle \mathbf{x}, \mathbf{y} \rangle, \text{ for all } \mathbf{x} \in \mathcal{H}$$

and $\parallel \mathbf{y} \parallel = \parallel \varphi \parallel$.

A complex valued function $\varphi(\mathbf{x}, \mathbf{y}), \mathbf{x}, \mathbf{y} \in \mathcal{H}$, which is linear in each variable, is called a bilinear functional and it is called bounded if there exists a positive number M such that

$$|\varphi(\mathbf{x}, \mathbf{y})| \leq M \|\mathbf{x}\| \|\mathbf{y}\|, \text{ for all } \mathbf{x}, \mathbf{y} \in \mathcal{H}.$$

Theorem 3.2 *For any bounded bilinear functional φ on $\mathcal{H}$ there exists a unique operator A on $\mathcal{H}$ such that*

$$\varphi(\mathbf{x}, \mathbf{y}) = \langle A\mathbf{x}, \mathbf{y} \rangle, \text{ for all } \mathbf{x} \in \mathcal{H}.$$

Proposition 3.4 *For any operator T on $\mathcal{H}$ there exists another operator T^* on $\mathcal{H}$ such that*

$$\langle T\mathbf{x}, \mathbf{y}\rangle = \langle \mathbf{x}, T^*\mathbf{y}\rangle, \ \text{for all } \mathbf{x}, \mathbf{y} \in \mathcal{H}.$$

The operator T^*, which is called the adjoint of T, has the same norm as T.

Proof. Let $\mathbf{y}$ be any fixed vector in $\mathcal{H}$. Consider the linear function φ defined by

$$\varphi(\mathbf{x}) = \langle T\mathbf{x}, \mathbf{y}\rangle, \ \text{for all } \in \mathcal{H}.$$

This functional φ is bounded because for any $\mathbf{x} \in \mathcal{H}$ we can write

$$|\varphi(\mathbf{x})| = |\langle T\mathbf{x}, \mathbf{y}\rangle| \leq \|T\mathbf{x}\|\|\mathbf{y}\| \leq \|T\|\|\mathbf{x}\|\|\mathbf{y}\|.$$

By Theorem 3.1 (Riesz), there exists $\mathbf{y}^* \in \mathcal{H}$ such that $\varphi(\mathbf{x}) = \langle \mathbf{x}, \mathbf{y}^*\rangle$ for every $\mathbf{x} \in \mathcal{H}$. Defining $T^* : \mathcal{H} \to \mathcal{H}$ by $T^*\mathbf{y} = \mathbf{y}^*$, one can check that T^* is linear, bounded, and satisfies $\langle T\mathbf{x}, \mathbf{y}\rangle = \varphi(\mathbf{x}) = \langle \mathbf{x}, \mathbf{y}^*\rangle = \langle \mathbf{x}, T^*\mathbf{y}\rangle$, for all $\mathbf{x}, \mathbf{y} \in \mathcal{H}$. ∎

Definition 3.14 *An operator T on the Hilbert space $\mathcal{H}$ is called normal if it commutes with T^*, Hermitian if it is equal to T^*, and unitary if $T^{-1} = T^*$.*

3.5 PROJECTION OPERATORS

For more about material covered in this section one can refer to [2, Sections 30–33, 35, 36].

Definition 3.15 *Let $\mathbf{x}$ be a vector in a Hilbert space $\mathcal{H}$ and M be a subset of $\mathcal{H}$. We say $\mathbf{x}$ is orthogonal to M and write $\mathbf{x} \perp M$ if $\mathbf{x} \perp \mathbf{y}$ for all $\mathbf{y}$ in M. In general, two sets are said to be orthogonal to each other if every vector in one set is orthogonal to the other set. The orthogonal complement $M^\perp$ of a subset M of a Hilbert space $\mathcal{H}$ is defined to be $M^\perp = \{\mathbf{x} \in H : \mathbf{x} \perp M\}$.*

Proposition 3.5 *Orthogonal complement of any subset M of a Hilbert space $\mathcal{H}$ is a subspace of $\mathcal{H}$.*

Proof. Let $\mathbf{y}$ and $\mathbf{z}$ be any two vectors in $M^\perp$, $\mathbf{x}$ be any vector in M, and α, β be any two complex numbers; then we have

$$\langle \mathbf{x}, \alpha\mathbf{y} + \beta\mathbf{z}\rangle = \alpha\langle \mathbf{x}, \mathbf{y}\rangle + \overline{\beta}\langle \mathbf{x}, \mathbf{z}\rangle = 0,$$

which shows $\alpha\mathbf{y} + \beta\mathbf{z} \in M^\perp$ and hence $M^\perp$ is a linear manifold. To show the remaining requirement, closedness of $M^\perp$, let $(\mathbf{y}_n)$ be a sequence in $M^\perp$

converging to some $\mathbf{y}$ in $\mathcal{H}$. Now for any $\mathbf{x}$ in M by continuity of the inner product (see Proposition 3.2) we can write

$$\langle \mathbf{x}, \mathbf{y} \rangle = \lim_{n\to\infty} \langle \mathbf{x}, \mathbf{y}_n \rangle = \lim_{n\to\infty} 0 = 0,$$

which shows $\mathbf{y}$ must be in $M^\perp$. ∎

Consider a subset M of a Hilbert space $\mathcal{H}$ and a vector $\mathbf{x}$ in $\mathcal{H}$ outside M. Is there in M a "closest" vector to $\mathbf{x}$? If there is such a vector is it unique? Considering $\delta(\mathbf{x}) = \inf_{\mathbf{y}\in M} \| \mathbf{x} - \mathbf{y} \|$, by closest element in M to $\mathbf{x}$ we mean any vector $\hat{\mathbf{x}} \in M$ with $\| \mathbf{x} - \hat{\mathbf{x}} \| = \delta(\mathbf{x})$. It is easy to see that for a general set M the answer is negative. The following important theorem, which can be found in any book on Hilbert spaces (e.g., see [202, Theorem 4.11]), provides a complete answer to this question.

Theorem 3.3 (Projection Theorem)) *If $\mathcal{M}$ is a subspace of the Hilbert space $\mathcal{H}$ and $\mathbf{x}$ is any vector in $\mathcal{H}$, then*

(a) *there is a unique vector $\widehat{\mathbf{x}} \in \mathcal{M}$, called orthogonal projection of $\mathbf{x}$ onto $\mathcal{M}$, such that*
$$\delta(\mathbf{x}) = \inf_{\mathbf{y}\in\mathcal{M}} \| \mathbf{x} - \mathbf{y} \| = \| \mathbf{x} - \widehat{\mathbf{x}} \|;$$

(b) *a vector $\widehat{\mathbf{x}} \in \mathcal{M}$ serves as the orthogonal projection of $\mathbf{x}$ onto $\mathcal{M}$ if and only if $\mathbf{x} - \widehat{\mathbf{x}} \in \mathcal{M}^\perp$.*

Proof. For (a), let $\delta(\mathbf{x}) = \inf_{\mathbf{y}\in\mathcal{M}} \| \mathbf{x}-\mathbf{y} \|$; then there is a sequence $\mathbf{x}_n \in \mathcal{M}$ with $\| \mathbf{x} - \mathbf{x}_n \| \to \delta(\mathbf{x})$. Since $\mathcal{M}$ is a subspace and hence $(\mathbf{x}_n + \mathbf{y}_n)/2 \in \mathcal{M}$, an application of the parallelogram law gives

$$\begin{aligned} 0 \le \| \mathbf{x}_n - \mathbf{x}_m \|^2 &= 2 \| \mathbf{x}_n - \mathbf{x} \|^2 + 2 \| \mathbf{x}_m - \mathbf{x} \|^2 - 4 \left\| \frac{\mathbf{x}_n + \mathbf{x}_m}{2} - \mathbf{x} \right\|^2 \\ &\le 2 \| \mathbf{x}_n - \mathbf{x} \|^2 + 2 \| \mathbf{x}_m - \mathbf{x} \|^2 - 4\delta(\mathbf{x}). \end{aligned}$$

Since the last quantity on the right has limit zero as $m, n \to \infty$, the sequence $\mathbf{x}_n$ is Cauchy and hence has a limit, $\widehat{\mathbf{x}}$, in $\mathcal{M}$. Now by continuity of norm given in Proposition 3.2(a),

$$\| \mathbf{x} - \widehat{\mathbf{x}} \| = \lim_{n\to\infty} \| \mathbf{x} - \mathbf{x}_n \| = \delta(\mathbf{x}),$$

which shows the existence in (a).

Assuming there are two different vectors $\widehat{\mathbf{x}}$ and $\widetilde{\mathbf{x}} \in \mathcal{M}$ with

$$\| \mathbf{x} - \widehat{\mathbf{x}} \| = \| \mathbf{x} - \widetilde{\mathbf{x}} \| = \delta(\mathbf{x}),$$

by the parallelogram law (Proposition 3.1), we get

$$0 \leq \| \widehat{\mathbf{x}} - \widetilde{\mathbf{x}} \| = 2 \| \widehat{\mathbf{x}} - \mathbf{x} \|^2 + 2 \| \widetilde{\mathbf{x}} - \mathbf{x} \|^2 - 4 \left\| \frac{\widehat{\mathbf{x}} + \widetilde{\mathbf{x}}}{2} - \mathbf{x} \right\|^2 \leq 4\delta(\mathbf{x}) - 4\delta(\mathbf{x}) = 0,$$

which contradicts $\widehat{\mathbf{x}} \neq \widetilde{\mathbf{x}}$.

For (b), if $\widehat{\mathbf{x}}$ is in $\mathcal{M}$ and $\mathbf{x} - \widehat{\mathbf{x}}$ is in $\mathcal{M}^\perp$, then $\widehat{\mathbf{x}}$ is the unique vector defined in (a). This is because for any $\mathbf{y} \in \mathcal{M}$

$$\| \mathbf{x} - \mathbf{y} \|^2 = \langle \mathbf{x} - \widehat{\mathbf{x}} + \widehat{\mathbf{x}} - \mathbf{y}, \mathbf{x} - \widehat{\mathbf{x}} + \widehat{\mathbf{x}} - \mathbf{y} \rangle = \| \mathbf{x} - \widehat{\mathbf{x}} \|^2 + \| \mathbf{y} - \widehat{\mathbf{x}} \|^2 \geq \| \mathbf{x} - \widehat{\mathbf{x}} \|^2 .$$

Conversely, if $\mathbf{x} \in \mathcal{M}$ but $\mathbf{x} - \widehat{\mathbf{x}}$ is not in $\mathcal{M}$, then $\widehat{\mathbf{x}}$ is not the closest vector of $\mathcal{M}$ to $\mathbf{x}$. One can easily check that, for any $\mathbf{y}$ in $\mathcal{M}$ with $\langle \mathbf{x} - \widehat{\mathbf{x}}, \mathbf{y} \rangle \neq 0$, the vector

$$\widetilde{\mathbf{x}} = \widehat{\mathbf{x}} + \frac{\langle \mathbf{x} - \widehat{\mathbf{x}}, \mathbf{y} \rangle}{\| \mathbf{y} \|^2} \mathbf{y}$$

is clearly in $\mathcal{M}$ and closer to $\mathbf{x}$ than $\widehat{\mathbf{x}}$. ∎

Part (a) of Theorem 3.3 establishes the existence of the orthogonal projection $\widehat{\mathbf{x}}$ of $\mathbf{x}$ while part (b), as we see in the following examples, helps us in identifying it.

Examples.

1. Let $\mathbf{x}_i, i = 1, 2, \ldots, n$, be n vectors in a Hilbert space $\mathcal{H}$, and $\mathcal{M}$ be the subspace $\mathcal{M} = \text{sp}\{\mathbf{x}_1, \mathbf{x}_2, \ldots, \mathbf{x}_n\}$ and $\mathbf{x}$ be any vector in $\mathcal{H}$ but not in $\mathcal{M}$. The closest vector in $\mathcal{M}$ to $\mathbf{x}$ is of the form

$$\widehat{\mathbf{x}} = \alpha_1 \mathbf{x}_1 + \alpha_2 \mathbf{x}_2 + \cdots + \alpha_n \mathbf{x}_n.$$

   Using part (b) of the Projection Theorem, for all $j = 1, 2, \ldots, n$ we must have $\mathbf{x} - \widehat{\mathbf{x}} \perp \mathbf{x}_j$ or $\langle \mathbf{x} - \Sigma \alpha_i \mathbf{x}_i, \mathbf{x}_j \rangle = 0$. So one can find α_i by solving the following system of n linear equations:

$$\Sigma \langle \mathbf{x}_i, \mathbf{x}_j \rangle \alpha_i = \langle \mathbf{x}, \mathbf{x}_j \rangle, \; j = 1, 2, \ldots, n.$$

2. As a special case when $\mathcal{M} = sp\{\mathbf{e}_1, \mathbf{e}_2, \ldots, \mathbf{e}_n\}$ with $\mathbf{e}_i \perp \mathbf{e}_j$ for all $i \neq j$ and $\| \mathbf{e}_i \| = 1$ for all i, then one can see that orthogonal projection of $\mathbf{x}$ on $\mathcal{M}$ turns out to be (see Problem 3.10 at the end of this chapter)

$$\widehat{\mathbf{x}} = \sum_{i=1}^{n} \langle \mathbf{x}, \mathbf{e}_i \rangle \mathbf{e}_i.$$

Definition 3.16 *Let $\mathcal{M}$ be a subspace of the Hilbert space $\mathcal{H}$; then the projection $P_{\mathcal{M}} : H \rightarrow \mathcal{M}$ is an operator defined by*

$$P_{\mathcal{M}}\mathbf{x} = \widehat{\mathbf{x}}, \text{ for all } \mathbf{x} \in \mathcal{H}.$$

Sometimes instead of $P_{\mathcal{M}}\mathbf{x}$ we use the notation $(\mathbf{x} \mid \mathcal{M})$.

The following Proposition gives some useful properties of projection operator.

Proposition 3.6 *If $\mathcal{M}$ and $\mathcal{N}$ are two subspaces of a Hilbert space $\mathcal{H}$, then*

(a) *$(I - P_{\mathcal{M}})$ is the projection from $\mathcal{H}$ onto $\mathcal{M}^{\perp}$;*

(b) *every $\mathbf{x} \in \mathcal{H}$ has a unique representation as the sum of two orthogonal vectors, one in $\mathcal{M}$ and the other one in $\mathcal{M}^{\perp}$*

$$\mathbf{x} = P_{\mathcal{M}}\mathbf{x} + (I - P_{\mathcal{M}})\mathbf{x};$$

(c) *$\| \mathbf{x} \|^2 = \| P_{\mathcal{M}}\mathbf{x} \|^2 + \| (I - P_{\mathcal{M}})\mathbf{x} \|^2$;*

(d) *$P_{\mathcal{M}}$ is a bounded operator with norm one (i.e. $\| P_{\mathcal{M}} \| = 1$);*

(e) *$\mathbf{x} \in \mathcal{M}$ if and only if $P_{\mathcal{M}}\mathbf{x} = \mathbf{x}$ and $\mathbf{x} \in \mathcal{M}^{\perp}$ if and only if $P_{\mathcal{M}}\mathbf{x} = 0$;*

(f) *$P_{\mathcal{M}}$ is idempotent (i.e., $P_{\mathcal{M}}^2 = P_{\mathcal{M}}$);*

(g) *$P_{\mathcal{M}}$ is self-adjoint;*

(h) *$P_{\mathcal{M}}$ is nonnegative definite (i.e., $\langle P_{\mathcal{M}}\mathbf{x}, \mathbf{x} \rangle \geq 0$ for all $\mathbf{x} \in \mathcal{H}$);*

(i) *$\mathcal{M} \subset \mathcal{N}$ if and only if $P_{\mathcal{M}}P_{\mathcal{N}} = P_{\mathcal{M}}$.*

Proof. Part (a) follows from the definition of projection. Part (b) follows from applying both sides of $I = P_{\mathcal{M}} + (I - P_{\mathcal{M}})$ to the vector $\mathbf{x}$. Part (c) follows from (b) and the fact that the summands $P_{\mathcal{M}}\mathbf{x}$ and $(I - P_{\mathcal{M}})\mathbf{x}$ are orthogonal. For part (d), consider any two vectors $\mathbf{x}, \mathbf{y} \in \mathcal{H}$ and any two complex numbers α, β. It is clear that

$$\alpha\widehat{\mathbf{x}} + \beta\widehat{\mathbf{y}} \in \mathcal{M}$$

and

$$\alpha\mathbf{x} + \beta\mathbf{y} - \alpha\widehat{\mathbf{x}} + \beta\widehat{\mathbf{y}} = (\alpha\mathbf{x} - \alpha\widehat{\mathbf{x}}) + (\beta\mathbf{y} - \beta\widehat{\mathbf{y}}) \perp \mathcal{M}$$

which implies $P_{\mathcal{M}}(\alpha\mathbf{x} + \beta\mathbf{y}) = \alpha P_{\mathcal{M}}\mathbf{x} + \beta P_{\mathcal{M}}\mathbf{y}$ and hence $P_{\mathcal{M}}$ is linear. Now from (c) we can write $\| P_{\mathcal{M}}\mathbf{x} \| \leq \| \mathbf{x} \|$, for all $\mathbf{x} \in \mathcal{H}$, which means $P_{\mathcal{M}}$ is bounded and its bound is one. Parts (e) and (f) are easy consequences of definition of projection. For (g), picking any two vectors $\mathbf{x}, \mathbf{y} \in \mathcal{H}$ we can write

$$\begin{aligned}\langle P_{\mathcal{M}}\mathbf{x}, \mathbf{y}\rangle &= \langle P_{\mathcal{M}}\mathbf{x}, P_{\mathcal{M}}\mathbf{y} + (1 - P_{\mathcal{M}})\mathbf{y}\rangle = \langle P_{\mathcal{M}}\mathbf{y}, P_{\mathcal{M}}\mathbf{y}\rangle \\ &= \langle P_{\mathcal{M}}\mathbf{x} + (1 - P_{\mathcal{M}})\mathbf{x}, P_{\mathcal{M}}\mathbf{y}\rangle = \langle \mathbf{x}, P_{\mathcal{M}}\mathbf{y}\rangle.\end{aligned}$$

This means $P_{\mathcal{M}}$ is self adjoint. Applying (f) and (g) we can write $\langle P_{\mathcal{M}}\mathbf{x}, \mathbf{x}\rangle = \langle P^2_{\mathcal{M}}\mathbf{x}, \mathbf{x}\rangle = \langle P_{\mathcal{M}}\mathbf{x}, P_{\mathcal{M}}\mathbf{x}\rangle = \| P_{\mathcal{M}}\mathbf{x} \|^2 \geq 0$, which completes the proof of (h). Proof of (i) is left to the reader as problem 3.7 at the end of this chapter. ∎

Proposition 3.7 *If P is a self-adjoint idempotent operator on a Hilbert space $\mathcal{H}$, then it is a projection on some closed subspace $\mathcal{M}$ of $\mathcal{H}$ (i.e., $P = P_{\mathcal{M}}$, for some $\mathcal{M}$).*

Proof. Let $\mathcal{M} = \{\mathbf{x} \in \mathcal{H} : P\mathbf{x} = \mathbf{x}\}$. It is easy to check that $\mathcal{M}$ is a linear manifold of $\mathcal{H}$. To show $\mathcal{M}$ is closed let $\mathbf{x}_n$ be a sequence in $\mathcal{H}$ that converges to some $\mathbf{x} \in \mathcal{H}$. So we have $P\mathbf{x}_n = \mathbf{x}_n$ for all $n \in \mathbb{N}$. Taking the limit of both sides and using the continuity of P, we get $P\mathbf{x} = \mathbf{x}$, which shows that $\mathbf{x}$ is in $\mathcal{M}$ and hence $\mathcal{M}$ is closed. Take any $\mathbf{x} \in \mathcal{H}$: first, $P\mathbf{x} \in \mathcal{M}$, because $P\mathbf{x} = P^2\mathbf{x} = P(P\mathbf{x})$ and second of all $\mathbf{x} - P\mathbf{x} \perp \mathcal{M}$, because for any $\mathbf{y} \in \mathcal{M}$ we have $\langle \mathbf{x} - P\mathbf{x}, \mathbf{y}\rangle = \langle \mathbf{x} - P\mathbf{x}, P\mathbf{y}\rangle = \langle (I - P)\mathbf{x}, P\mathbf{y}\rangle = \langle P(I - P)\mathbf{x}, \mathbf{y}\rangle = \langle (P - P^2)\mathbf{x}, \mathbf{y}\rangle = \langle 0, \mathbf{y}\rangle = 0$. So for any $\mathbf{x} \in \mathcal{H}$, $P\mathbf{x} = \widehat{\mathbf{x}} = P_{\mathcal{M}}\mathbf{x}$, by definitions of $\widehat{\mathbf{x}}$ and $P_{\mathcal{M}}$. ∎

Projections have the following additional properties, the proofs of which can be found in [2, Chapter III].

Proposition 3.8 *If $\mathcal{M}_j, j = 1, 2, \ldots, n$ are subspaces of $\mathcal{H}$, then:*

(a) *The composition $P_{\mathcal{M}}P_{\mathcal{N}}$ is a projection if and only if $P_{\mathcal{M}_1}P_{\mathcal{M}_2} = P_{\mathcal{M}_2}P_{\mathcal{M}_1}$. If this is the case, then $P_{\mathcal{M}_1}P_{\mathcal{M}_2} = P_{\mathcal{M}_1 \cap \mathcal{M}_2}$.*

(b) *The operator*

$$Q = P_{\mathcal{M}_1} + P_{\mathcal{M}_2} + \ldots + P_{\mathcal{M}_n}$$

is a projection if and only if $P_{\mathcal{M}_j}P_{\mathcal{M}_k} = 0$, whenever $j \neq k$ and if this is the case then $Q = P_{\mathcal{M}}$, where $\mathcal{M} = \mathcal{M}_1 \oplus \mathcal{M}_2 \oplus \cdots \oplus \mathcal{M}_n$.

(c) *The operator $P_{\mathcal{M}_1} - P_{\mathcal{M}_2}$ is a projection if and only if $\mathcal{M}_2 \subset \mathcal{M}_1$ and if this the case then $P_{\mathcal{M}_1} - P_{\mathcal{M}_2} = P_{\mathcal{M}}$, where*

$$\mathcal{M} = \mathcal{M}_1 \ominus \mathcal{M}_2 = \{\mathbf{x} \in \mathcal{M}_1 : \mathbf{x} \perp \mathcal{M}_2\}$$

.

Proposition 3.9 *Suppose $(P_{\mathcal{M}_n}), n \in N$, is a nondecreasing sequence of projections; that is, we have $P_{\mathcal{M}_n} \leq P_{\mathcal{M}_{n+1}}$ for all n. Then $P = \lim_{n\to\infty} P_{\mathcal{M}_n}$ exists and is a projection. The limit here is in the strong sense,*

$$\lim_{k\to\infty} \| P\mathbf{x} - P_{\mathcal{M}_k}\mathbf{x} \| = 0, \text{ for all } \mathbf{x} \in \mathcal{H}.$$

The next proposition gives some properties of orthogonal complements.

Proposition 3.10 *Let $\mathcal{M}$ be a subspace and L be a subset of $\mathcal{H}$, then*

(a) $L \cap L^{\perp} = 0$;

(b) $\overline{sp}(L) = (L^{\perp})^{\perp}$;

(c) $L \subseteq (L^{\perp})^{\perp}$ *and* $\mathcal{M} = (\mathcal{M}^{\perp})^{\perp}$;

(d) $[\overline{sp}(\cup_n \mathcal{M}_n)]^{\perp} = \cap_n \mathcal{M}_n^{\perp}$.

Proof. For (a), suppose $\mathbf{x} \in L \cap L^{\perp}$, then we must have $\mathbf{x} \perp \mathbf{x}$ and $\|\mathbf{x}\|^2 = \langle \mathbf{x}, \mathbf{x} \rangle = 0$, which implies $\mathbf{x} = 0$. For (b), let $\mathbf{x} \in L$, then $\mathbf{x} \perp \mathbf{y}$ for every $\mathbf{y} \in L^{\perp}$. This means $\mathbf{x} \perp L^{\perp}$ or $\mathbf{x} \in (L^{\perp})^{\perp}$. So $L \subseteq (L^{\perp})^{\perp}$. Now since $(L^{\perp})^{\perp}$ is a subspace we get the inclusion $\overline{\text{sp}}(L) \subseteq (L^{\perp})^{\perp}$. To show the other inclusion, suppose there is a vector $\mathbf{x} \in (L^{\perp})^{\perp}$ not in $\overline{\text{sp}}(L)$ and let $\widehat{\mathbf{x}}$ be the projection of $\mathbf{x}$ on $\overline{\text{sp}}(L)$. Then $\mathbf{x} - \widehat{\mathbf{x}}$ is in both $(L^{\perp})^{\perp}$ and $L^{\perp}$, so $\|\mathbf{x}\|^2 = \langle \mathbf{x}, \mathbf{x} \rangle = 0$, which is a contradiction. Part (c) is an immediate consequence of (b). Proof of (d) is left to the reader as Problem 3.8 at the end of this chapter. ∎

Definition 3.17 *For a finite or countable family $(\mathcal{H}_n)$ of Hilbert spaces, their direct sum $\mathcal{H} = \bigoplus \sum_n \mathcal{H}_n$ is defined to be the collection of all sequences $\mathbf{X} = (\mathbf{x}_n)$, where each $\mathbf{x}_n \in \mathcal{H}_n$ and $\sum \|\mathbf{x}_n\|^2 < \infty$. Addition and scalar multiplication on $\mathcal{H}$ are defined coordinate-wise and inner product is defined by*

$$\langle \mathbf{X}, \mathbf{Y} \rangle = \sum_n \langle \mathbf{x}_n, \mathbf{x}_n \rangle.$$

One can easily check that $\mathcal{H} = \bigoplus \sum_n \mathcal{H}_n$ together with this inner product becomes a Hilbert space. Sometimes an arbitrary $\mathbf{x} \in \mathcal{H}$ is written as $\mathbf{X} = \bigoplus \sum_n \mathbf{x}_n$. If these $\mathcal{H}_n$ are mutually orthogonal subspaces of some Hilbert space $\mathcal{H}$, then $\mathcal{H} = \bigoplus \sum_n \mathcal{H}_n$ is called the orthogonal direct sum of the $\{\mathcal{H}_n\}$. If we have finitely many $\mathcal{H}_n, n = 1, 2, \ldots, k$, then $\mathcal{H} = \bigoplus \sum_1^k \mathcal{H}_n$ can be identified with the space of all vectors $\mathbf{X} = [\mathbf{x}^1, \mathbf{x}^2, \ldots, \mathbf{x}^k]$ with $\mathbf{x}^i \in \mathcal{H}_i$.

Proposition 3.11 *Suppose $\{\mathcal{H}_n; n = 1, 2, \ldots, k\}$ is a collection of mutually orthogonal subspaces of a Hilbert $\mathcal{H}$ such that $\cap_{n=1}^k (\mathcal{H}_n)^{\perp} = \mathbf{0}$. Then $\mathcal{H}$ is isomorphic to the orthogonal direct sum $\bigoplus \sum_{n=1}^k \mathcal{H}_n$.*

The straightforward proof is omitted.

Remark. If $\mathcal{M}$ is a subspace of $\mathcal{H}$, then by the preceding Proposition, $\mathcal{H}$ turns out to be isomorphic to the orthogonal sum of $\mathcal{M} \bigoplus \mathcal{M}^{\perp}$ but we usually say $\mathcal{H}$ is the same as the orthogonal sum of $\mathcal{M} \bigoplus \mathcal{M}^{\perp}$ and write $\mathcal{H} = \mathcal{M} \bigoplus \mathcal{M}^{\perp}$. If this is the case, any $\mathbf{x} \in \mathcal{H}$ can be expressed as the sum

$$\mathbf{x} = \mathbf{y} + \mathbf{z}$$

of some $\mathbf{y} \in \mathcal{M}$ and some $\mathbf{z} \in \mathcal{M}^{\perp}$.

Definition 3.18 *A set $E \subset \mathcal{H}$ is called normalized if $\|\mathbf{x}\| = 1$ for all $\mathbf{x} \in E$. If $\langle \mathbf{x}, \mathbf{y} \rangle = 0$ for any distinct pair $\mathbf{x}, \mathbf{y} \in E$, then E is called an orthogonal set. A set E that is both normalized and orthogonal is called orthonormal. A complete orthonormal set in $\mathcal{H}$ is a dense orthonormal subset E of $\mathcal{H}$.*

Gram–Schmidt Orthogonalization Process. From any linearly independent sequence $(\mathbf{x}_n) \subset \mathcal{H}$, using the Gram–Schmidt orthogonalization process, one can construct an orthonormal set $\{\mathbf{e}_n\}$ that is equivalent to it, namely,

$$\text{sp}\{\mathbf{e}_1, \mathbf{e}_2, \ldots, \mathbf{e}_k\} = \text{sp}\{\mathbf{x}_1, \mathbf{x}_2, \ldots, \mathbf{x}_k\}, \text{for any } k \geq 1.$$

The process is based on induction and starts with setting

$$\mathbf{e}_1 = \frac{\mathbf{x}_1}{\| \mathbf{x}_1 \|}.$$

After $\mathbf{e}_1, \mathbf{e}_2, \ldots, \mathbf{e}_k$ have been constructed, $\mathbf{e}_{k+1}$ is taken to be

$$\mathbf{e}_{k+1} = \frac{\mathbf{c}_{k+1}}{\| \mathbf{c}_{k+1} \|},$$

with $\mathbf{c}_{k+1} = \mathbf{x}_{k+1} - \sum_{n=1}^{k} \langle \mathbf{x}_{k+1} - \mathbf{e}_n \rangle \mathbf{e}_n$.

For proofs and a more complete discussion of the Gram–Schmidt process, see [2, Sections 8 and 9].

Remark. The original sequence need not be linearly independent because we can always extract a linearly independent subsequence from one that is not.

Definition 3.19 *A Hilbert space $\mathcal{H}$ is called separable if it has a countable dense subset E.*

One can easily prove that Gram–Schmidt orthogonalization process gives the following important result (see Problem 3.9 at the end of this Chapter).

Proposition 3.12 *Every separable Hilbert space has a complete orthonormal sequence.*

Proposition 3.13 (Fourier Series) *If* $\mathbf{E} = \{\mathbf{e}_n : n \in \mathbb{N}\}$ *is a complete orthonormal set in* $\mathcal{H}$*, then any vector* $\mathbf{x} \in \mathcal{H}$ *can be expressed as*

$$\mathbf{x} = \sum_{n=1}^{\infty} \langle \mathbf{x}, \mathbf{e}_n \rangle \mathbf{e}_n.$$

One can then check that a countable orthonormal set $\mathbf{E} = \{\mathbf{e}_n : n \in \mathbb{N}\} \subset \mathcal{H}$ is complete if and only if

$$\| \mathbf{x} \|^2 = \sum_{n=1}^{\infty} |\langle \mathbf{x}, \mathbf{e}_n \rangle|^2, \text{ for all } \mathbf{x} \in \mathcal{H}.$$

That is why a complete orthonormal set in $\mathcal{H}$ is also called an orthonormal basis for $\mathcal{H}$.

3.6 SPECTRAL THEORY OF UNITARY OPERATORS

One of the classical results of operator theory is the spectral theorem for normal operators. In this section we state this result for the case of unitary operators so we can use it in subsequent chapters. For more detail on material presented here one can refer to [2, Sections 62 and 63].

3.6.1 Spectral Measures

Here we first introduce the notion of spectral measures and then briefly discuss their properties. Throughout this section we work with a measure space $(X, \mathcal{F})$ consisting of a set X and a σ-algebra $\mathcal{F}$ of its subsets.

Definition 3.20 *A function* E *defined on* σ*-algebra* $\mathcal{F}$ *of subsets of* X *whose values are orthogonal projections in a Hilbert space* $\mathcal{H}$ *is called a spectral measure if* $E(X) = I$ *and for any sequence* (M_n) *of disjoint subsets in* $\mathcal{F}$

$$E(\cup_{n=1}^{\infty} M_n) = \sum_{n=1}^{\infty} E(M_n).$$

For an elementary and yet interesting example of spectral measures see Problem 3.11 at the end of this chapter.

The following proposition lists several useful properties of spectral measures. For proofs the reader can refer to [87, Section 36].

Proposition 3.14 *If* E *is a spectral measure on* $\mathcal{F}$ *then* $E(\emptyset) = \mathbf{0}$ *and it is*

(a) finitely additive: *for any finite disjoint family of sets* M_n *in* $\mathcal{F}$*,*

$$E(\cup_{n=1} M_n) = \sum_{n=1} E(M_n);$$

(b) modular: *for any two sets* M, N *in* $\mathcal{F}$,

$$E(M \cup N) = E(M) + E(N) - E(M \cap N);$$

(c) multiplicative: *for any two sets* M, N *in* $\mathcal{F}$,

$$E(M \cap N) = E(M)E(N);$$

(d) subtractive: *for any two sets* M, N *in* $\mathcal{F}$ *with* $M \subseteq N$,

$$E(N - M) \leq E(N) - E(M);$$

(e) monotone: *for any two sets* M, N *in* $\mathcal{F}$ *with* $M \subseteq N$,

$$E(M) \leq E(N);$$

(f) commutative: *for any two sets* M, N *in* $\mathcal{F}$

$$E(M)E(N) = E(N)E(M);$$

(g) orthogonally scattered: *for any two disjoint sets* M, N *in* $\mathcal{F}$,

$$E(M) \perp E(N).$$

The following theorem, proved in [87, Section 36], reveals a very close and useful tie that exists between spectral measures and scalar measures.

Theorem 3.4 *Let* $(X, \mathcal{F})$ *be a measure space. A projection valued set function* E *on* $\mathcal{F}$ *is a spectral measure if and only if* $E(X) = I$ *and for each pair of vectors* $\mathbf{x}$ *and* $\mathbf{y}$ *in* $\mathcal{H}$ *the scalar valued set function* $\mu_{\mathbf{x},\mathbf{y}}(M) = \langle E(M)\mathbf{x}, \mathbf{y}\rangle$ *is a countably additive measure.*

3.6.2 Spectral Integrals

Our presentation here is in a manner found in [87] and [187]. Let f be a bounded measurable function on $(X, \mathcal{F})$ and E be a spectral measure on $\mathcal{F}$. For each pair of vectors $\mathbf{x}$ and $\mathbf{y}$ in $\mathcal{H}$, the familiar integral $\int f(\lambda) d\mu_{x,y}$ of f with respect to scalar valued measure $\mu_{x,y}(M) = \langle E(M)\mathbf{x}, \mathbf{y}\rangle$ can be formed. It is easy to see that φ defined by $\varphi(\mathbf{x}, \mathbf{y}) = \int f(\lambda) d\mu_{x,y}$ is a bilinear functional. We claim this is also bounded. In fact, for any vector $\mathbf{x} \in \mathcal{H}$,

$$\begin{aligned}
|\varphi(\mathbf{x}, \mathbf{x})| &\leq \int |f(\lambda)|\, |\langle E(d\lambda)\mathbf{x}, \mathbf{x}\rangle| \\
&\leq \int |f(\lambda)|\, |\langle E(d\lambda)\mathbf{x}, E(d\lambda)\mathbf{x}\rangle| \\
&\leq |f|_{\sup} \|E(X)\mathbf{x}\|^2 \\
&\leq |f|_{\sup} \|\mathbf{x}\|^2,
\end{aligned}$$

which in conjunction with the parallelogram law (part (e) of Proposition 3.1) gives

$$|\varphi(\mathbf{x},\mathbf{y})| \le 2|f|_{\sup}\|\mathbf{x}\|\,\|\mathbf{y}\|.$$

Therefore by Theorem 3.2 there is a unique operator $A(f)$ on $\mathcal{H}$ such that

$$\langle A(f)\mathbf{x},\mathbf{y}\rangle = \varphi(\mathbf{x},\mathbf{y}) = \int f(\lambda)\langle E(d\lambda)\mathbf{x},\mathbf{y}\rangle.$$

The dependence of $A(f)$ on f and E will be denoted by

$$\int f(\lambda)E(d\lambda) = A(f),$$

and this defines the spectral integral $\int f(\lambda)E(d\lambda)$ as $A(f)$. The following properties of the spectral integral we just introduced are proved in [87, Section 37] and [187, Section 1.4].

Theorem 3.5 *If E is a spectral measure on $(X,\mathcal{F})$, f and g are bounded complex valued measurable functions on X, and α is a complex number, then*

(a) $\int(\alpha f(\lambda))E(d\lambda) = \alpha\int f(\lambda)E(d\lambda)$;

(b) $\int(f(\lambda)+g(\lambda))E(d\lambda) = \int f(\lambda)E(d\lambda) + \int g(\lambda)E(d\lambda)$;

(c) $\int \overline{f(\lambda)}E(d\lambda) = [\int f(\lambda)E(d\lambda)]^*$;

(d) $\int f(\lambda)g(\lambda)E(d\lambda) = \int f(\lambda)E(d\lambda)\int g(\lambda)E(d\lambda)$.

The dependence of $A(f)$ and $\int f(\lambda)E(d\lambda)$ mentioned above is often expressed in the more application oriented form of

$$\int f(\lambda)E(d\lambda)\mathbf{x} = A(f)\mathbf{x}, \text{ for every } \mathbf{x}\in\mathcal{H}. \tag{3.2}$$

The integral $\int f(\lambda)E(d\lambda)\mathbf{x}$ here can also be defined in the customary measure theory way. First, we define the integral of a simple function $f = \sum_{j=1}^{n}\alpha_j 1_{\Delta_j}$ to be

$$\int f(\lambda)E(d\lambda)\mathbf{x} = \sum_{j=1}^{n}\alpha_j E(\Delta_j)\mathbf{x}.$$

Second, using the usual techniques of scalar measures we check that this integral, defined for simple functions, is well defined, linear, and additive. Third, we show that for any simple function f

$$\left\|\int f(\lambda)E(d\lambda)\mathbf{x}\right\| = \|f\|_{L^2(\mu_x)}, \tag{3.3}$$

where $\mu_x(\Delta) = \mu_{x,x}(\Delta) = \langle E(\Delta)\mathbf{x}, \mathbf{x}\rangle$. In fact, if we express the simple function f in the form $f = \sum_{j=1}^n \alpha_j 1_{\Delta_j}$, with mutually disjoint sets Δ_j, then we can write

$$\begin{aligned}
\Big\| \int f(\lambda)E(d\lambda)\mathbf{x}\Big\|^2 &= \Big\|\sum_{j=1}^n \alpha_j E(\Delta_j)\mathbf{x}\Big\|^2 \\
&= \sum_{j=1}^n |\alpha_j|^2 \|E(\Delta_j)\mathbf{x}\|^2 \\
&= \sum_{j=1}^n |\alpha_j|^2 \langle E(\Delta_j)\mathbf{x}, \mathbf{x}\rangle \\
&= \int |f|^2 d\mu_x,
\end{aligned}$$

where the second equality follows from the orthogonality property of spectral measures stated in part (g) of Proposition 3.14.

Fourth, take an arbitrary complex valued function f in $L^2(\mu_x)$ and let (f_n) be any sequence of simple functions converging to f in $L^2(\mu_x)$-norm. This implies that the sequence (f_n) is Cauchy $\in L^2(\mu_x)$. Therefore by (3.3) the sequence $\int f_n(\lambda)E(d\lambda)\mathbf{x}$ must be Cauchy in $\mathcal{H}$ and thus has a limit $B_\mathbf{x}$ in $\mathcal{H}$. We define

$$\int f(\lambda)E(d\lambda)\mathbf{x} = B_\mathbf{x}(f). \tag{3.4}$$

Finally, we check that this integral retains linearity and additivity as well as (3.3).

Now we have two seemingly different definitions for the spectral integral $\int f(\lambda)E(d\lambda)\mathbf{x}$: namely, $A(f)\mathbf{x}$ given through (3.2) and $B_\mathbf{x}(f)$ given through (3.4). We claim these two are the same, that is, $A(f)\mathbf{x} = B_\mathbf{x}$, and hence there is no confusion. To verify our claim for a simple $f = \sum_{j=1}^n \alpha_j 1_{\Delta_j}$ and any $\mathbf{y} \in \mathcal{H}$ we can write

$$\begin{aligned}
\langle B_\mathbf{x}(f), \mathbf{y}\rangle &= \Big\langle \int f(\lambda)E(d\lambda)\mathbf{x}, \mathbf{y}\Big\rangle \\
&= \Big\langle \sum_{j=1}^n \alpha_j E\Delta_j \mathbf{x}, \mathbf{y}\Big\rangle \\
&= \sum_{j=1}^n \alpha_j \langle E\Delta_j \mathbf{x}, \mathbf{y}\rangle \\
&= \int f(\lambda)\langle E(d\lambda)\mathbf{x}, \mathbf{y}\rangle \\
&= \langle A(f)\mathbf{x}, \mathbf{y}\rangle,
\end{aligned}$$

which in virtue of the uniqueness part of Riesz Theorem 3.1 completes the proof. The argument for a general function f is the standard limiting argument of passing from the simple functions case to general ones.

3.6.3 Spectral Theorems

In the last two subsections we introduced the notion of spectral measure and then, starting with a given spectral measure on a measure space $(X, \mathcal{F})$ with values in a Hilbert space $\mathcal{H}$, we proceed to define the spectral integral $\int f(\lambda)E(d\lambda)$ as an operator $A(f)$ such that

$$\int f(\lambda)d\langle E(\lambda)\mathbf{x}, \mathbf{y}\rangle = \langle A(f)\mathbf{x}, \mathbf{y}\rangle, \quad \text{for any pair } \mathbf{x}, \mathbf{y} \in \mathcal{H}.$$

However, in most applications we have an operator $A : \mathcal{H} \to \mathcal{H}$ and we would like to represent it as a spectral integral. The question is: Given an operator $A : \mathcal{H} \to \mathcal{H}$, does there exist a spectral measure E on some measure space $(X, \mathcal{F})$ and a function f such that $A = \int f(\lambda)E(d\lambda)$? The answer for a large class of operators, known as normal operators, is affirmative. This important result, which is called *the spectral theorem for normal operators*, says that every bounded normal operator A has a spectral measure E on Borel subsets of complex plane $\mathbb{C}$ for which $A = \int_{\mathbb{C}} \lambda E(d\lambda)$. This result can be found,for example, in [87], [187, Section 1.4] and [2, Section 62].

In this book we are interested in the special case of this result for unitary operators: If U is a unitary operator on a Hilbert space $\mathcal{H}$, then there exists a unique spectral measure E on Borel subsets of the unit circle T such that

$$U = \int_T \lambda E(d\lambda).$$

If we identify T with $[0, 2\pi)$ in the usual way one can state this theorem as follows.

Theorem 3.6 (Spectral Theorem for Unitary operators) *For any unitary operator U on a Hilbert space $\mathcal{H}$ there exists a unique spectral measure E on the Borel subsets of $[0, 2\pi)$ such that*

$$U = \int_0^{2\pi} e^{i\lambda} E(d\lambda). \tag{3.5}$$

Finally this last theorem and with part (c) and (d) of Theorem 3.5 establish the following result.

Theorem 3.7 *For any unitary operator U on a Hilbert space $\mathcal{H}$ there exists a unique spectral measure E on Borel subsets of $[0, 2\pi)$ such that*

$$U^t = \int_0^{2\pi} e^{it\lambda} E(d\lambda), \quad \textit{for any integer } t. \tag{3.6}$$

PROBLEMS AND SUPPLEMENTS

3.1 Show that for any two complex numbers α and β,

$$|\alpha + \beta|^2 \leq 2|\alpha|^2 + 2|\beta|^2 .$$

3.2 Show that the Hilbert space l^2 as in Section 3.3 is separable.

3.3 Show that any separable Hilbert space $\mathcal{H}$ is isomorphic to l^2. Hint: Take a countable dense set $\{\mathbf{x}_n\}$ in $\mathcal{H}$, denote its Gram–Schmidt orthogonalized sequence by $\mathbf{e}_n$, and then show that the mapping $T : \mathcal{H} \to l^2$ defined by $T\mathbf{x} = (\langle \mathbf{x}, \mathbf{e}_n \rangle)$ is an isomorphism.

3.4 Let U be a unitary operator on some Hilbert space $\mathcal{H}$.

(a) Show that for any two orthogonal subspaces $\mathcal{M}$ and $\mathcal{N}$ of $\mathcal{H}$ we have

$$U(\mathcal{M} \oplus \mathcal{N}) = U\mathcal{M} \oplus U\mathcal{N}.$$

(b) Show that if $\mathcal{M}$ and $\mathcal{N}$ are subspaces of $\mathcal{H}$ then

$$U(\mathcal{M} \ominus \mathcal{N}) = U\mathcal{M} \ominus U\mathcal{N}.$$

3.5 Suppose $\mathcal{M}$ is a subspace of a Hilbert space $\mathcal{H}$ and U is a unitary operator on $\mathcal{H}$. Show that $UP_{\mathcal{M}} = P_{\mathcal{M}}U$ if and only if $U\mathcal{M} = \mathcal{M}$.

3.6 Show that the "subspace" assumption in Theorem 3.3 is essential.

3.7 Let $\mathcal{M}$ and $\mathcal{N}$ be two subspaces of a Hilbert space $\mathcal{H}$ and show that $\mathcal{M} \subset \mathcal{N}$ if and only if $P_{\mathcal{M}} P_{\mathcal{N}} = P_{\mathcal{M}}$.

3.8 Let $(\mathcal{M}_n)$ be a sequence of subspaces of a Hilbert space $\mathcal{H}$ and show that

$$[\overline{\text{sp}}(\cup_n \mathcal{M}_n)]^{\perp} = \cap_n \mathcal{M}_n^{\perp}.$$

3.9 Show that any separable Hilbert space has a complete orthonormal sequence. Hint: Start with a countable dense subset $\{\mathbf{x}_n : n \in \mathbb{N}\}$ of $\mathcal{H}$ and consider its corresponding Gram–Schmidt orthogonalized set.

3.10 Consider a Hilbert space $\mathcal{H}$ and its subspace $\mathcal{M}$ spanned by an orthonormal set $\{\mathbf{e}_1, \mathbf{e}_2, \ldots, \mathbf{e}_n\}$. Show that the projection of a vector $\mathbf{x} \in \mathcal{H}$ on $\mathcal{M}$ is given by

$$P_{\mathcal{M}}\mathbf{x} = \sum_{i=1}^{n} \langle \mathbf{x}, \mathbf{e}_i \rangle \mathbf{e}_i .$$

3.11 Define $E : \mathcal{F} \to L^2(X, \mathcal{F}, \mu)$ by taking $E(A) : L^2(X, \mathcal{F}, \mu) \to L^2(X, \mathcal{F}, \mu)$ for each set $A \in \mathcal{F}$, to be the multiplication by the characteristic function of A. That is,

$$E(A)g = g1_A \text{ or } (E(A)g)(\lambda) = g(\lambda)1_A(\lambda).$$

Show that E is a spectral measure (see Definition 3.20).

CHAPTER 4

STATIONARY RANDOM SEQUENCES

Here we review the pertinent facts needed about stationary random sequences, both univariate and multivariate. Half the chapter is devoted to univariate and the other half to the multivariate case. For a good introduction to stationary univariate sequences, see Pourahmadi [183] and Brockwell and Davis [28]; for multivariate stationary sequences, see Rozanov [201], Wiener and Masani [224], [225] and Masani [152].

We recall from Chapter 1 that (weak) stationarity is defined by the conditions $EX_t = m$ (constant) for all t and $R(s,t) = \text{Cov}\,(X_s, X_t) = R(s-t)$ for all $s, t \in \mathbb{Z}$. Here $R(\tau) = \text{Cov}\,(X_\tau, X_0)$ is called the *covariance function* of X_t. Throughout this chapter we assume that all our random variables have zero mean and finite variance and we equip the set of all these random variables with the inner product

$$\langle X, Y \rangle = \text{Cov}\,(X, Y).$$

One of the most important properties of the covariance function of a stationary random process is its nonnegative definiteness. We omit its easy proof.

Periodically Correlated Random Sequences:Spectral Theory and Practice. By H.L. Hurd and A.G. Miamee

Definition 4.1 *A complex valued function $K(\tau)$ defined on integers is said to be* nonnegative definite *if $K(-\tau) = \overline{K(\tau)}$ for all $\tau \in \mathbb{Z}$ and*

$$\sum_{i,j=1}^{n} a_i \overline{a_j} K(t_i - t_j) \geq 0$$

for any positive integer n, scalars $a_1, a_2, \ldots, a_n$ and integers $t_1, t_2, \ldots, t_n$.

Its importance stems from Herglotz' theorem(see [28]), which says any nonnegative definite function has a spectral representation.

Theorem 4.1 (Herglotz) *A complex valued function $K(\cdot)$ defined on integers is nonnegative definite if and only if there exists a bounded, nondecreasing and left continuous function F_λ on $[0, 2\pi)$, which vanishes at 0, and for which*

$$K(\tau) = \int_0^{2\pi} e^{i\lambda\tau} dF_\lambda, \quad \text{for any } \tau \in \mathbb{Z}.$$

The function F_λ is called the spectral distribution function function of X_t.

Some authors develop the spectral representation of stationary random processes starting with this theorem. However, because we wish to emphasize the role of unitary operators in the spectral theory for periodically correlated processes, we will start with their role in the spectral theory of univariate stationary sequences.

4.1 UNIVARIATE SPECTRAL THEORY

In this section the spectral theorem for unitary operators is used to obtain spectral representations of stationary sequences and their covariance functions. Then the isomorphism between the time and spectral domain is studied and its importance in prediction theory of stationary random sequences is discussed.

4.1.1 Unitary Shift

Although much of the spectral theory of stationary sequences can be derived, and historically was derived, without the explicit use of the unitary shift, we believe it is the most fundamental idea present. Hence we will use it here as the basis for discussing the spectral theory of stationary sequences; then later we will use it in discussing the spectral theory for periodically correlated random sequences.

For any second order random sequences X_t, there is a naturally defined smallest subspace on which we can focus our attention. This subspace, called *time domain* of X_t, is defined next.

Definition 4.2 *The time domain $\mathcal{H}_X$ of a second order process X_t is the closed subspace spanned by all the vectors X_t with $t \in \mathbb{Z}$,*

$$\mathcal{H}_X = \overline{\text{sp}}\{X_t : t \in \mathbb{Z}\}, \tag{4.1}$$

where the closure is in the mean-square sense.

In the following we take $m = E\{X_t\}$ to be zero but will remark later on the consequence of assuming $m \neq 0$. We now show that the condition of stationarity is equivalent to the existence of a unitary shift operator.

Proposition 4.1 *A second order stochastic sequence X_t is stationary if and only if there exists a unitary operator U defined on $\mathcal{H}_X$ for which*

$$X_{t+1} = UX_t \tag{4.2}$$

for every $t \in \mathbb{Z}$.

Proof. If there exists a unitary operator U for which (4.2) holds, then it is easy to see

$$\langle X_t, X_s \rangle = \langle UX_t, UX_s \rangle = \langle X_{t+1}, X_{s+1} \rangle,$$

which means X_t is stationary.

Conversely, suppose X_t is stationary; then we define $U : L_X = \text{sp}\{X_t : t \in \mathbb{Z}\} \to L_X$ by $U\mathbf{z} = \sum_{j=1}^n a_j X_{t_j+1}$, for any $\mathbf{z} = \sum_{j=1}^n a_j X_{t_j}$ in L_X. We will show that U is well defined, linear, and onto. To show it is well defined, suppose a vector $\mathbf{z}$ in L_X has two different expressions, which, without loss of generality, can be taken to be $\mathbf{z} = \sum_{j=1}^n a_j X_{t_j}$ and $\mathbf{z} = \sum_{j=1}^n b_j X_{t_j}$. Then we can write

$$\begin{aligned}
\left\| \sum_{j=1}^n a_j X_{t_j+1} - \sum_{j=1}^n b_j X_{t_j+1} \right\|^2 &= \left\| \sum_{j=1}^n (a_j - b_j) X_{t_j+1} \right\|^2 \\
&= \sum_{j,k=1}^n (a_j - b_j)\overline{(a_k - b_k)} \langle X_{t_j+1}, X_{t_k+1} \rangle \\
&= \sum_{j,k=1}^n (a_j - b_j)\overline{(a_k - b_k)} \langle X_{t_j}, X_{t_k} \rangle \\
&= \left\| \sum_{j=1}^n a_j X_{t_j} - \sum_{j=1}^n b_j X_{t_j} \right\|^2 = 0,
\end{aligned}$$

which means $\sum_{j=1}^n a_j X_{t_j+1} = \sum_{j=1}^n b_j X_{t_j+1}$. For linearity, suppose $\mathbf{z} = \sum_{j=1}^n a_j X_{t_j}$ and $\mathbf{z}' = \sum_{j=1}^n b_j X_{t_j}$ are two members of $\in L_X$; then for any scalar α we have

$$\begin{aligned}
U(\alpha \mathbf{z} + \mathbf{z}') &= U\Big(\alpha \sum_{j=1}^{n} a_j X_{t_j} + \sum_{j=1}^{n} b_j X_{t_j}\Big) \\
&= U\Big(\sum_{j=1}^{n} (\alpha a_j + b_j) X_{t_j}\Big) \\
&= \sum_{j=1}^{n} (\alpha a_j + b_j) X_{t_{j+1}} \\
&= \alpha \sum_{j=1}^{n} a_j X_{t_{j+1}} + \sum_{j=1}^{n} b_j X_{t_{j+1}} = \alpha U\mathbf{z} + U\mathbf{z}'.
\end{aligned}$$

To show U is onto, pick any $\mathbf{z} = \sum_{j=1}^{n} a_j X_{t_j} \in L^X$, and then we have $U(\sum_{j=1}^{n} a_j X_{t_j - 1}) = \sum_{j=1}^{n} a_j X_{t_j} = \mathbf{z}$. Finally, U is an isometry because for any two vectors $\mathbf{z} = \sum_{j=1}^{n} a_j X_{t_j}$ and $\mathbf{z}' = \sum_{j=1}^{n} b_j X_{t_j}$ in L^X we can write

$$\langle \mathbf{z}, \mathbf{z}' \rangle = \sum_{j,k=1}^{n} a_j \overline{b_k} \langle X_{t_j}, X_{t_k} \rangle = \sum_{j,k=1}^{n} a_j \overline{b_k} \langle X_{t_j+1}, X_{t_k+1} \rangle = \langle U\mathbf{z}, U\mathbf{z}' \rangle,$$

which shows U preserves inner product as well as norm. Therefore U can be extended by continuity (see [2]) to a unitary operator U on $\mathcal{H}_X = \overline{L_X}$ and (4.2) holds for U. ∎

Now we address the issue of $m = EX \neq 0$. Since $\mathcal{H}_X \subset L^2$ and the constant random variable is also in L^2, then

$$m = \int X_t(\omega) P(d\omega) = \langle X_t, 1 \rangle.$$

Thus a constant mean for a process says that every X_t has the same projection onto 1. But it is not necessarily true that $1 \in \mathcal{H}_X$. If $1 \in \mathcal{H}$ then $U1 = 1$, meaning 1 is an eigenvector of U with eigenvalue of 1. See the supplements for a sketch. The fuss over the case of nonzero mean can be avoided by always including the vector 1 in forming L_X and H_X; that is, define

$$L_X = \text{sp}\{1, X_t : t \in \mathbb{Z}\}\}$$

and then $\mathcal{H}_X = \overline{L_X}$.

4.1.2 Spectral Representation

Let X_t be a stationary sequence with unitary shift U as defined in Section 4.1.1. Due to the Spectral Theorem for Unitary Operators (Theorem 3.7) there exists a spectral measure E on Borel subsets of $[0, 2\pi)$ for which

$$U^t = \int_0^{2\pi} e^{it\lambda} E(d\lambda). \tag{4.3}$$

Therefore we can write

$$X_t = U^t X_0 = \int_0^{2\pi} e^{it\lambda} E(d\lambda) X_0 = \int_0^{2\pi} e^{it\lambda} \xi(d\lambda),$$

where the set function $\xi(E) = E(A)X_0$, for A any Borel subset of $[0, 2\pi)$, turns out to be a countably additive vector measure called the *random spectral measure* or *random measure* of X_t. Furthermore, $\xi(\cdot)$ is orthogonally scattered in the sense that $\langle \xi(A), \xi(B) \rangle = 0$, whenever $A \cap B = \emptyset$. These properties are inherited from the corresponding properties of the spectral measure E explained in Section 3.6.

We have just proved the necessity part of the following spectral representation theorem for stationary random processes.

Theorem 4.2 *For a second order sequence X_t to be stationary it is necessary and sufficient that there exists a countably additive orthogonally scattered measure $\xi(\cdot)$ on the Borel subsets of $[0, 2\pi)$ such that*

$$X_t = \int_0^{2\pi} e^{it\lambda} \xi(d\lambda). \tag{4.4}$$

Proof. To prove sufficiency suppose X_t has representation (4.4). Formally, we then have

$$\begin{aligned} \langle X_{t+\tau}, X_t \rangle &= \langle \int_0^{2\pi} e^{i(t+\tau)\lambda} \xi(d\lambda), \int_0^{2\pi} e^{it\lambda} \xi(d\lambda) \rangle \\ &= \int_0^{2\pi} \int_0^{2\pi} e^{i(t+\tau)\lambda} e^{-it\theta} \langle \xi(d\lambda), \xi(d\theta) \rangle \\ &= \int_0^{2\pi} e^{i\tau\lambda} F(d\lambda) = R(\tau), \end{aligned} \tag{4.5}$$

which shows X_t is stationary. Here F is the scalar valued *spectral measure* of X_t defined on Borel subsets of $[0, 2\pi)$ by

$$F(\Delta_1 \cap \Delta_2) = \langle \xi(\Delta_1), \xi(\Delta_2) \rangle = ||\xi(\Delta_1 \cap \Delta_2)||^2.$$

This can be made precise by interpreting (4.3) and hence (4.4) in the Riemann–Stieltjes sense. Or it can be argued using Proposition 5.8. ∎

Corresponding to the Riemann-Stieltjes interpretation of (4.3) and (4.4), the expression (4.5) can also be expressed as a Riemann–Stieltjes integral,

$$R(\tau) = \int_0^{2\pi} e^{i\tau\lambda} d\tilde{F}(\lambda) \tag{4.6}$$

with respect to a distribution function $\tilde{F}$ defined on $[0, 2\pi)$ by

$$\tilde{F}(\lambda) = \begin{cases} 0 & \text{if } \lambda = 0 \\ F([0, \lambda)) & \text{if } 0 < \lambda < 2\pi. \end{cases}$$

The function $\tilde{F}$, which is called the *spectral distribution function* of X_t, is nondecreasing and left continuous.

Henceforth we will use the notation F for both the spectral measure and the spectral distribution function of X_t. The integration with respect to F, the measure, is in the Lebesgue sense and is identified by using $F(d\lambda)$ while the integration with respect to F, the distribution, is in the Riemann–Stieltjes sense and is identified by using $dF(\lambda)$.

Nonzero Mean. Only a slight modification must be made if X_t has a nonzero mean m. Writing $X_t = X'_t + m$, we see that X'_t has zero mean. This leads us to see that $\xi(\cdot)$ must have an atom of weight m at $\lambda = 0$, or $\xi(A) = \xi'(A) + m$ whenever $\{0\} \in A$ and otherwise $\xi(A) = \xi'(A)$, where $\xi(\cdot)$ is the random measure associated with X'_t. This leads to $\langle X_{t+\tau}, X_t \rangle = \langle X'_{t+\tau}, X'_t \rangle + |m|^2$ and $F_X(A) = F_{X'}(A) + |m|^2$ whenever $\{0\} \in A$ and otherwise $F_X(A) = F_{X'}(A)$. We emphasize that the $\xi(\cdot)$ measure of a single point is a random variable, and a constant is a special random variable. So it is possible for $\xi'(\{0\})$ to be a nonzero random variable, of mean zero but positive variance, and still X'_t would have zero mean. This comment is very closely connected to the following topic.

4.1.3 Mean Ergodic Theorem

For stationary sequences, the mean ergodic theorem addresses mean square convergence, that is, convergence in $L^2(\Omega, \mathcal{F}, P)$, of

$$S_N = \frac{1}{2N+1} \sum_{t=-N}^{N} X_t. \tag{4.7}$$

More generally, we can address the convergence of

$$S_N(\lambda) = \frac{1}{2N+1} \sum_{t=-N}^{N} X_t e^{-i\lambda t} \tag{4.8}$$

from the spectral representation (4.4).

Proposition 4.2 *If X_t is a stationary sequence, then in the L^2 sense*

$$\lim_{N\to\infty} S_N(\lambda) = \xi(\{\lambda\}). \tag{4.9}$$

Proof. This follows immediately from

$$\begin{aligned} S_N(\lambda) &= \frac{1}{2N+1}\sum_{t=-N}^{N}\int_0^{2\pi} e^{i(\gamma-\lambda)t}\xi(d\gamma) \\ &= \int_0^{2\pi} d_N(\gamma-\lambda)\xi(d\gamma), \end{aligned}$$

where the Dirichlet kernel $d_N(x)$ is bounded, $|d_N(x)| \leq 1$, and is continuous for every N and converges to $1_{\{0\}}(x)$. $S_N(\lambda) \xrightarrow{L^2} \xi(\{\lambda\})$ follows from

$$\langle S_N(\lambda), \xi(\lambda)\rangle = \int_0^{2\pi} d_N(\gamma-\lambda)F(d\gamma\cap\{\lambda\}) \longrightarrow F(\{\lambda\})$$

$$\begin{aligned} \| S_N(\lambda) \|^2 &= \frac{1}{(2N+1)^2}\sum_{s=-N}^{N}\sum_{t=-N}^{N} R(s-t)e^{-i\lambda(s-t)} \\ &= \int_0^{2\pi} |d_N(\gamma-\lambda)|^2 F(d\gamma) \longrightarrow F(\{\lambda\}). \qquad \blacksquare \end{aligned}$$

Definition 4.3 *Any stationary sequence for which*

$$\lim_{N\to\infty} S_N(0) = m, \tag{4.10}$$

in mean square or $L^2(\Omega, \mathcal{F}, P)$ sense, is called mean ergodic.

Thus from Proposition 4.2 and the discussion in the preceding section, we can see that $S_N(0) \to m$ if and only if the atom of $\xi(\cdot)$ at $\{0\}$ is the constant random variable m. This also means that $F(\{0\}) = |m|^2$, for if $F(\{0\}) > |m|^2$, then $\xi(\{0\}) = X + m$, where X has mean zero but positive variance. In other words, the sequence is mean ergodic ($S_N(0) \to m$) if and only if $F(0) = |m|^2$, meaning the atom at $\lambda = 0$ is only large enough to account for the mean; there is no random component having positive variance at $\lambda = 0$.

4.1.4 Spectral Domain

Let X_t be a stationary sequence with spectral measure F and let $L^2(F)$ denote the set of all complex valued Borel measurable functions on $[0, 2\pi)$ that are square integrable with respect to the measure F. That is,

$$L^2(F) = \Big\{ f : [0, 2\pi) \to \mathbb{C} \text{ with } \int_0^{2\pi} | f(\lambda) |^2 \, F(d\lambda) < \infty \Big\}. \tag{4.11}$$

$L^2(F)$ equipped with the inner product

$$\langle f, g \rangle = \int_0^{2\pi} f(\lambda)\overline{g(\lambda)}F(d\lambda) \tag{4.12}$$

becomes a Hilbert space, which is called the *spectral domain* of X_t. The spectral representation of Theorem 4.2 establishes a bridge between the time and spectral domains of X_t. To see this, from the spectral representation of X_t it is natural to consider the transformation V, which maps each finite linear combination $\sum c_j X_{t_j}$ in L^X to $\sum c_j e^{i\lambda t_j}$ in its spectral domain $L^2(F)$,

$$V\Big(\sum_{j\in J} c_j X_{t_j}\Big) = \sum_{j \in J} c_j e^{i\lambda t_j}. \tag{4.13}$$

It is easy to see from

$$\begin{aligned}
\Big\langle V\Big(\sum_{j\in J} c_j X_{t_j}\Big), V\Big(\sum_{k\in K} d_k X_{t_k}\Big)\Big\rangle &= \Big\langle \sum_{j\in J} c_j e^{i\lambda t_j}, \sum_{k\in K} d_k e^{i\lambda t_k} \Big\rangle \\
&= \int_0^{2\pi} \sum_{j\in J}\sum_{k\in K} c_j \overline{d_k} e^{i\lambda(t_j - t_k)} F(d\lambda) \\
&= \sum_{j\in J}\sum_{k\in K} c_j \overline{d_k} \int_0^{2\pi} e^{i\lambda(t_j - t_k)} F(d\lambda) \\
&= \sum_{j\in J}\sum_{k\in K} c_j \overline{d_k} R(t_i, t_j) \\
&= \Big\langle \sum_{j\in J} c_j X_{t_j}, \sum_{k\in K} d_k X_{t_k} \Big\rangle
\end{aligned}$$

that V is a well defined isometry. Hence V can be extended to an isometric isomorphism, called the *Kolmogorov isomorphism*, from $\mathcal{H}_X$ onto $L^2(F)$. This Kolmogorov isomorphism, as we see later in this chapter, allows us to transfer questions regarding a stationary sequence from the time domain to corresponding questions in the spectral domain. Then using Fourier analysis, we analyze the question in the spectral domain and then transfer our findings back to the time domain.

4.2 UNIVARIATE PREDICTION THEORY

A predictor[1] $\widetilde{X}_t$ for X_t, based on some subset $S = \{X_{j_1}, X_{j_2}, \dots\}$, is a random variable that is close to X_t in some acceptable sense. Typically, we make the error, $X_t - \widetilde{X}_t$, as small as possible in the chosen sense. For example, we can choose $\widetilde{X}_t$ so that either $P(|X_t - \widetilde{X}_t| \geq \epsilon)$ or $E|X_t - \widetilde{X}_t|^2$ is smallest. Here we will address only *linear least-squares prediction*, so that $\widetilde{X}_t$ is a linear function of the elements of S that minimizes $E|X_t - \widetilde{X}_t|^2$. The problem is solved by the projection of X_t onto $\mathcal{M}(S)$, the Hilbert subspace spanned by the elements of S. Here we examine two cases that have conceptual and practical significance: first, when S is the infinite set $\{X_s : s \leq t\}$ and we wish to predict the element X_{t+1}, and second, when S is a finite contiguous sequence of n elements $\{X_{t-n+1}, X_{t-n+2}, \dots, X_t\}$ and we wish to predict X_{t-n} (backward) and X_{t+1} (forward). There are other cases of theoretical and practical interest, but these two permit us to discuss the main ideas and provide the basis for extending it to the case of PC random sequences.

We do not pursue finding the complete solution to the case when S is the infinite past, as to do so would take us a little too far from our main direction. However, we will give appropriate references for interested readers.

4.2.1 Infinite Past, Regularity and Singularity

We begin with some subspaces of the time domain $\mathcal{H}_X$ that are important for discussing prediction on an infinite past. In the following discussion, the closure is with respect to the inner product (4.1) defined on $\mathcal{H}_X$.

Definition 4.4 *Let X_t be any second order random sequence. Its (linear) past up to and including time t is defined to be the subspace*

$$\mathcal{H}(t) = \overline{\text{sp}}\{X_s : s \leq t\} \tag{4.14}$$

generated by all the vectors X_s with $s \leq t$ and its remote past is defined to be the subspace

$$\mathcal{H}(-\infty) = \bigcap_{t \in \mathbb{Z}} \mathcal{H}(t). \tag{4.15}$$

When the context requires it, we will add additional notation $\mathcal{H}_X$, $\mathcal{H}_X(t)$, and $\mathcal{H}_X(-\infty)$, to signify the random sequence X_t in question.

We note that $\mathcal{H}(t)$ is nondecreasing with respect to t, that is, $\mathcal{H}(t) \supseteq \mathcal{H}(s)$ for $t \geq s$, and for this reason we can also express

$$\mathcal{H}(-\infty) = \bigcap_{t<0} \mathcal{H}(t) = \bigcap_{t_k<0} \mathcal{H}(t_k),$$

[1]In some disciplines, $\widetilde{X}_t$ is called the *predictand* and S the predictors.

where $\{t_k\}$ is any sequence of integers that converges to $-\infty$.

Definition 4.5 *A stationary random sequence is called* purely nondeterministic *or* regular *if*

$$\mathcal{H}(-\infty) = \{\mathbf{0}\}$$

and is called deterministic *or* singular *if*

$$\mathcal{H}(-\infty) = \mathcal{H},$$

or equivalently,

$$\mathcal{H}(s) = \mathcal{H}(t); \quad \textit{for all} \quad s, t \in \mathbb{Z}.$$

When we are working with a second order random sequence X_t, it is natural to take predictor $\widetilde{X}_{t+1}$ of X_{t+1} based on the past of the process up to and including time t to be that random variable in $\mathcal{H}(t)$ which generates the least error. It is natural to work with the linear past as the set of acceptable predictors and with mean-square error as the criterion for goodness because the solution to the prediction problem is then given by the Projection Theorem 3.3: that is, $\widetilde{X}_{t+1}$ is simply the projection $(X_{t+1} \mid \mathcal{H}(t))$ of X_{t+1} on $\mathcal{H}(t)$. The Projection Theorem also ensures that the predictor $\widetilde{X}_{t+1}$ (as a vector) is unique, although its implementation may not be unique, especially in nonstationary situations to be addressed later. Another important reason for considering the linear least-square predictor is the fact that in the very important Gaussian case, the nonlinear and linear predictors become identical. See [183] for a good introductory discussion of nonlinear prediction.

4.2.2 Wold Decomposition

Some important results in prediction on the infinite past arise from the relationship between the propagating unitary operator U defined in (4.2) and the subspaces we have just defined. At this time we give the following relationships.

Lemma 4.1 *If X_t is second order stationary with unitary shift U, then*

(a) $\mathcal{H}(t+1) = U\mathcal{H}(t)$;

(b) $\mathcal{H} = U\mathcal{H}$;

(c) $\mathcal{H}(-\infty) = U\mathcal{H}(-\infty)$.

Proof. For (a), recall that for any mapping $A : \mathcal{H} \mapsto \mathcal{H}$ and subset M of $\subset \mathcal{H}$, $AM = \{\mathbf{y} \in \mathcal{H} : \mathbf{y} = A\mathbf{x}, \mathbf{x} \in M\}$. Thus taking M to be $L(t) = \text{sp}\{X_s : s \leq t\}$ we have $L(t+1) = UL(t)$; for if $\mathbf{z} \in L(t)$, then $\mathbf{z} = \sum_{j=1}^{n} \alpha_j X_{t_j} : t_j \leq t$ so that $U\mathbf{z} = \sum_{j=1}^{n} \alpha_j X_{t_j+1} \in L(t+1)$, which means $UL(t) \subset L(t+1)$. Now let $\mathbf{z} \in U\mathcal{H}(t)$, so $\mathbf{z} = U\mathbf{w}$ for some $\mathbf{w} \in \mathcal{H}(t)$. Then $\mathbf{w} = \lim \mathbf{w}_n$ for $\mathbf{w}_n \in L(t)$. By continuity of U we can write $\mathbf{z} = U\mathbf{w} = U(\lim \mathbf{w}_n) = \lim U\mathbf{w}_n$. But since $U\mathbf{w}_n \in L(t+1)$ for all n, we conclude that $\mathbf{z} \in \overline{L(t+1)} = \mathcal{H}(t+1)$ and so $U\mathcal{H}(t) \subseteq \mathcal{H}(t+1)$. A similar argument using the continuity of U^{-1} produces $\mathcal{H}(t+1) \subseteq U\mathcal{H}(t)$. The proof of (b) is essentially the same. For (c), first suppose $\mathbf{z} \in \mathcal{H}(-\infty)$, which means $\mathbf{z} \in \mathcal{H}(t)$ for all t. But then $\mathbf{z} \in \mathcal{H}(t+1) = U\mathcal{H}(t)$, so for each t there is some $\mathbf{y}_t \in \mathcal{H}(t)$ with $\mathbf{z} = U\mathbf{y}_t$. But since U is one-to-one, $\mathbf{y}_t = \mathbf{y}$ for some $\mathbf{y}$ and all $t \in \mathbb{Z}$. It is clear that $\mathbf{y} \in \mathcal{H}(-\infty)$ and $\mathbf{z} = U\mathbf{y}$, which implies $\mathbf{z} \in U\mathcal{H}(-\infty)$. ∎

Theorem 4.3 (Wold Decomposition Theorem) *Any second order stationary sequence X_t has a unique decomposition*

$$X_t = Y_t + Z_t, \tag{4.16}$$

in terms of two orthogonal stationary sequences Y_t and Z_t such that

(a) $\mathcal{H}_X(t) = \mathcal{H}_Y(t) \oplus \mathcal{H}_Z(t)$;

(b) $\mathcal{H}_X(t) = \mathcal{H}_Y(-\infty) \oplus \mathcal{H}_Z(t)$;

(c) Y_t *is deterministic and* Z_t *is purely nondeterministic;*

(d) $U_Y = U_X|_{\mathcal{H}(-\infty)}$; *and* $U_Z = U_X|_{\mathcal{H}(-\infty)^\perp}$.

Proof. For each $t \in \mathbb{Z}$ set

$$Y_t = (X_t|\mathcal{H}(-\infty)) \text{ and } Z_t = X_t - Y_t. \tag{4.17}$$

On one hand, for each s, $Y_s \in \mathcal{H}_X(-\infty)$, and on the other hand, $Z_t = X_t - Y_t = X_t - (X_t|\mathcal{H}_X(-\infty)) \perp \mathcal{H}_X(-\infty)$ for each t. Hence $Y_s \perp Z_t$ for all $s, t \in \mathbb{Z}$, which means $\mathcal{H}_Y \perp \mathcal{H}_Z$. For each $t \in \mathbb{Z}, Y_t \in \mathcal{H}_X(-\infty) \subseteq \mathcal{H}_X(t)$ and hence $Z_t = X_t - Y_t \in \mathcal{H}_X(t)$ and therefore $\mathcal{H}_Y(t) \oplus \mathcal{H}_Z(t) \subseteq \mathcal{H}_X(t)$. In order to complete the proof of (a), it suffices to show that $\mathcal{H}_X(t) \subseteq \mathcal{H}_Y(t) \oplus \mathcal{H}_Z(t)$. To do this, pick some $\mathbf{w} \in \mathcal{H}_X(t)$; then $\mathbf{w} = \lim \mathbf{w}_n$, where each vector $\mathbf{w}_n \in \mathcal{M}_X(t)$ and hence $\mathbf{w}_n = \mathbf{u}_n + \mathbf{v}_n$, $\mathbf{u}_n \in L^Y(t)$, $\mathbf{v}_n \in L^Z(t)$. But since $\mathbf{w}_n$ must be Cauchy and $\|\mathbf{u}_n - \mathbf{u}_m\| \leq \|\mathbf{w}_n - \mathbf{w}_m\|$, and similarly for $\mathbf{v}_n$, then $\mathbf{u}_n \to \mathbf{u} \in \mathcal{H}_Y(t)$, $\mathbf{v}_n \to \mathbf{v} \in \mathcal{H}_Z(t)$. Taking the limit of $\mathbf{w}_n = \mathbf{u}_n + \mathbf{v}_n$, we get $\mathbf{w} = \mathbf{u} + \mathbf{v}$, which shows $\mathcal{H}_X(t) \subset \mathcal{H}_Y(t) \oplus \mathcal{H}_Z(t)$. This completes the proof of (a). For (b), it is sufficient to show

$$\mathcal{H}_X(-\infty) = \mathcal{H}_Y(t), t \in \mathbb{Z}. \tag{4.18}$$

From the definition of Y_t it is clear that $\mathcal{H}_Y(t) \subset \mathcal{H}_X(-\infty)$. If this subset was proper for some t, then there must exist a nonzero vector $\mathbf{u} \in \mathcal{H}_X(-\infty) \ominus$

$\mathcal{H}_Y(t)$. In that case, for each $s \le t$, we have $\mathbf{u} \perp Y_s$ and $\mathbf{u} \perp Z_s$ so that $\mathbf{u} \perp X_s = Y_s + Z_s$ and hence $\mathbf{u} \perp \mathcal{H}_X(t)$, and therefore $\mathbf{u} \perp \mathcal{H}_X(-\infty)$. This implies $\mathbf{u} = \mathbf{0}$, which is a contradiction. Singularity of Y_t in (c) is an immediate consequence of (4.18). Taking the intersection, over all integers t, of both sides of the equality in item (b) implies $\mathcal{H}_X(-\infty) = \mathcal{H}_Y(-\infty) \oplus \mathcal{H}_Z(-\infty)$. From this and (4.18) one gets $\mathcal{H}_Z(-\infty) = 0$, which means Z_t is regular. For (d), using the definition (4.17) of Y_t yields

$$\begin{aligned} Y_{t+1} &= (X_{t+1}|\mathcal{H}_X(-\infty)) = (U_X(X_t)|\mathcal{H}_{\mathcal{X}}(-\infty)) \\ &= U_X(X_t|\mathcal{H}_X(-\infty)) = U_X\, Y_t \end{aligned}$$

for any $t \in \mathbb{Z}$. This and Proposition 4.1 imply that Y_t is stationary. This also shows that U_X and U_Y both act on $\mathcal{H}_Y$, which together with the fact that $\mathcal{H}_X(-\infty) = \mathcal{H}_Y$, proved above, gives $U_Y = U_X|_{\mathcal{H}_X(-\infty)}$. The statements about Z_t follow easily. ∎

4.2.3 Innovation Subspaces

Since innovation refers to something entirely new, it is both natural and important to consider the innovation subspace $\mathcal{I}_X(t)$ at each time t. We start with defining several notions that we will need later on in this chapter.

Definition 4.6 (Innovations and their subspaces)

(a) *An uncorrelated sequence ε_t of random variables with mean m and variance σ^2 is called a white noise and is denoted by $WN(m, \sigma^2)$. If $m = 0$ and $\sigma^2 = 1$, then the sequence is called a normalized white noise.*

(b) *A sequence $\mathbf{e}_t$ of vectors in a Hilbert space is called orthogonal if $\langle \mathbf{e}_t, \mathbf{e}_s \rangle = 0$ whenever $s \ne t$ and it is called an orthonormal process*orthonormal *if $\langle \mathbf{e}_t, \mathbf{e}_s \rangle = \delta_{s-t}$.*

(c) *The innovation of a second order process X_t at time t is defined to be $\zeta_t = X_t - P_{\mathcal{H}(t-1)} X_t$.*

(d) *The innovation space of a second order sequence X_t at time t is defined to be*

$$\begin{aligned} \mathcal{I}_X(t) &= \mathcal{H}_X(t) \ominus \mathcal{H}_X(t-1) \\ &= \{x \in \mathcal{H}_X(t) : x \perp \mathcal{H}_X(t-1)\}. \end{aligned} \tag{4.19}$$

The preceding expression can be written as

$$\mathcal{H}_X(t) = \mathcal{I}_X(t) \oplus \mathcal{H}_X(t-1)$$

which when iterated seems to suggest that we can express the entire history of X_t as

$$\mathcal{H}_X(t) = \oplus \sum_{j=0}^{\infty} \mathcal{I}_X(t-j).$$

But this is not quite correct because some nonzero vectors $x \in \mathcal{H}_X(-\infty)$ do not necessarily belong to $\oplus \sum_{j \leq t} \mathcal{I}_X(j)$. This suggests the following modification:

$$\mathcal{H}_X(t) = \mathcal{H}_X(-\infty) \oplus \sum_{j=0}^{\infty} \mathcal{I}_X(t-j),$$

which turns out to be true. We will now make this a bit more precise while presenting some basic facts about $\mathcal{I}_X(t)$ that are important for understanding its role in prediction.

Lemma 4.2 *If X_t is a second order stationary sequence with unitary shift U, then for every integer t,*

(a) $\mathcal{I}_X(t+1) = U\mathcal{I}_X(t)$;

(b) $\mathcal{I}_X(t) \perp \mathcal{H}_X(-\infty,),\ \mathcal{I}_X(t) = \mathcal{I}_Z(t)$;

(c) $\mathcal{H}_X(t) = \mathcal{H}_X(-\infty) \oplus \sum_{j=0}^{\infty} \mathcal{I}_X(t-j)$;

(d) $\dim \mathcal{I}_X(t) = \begin{cases} 0 & \textit{if } X_t \textit{ is singular} \\ 1 & \textit{otherwise} \end{cases}$.

Proof. Part (a) is clear from

$$\begin{aligned} \mathcal{I}_X(t+1) &= \mathcal{H}_X(t+1) \ominus \mathcal{H}_X(t) \\ &= U\mathcal{H}_X(t) \ominus U\mathcal{H}_X(t-1) \\ &= U[\mathcal{H}_X(t) \ominus \mathcal{H}_X(t-1)] = U\mathcal{I}_X(t) \end{aligned}$$

where the first equality in the last row follows from Problem 3.4 at the end of Chapter 3. The first part of (b) is obvious. For the second part, applying part (a) of the Wold Decomposition Theorem 4.3 followed by its part (c), we get

$$\begin{aligned} \mathcal{I}_X(t) &= \mathcal{H}_X(t) \ominus \mathcal{H}_X(t-1) \\ &= [\mathcal{H}_Y(t) \oplus cH_Z(t)] \ominus [\mathcal{H}_Y(t-1) \oplus \mathcal{H}_Z(t-1)] \\ &= [\mathcal{H}_Y(t) \ominus \mathcal{H}_Y(t-1)] \oplus [\mathcal{H}_Z(t) \ominus \mathcal{H}_Z(t-1)] \\ &= \{0\} \oplus [\mathcal{H}_Z(t) \ominus \mathcal{H}_Z(t-1)] = \mathcal{I}_Z(t). \end{aligned}$$

Proof of (c) is left to reader as an exercise. For (d), if X_t is singular, then $X_t \in \mathcal{H}_X(t-1)$ and hence $\dim \mathcal{I}_X(t) \equiv 0$. Now if X_t is not singular, then the regular component Z_t of the Wold decomposition (4.16) must have positive length, that is, $\| Z_t \| > 0$ for all $t \in \mathbb{Z}$. Furthermore, the innovation vector

$$\begin{aligned} \zeta_t &= X_t - P_{\mathcal{H}_X(t-1)} X_t \in \mathcal{I}_X(t) \\ &= Z_t - P_{\mathcal{H}_Z(t-1)} Z_t \in \mathcal{I}_Z(t) \end{aligned} \tag{4.20}$$

cannot be null. Thus $\mathcal{I}_X(t) = \mathcal{I}_Z(t)$ contains at least one nonzero vector. That means $\dim \mathcal{I}_X(t) \geq 1$. Suppose $\dim \mathcal{I}_X(t) > 1$; then there should be a vector $\mathbf{w} \in \mathcal{I}_X(t)$ that is orthogonal to ζ_t. Hence $\mathbf{w} \perp X_t - P_{\mathcal{H}_X(t-1)} X_t$ and also $\mathbf{w} \perp \mathcal{H}_X(t-1)$, which leads to $\mathbf{w} \perp H_X(t)$, a contradiction. Note that it is also clear from part (a) that $d_X(t)$ is constant with respect to t. ∎

It is clear from the preceding that $\mathcal{I}_X(t) = \mathrm{sp}\{\zeta_t\}$ and we denote the prediction error variance as $\sigma_X^2(t) = \mathrm{Var}(\zeta_t)$. Note that $\sigma_X(t) > 0$ if and only if X_t has a nontrivial regular part. Since the least square predictor $\widetilde{X}_t$ of X_t based on its past $\mathcal{H}_X(t-1)$ is defined to be the orthogonal projection of X_t on $\mathcal{H}_X(t-1)$. The following additional facts about innovations are true in a trivial way (i.e., with $\sigma_\zeta(t) = 0$) if X_t is deterministic or singular. However, the more interesting situation for prediction purposes is when the sequence is nondeterministic, which is equivalent to $\sigma_X(t) \neq 0$.

Lemma 4.3 *Let ζ_t be the innovation sequence of a stationary sequence X_t. Then*

(a) *ζ_t is stationary and has the same shift as X_t and hence X_t and ζ_t are jointly stationary and $\sigma_X(t) \equiv \sigma$;*

(b) *ζ_t is an orthogonal sequence with constant variance, that is,*

$$\langle \zeta_s, \zeta_t \rangle = \sigma^2 \delta_{s-t}, \text{ for any } s, t \in \mathbb{Z};$$

(c) *for any integer t and any positive integer k,*

$$\langle \zeta_t, X_{t-k} \rangle = 0 \text{ and } \langle \zeta_t, X_t \rangle = \sigma^2,$$

which in particular shows that any future innovation is orthogonal to the past of the sequence up to that point.

Proof. For (a), let U denote the shift operator of the stationary sequence X_t. Since U commutes with the projection onto $\mathcal{H}(t)$, for every t

$$\begin{aligned} \zeta_{t+1} &= X_{t+1} - P_{\mathcal{H}(t)} X_{t+1} = U X_t - P_{\mathcal{H}(t)} U X_t \\ &= U\Big(X_t - P_{\mathcal{H}(t-1)} X_t\Big) = U \zeta_t, \end{aligned}$$

showing that U serves as a unitary shift for ζ_t and therefore ζ_t is stationary with variance σ_X^2. For (b), if $s < t$, then $\zeta_s \in \mathcal{H}_X(s)$ and $\zeta_t = X_t - \widetilde{X}_t \perp \mathcal{H}_X(s)$. Therefore $\zeta_s \perp \zeta_t$. For (c), note that for any $k \geq 1$, $X_{t-k} \in \mathcal{H}_X(t-1)$ and $\zeta_t = X_t - \widetilde{X}_t$ is orthogonal to $\mathcal{H}_X(t-1)$. Therefore we have $\langle \zeta_t, X_{t-k} \rangle = 0$. Furthermore, from the definition of σ_X^2 in part (a), we have

$$\sigma^2 = \langle \zeta_t, \zeta_t \rangle = \langle \zeta_t, X_t - \widetilde{X}_t \rangle = \langle \zeta_t, X_t \rangle. \quad \blacksquare$$

Next, we study the relationship between regular stationary sequences, white noise, orthonormal sequences, and innovation sequences. The very elementary, yet important, result is that any white noise sequence is regular. We begin with a lemma.

Lemma 4.4

(a) *Any orthogonal sequence* $\mathbf{e}_t$ *is regular.*

(b) *Any white noise* ε_t *is regular.*

(c) *The innovation* ζ_t *of any stationary sequence* X_t *is regular.*

Proof. Since all white noises and all innovation sequences are orthogonal, we need only prove the first statement. Take any vector Y in $\mathcal{H}_{\mathbf{e}}(-\infty) \subset \mathcal{H}_{\mathbf{e}}$. Since $\{\mathbf{e}_t\}$ forms an orthonormal basis for $\mathcal{H}_{\mathbf{e}}$, we have the expansion $Y = \sum_{j=-\infty}^{\infty} a_t \mathbf{e}_t$, with $a_t = \langle Y, \mathbf{e}_t \rangle$. On the other hand, for each integer t, the expression

$$Y \in \mathcal{H}_{\mathbf{e}}(-\infty) \subset \mathcal{H}_{\mathbf{e}}(t-1) \subset \mathcal{H}_{\mathbf{e}}(t),$$

in conjunction with $\mathbf{e}_t \perp \mathcal{H}_{\mathbf{e}}(t-1)$, implies that $a_t = \langle Y, \mathbf{e}_t \rangle = 0$, for each integer t. Therefore $Y = 0$. $\blacksquare$

Using this lemma one can characterize regular sequences.

Proposition 4.3 (Moving Average Representation) *A second order random sequence* X_t *is stationary and regular if and only if it has a one sided moving average expansion*

$$X_t = \sum_{j \geq 0} a_j \mathbf{e}_{t-j}, \ t \in \mathbb{Z}, \quad \textit{with} \ \sum_{j \geq 0} |a_j|^2 < \infty, \tag{4.21}$$

with respect to an orthonormal sequence $\mathbf{e}_t$.

Proof. If X_t satisfies (4.21) then it is clear that X_t is an L^2 random variable for every t. Taking $t \geq s$ and observing that

$$\begin{aligned} R(t,s) &= \langle X_t, X_s \rangle = \sum_{j \geq 0} \sum_{k \geq 0} a_j \overline{a_k} \langle \mathbf{e}_{t-j}, \mathbf{e}_{s-k} \rangle \\ &= \sum_{k \geq 0} a_{k+t-s} \overline{a_k} = \sum_{k \geq 0} a_{k+t+1-s-1} \overline{a_k} \\ &= R(t+1, s+1) \end{aligned} \tag{4.22}$$

shows that X_t is stationary. Since clearly

$$\mathcal{H}_X(t) \subset \mathcal{H}_{\mathbf{e}}(t) = \overline{\mathrm{sp}}\{\mathbf{e}_s : s \leq t\},$$

we can write

$$\bigcap_t \mathcal{H}_X(t) \subset \bigcap_t \mathcal{H}_{\mathbf{e}}(t) = \{0\},$$

showing that X_t is regular.

Conversely, suppose X_t is a regular stationary sequence. The innovation spaces $\mathcal{I}_X(t), t \in \mathbb{Z}$ defined in (4.19) are one dimensional and hence $\mathcal{I}_X(t) = \mathrm{sp}\{\zeta_t\}$, where $\zeta_t = X_t - P_{\mathcal{H}(t-1)} X_t$ is the innovation of X_t at time t. Since X_t is regular, part (c) of Lemma 4.2 reduces to $\mathcal{H}_X(t) = \oplus \sum_{j=0}^{\infty} \mathcal{I}_X(t-j)$. Therefore any vector $\mathbf{Y} \in \mathcal{H}_X(t)$ can be represented as

$$\mathbf{Y} = \sum_{j \geq 0} a_j(\mathbf{Y}) \zeta_{t-j}.$$

In particular, since $X_t \in \mathcal{H}_X(t)$, it must have a representation

$$X_t = \sum_{j \geq 0} a_j(t) \zeta_{t-j}, \text{ with } \sum_{j \geq 0} |a_j(t)|^2 < \infty.$$

Now on the one hand,

$$X_{t+1} = \sum_{j \geq 0} a_j(t+1) \zeta_{t+1-j}$$

and on the other hand,

$$X_{t+1} = U X_t = \sum_{j \geq 0} a_j(t) U \zeta_{t-j} = \sum_{j \geq 0} a_j(t) \zeta_{t+1-j}.$$

In the last equation, U moved inside the summation because of mean-square convergence of the partial sums and continuity of U. In virtue of uniqueness of such representations for each j, $a_j(t)$ must be constant in t. That is, $a_j(t) = a_j$ and we get the one sided moving average representation

$$X_t = \sum_{j \geq 0} a_j \zeta_{t-j}, \text{ with } \sum_{j \geq 0} |a_j|^2 < \infty. \quad \blacksquare \tag{4.23}$$

For another characterization of regular stationary processes see problem 4.9.

Since we usually observe the process X_t itself and not its innovations ζ_t, it seems pointless, from a practical viewpoint, to express the predictor $\widetilde{X}_\delta$ in terms of innovations, and that is true. However, expressing the predictor this way permits us to evaluate the error variance $\| X_\delta - \widetilde{X}_\delta \|^2$ as a function of δ. We begin with expressing the predictor for a purely nondeterministic sequence.

Proposition 4.4 *If X_t is a regular stationary sequence with one sided moving average representation (4.23) in terms of its innovations ζ_t, then the δ-step ahead predictor of X_δ based on its past, namely, $\widetilde{X}_\delta = (X_\delta | \mathcal{H}_X(0))$, is given by*

$$\widetilde{X}_\delta = \sum_{j=\delta}^{\infty} a_j \zeta_{\delta-j}, \quad \delta \geq 1, \tag{4.24}$$

and the resulting prediction error given by

$$X_\delta - \widetilde{X}_\delta = \sum_{j=0}^{\delta-1} a_j \zeta_{\delta-j}, \tag{4.25}$$

has variance

$$\| X_\delta - \widetilde{X}_\delta \|^2 = \sigma^2 \sum_{j=0}^{\delta-1} |a_j|^2. \tag{4.26}$$

Proof. It is immediate from (4.23) that $\mathcal{H}_X(0) \subset \mathcal{H}_\zeta(0)$ and it is equally immediate from $\zeta_t = X_t - P_{\mathcal{M}(t-1)}$ that $\mathcal{H}_\zeta(0) \subset \mathcal{H}_X(0)$. Thus we have $\mathcal{H}_\zeta(0) = \mathcal{H}_X(0)$. Therefore $\sum_{j=\delta}^{\infty} a_j \zeta_{\delta-j}$, which clearly belongs to $\mathcal{H}_\zeta(0)$, is in $\mathcal{H}_X(0)$. On the other hand, $\sum_{j=0}^{\delta-1} a_j \zeta_{\delta-j}$, which is clearly orthogonal to $\mathcal{H}_\zeta(0)$, is orthogonal to $\mathcal{H}_X(0)$. Therefore by the projection theorem we have

$$\widetilde{X}_\delta = (X_\delta | \mathcal{H}_X(0)) = \sum_{j=\delta}^{\infty} a_j \zeta_{\delta-j}.$$

The error formulas are now immediate. $\blacksquare$

Corollary 4.4.1 *If X_t is a stationary sequence having Wold decomposition $X_t = Y_t + Z_t$ of Proposition 4.3, with Y_t its deterministic and $Z_t = \sum_{j\geq 0} a_j \zeta_{t-j}$ its regular component, then the δ-step ahead predictor of X_δ based on its past $\ldots, X_{-1}, X_0$ is given by*

$$\widetilde{X}_\delta = \sum_{j=\delta}^{\infty} a_j \zeta_{\delta-j} + Y_\delta, \quad \delta \geq 1 \tag{4.27}$$

with (4.25) and (4.26) remaining to hold.

Proof. In this case, $\mathcal{H}_X(0) = \mathcal{H}_X(-\infty) \oplus \mathcal{H}_Z(0)$ and so $Y_\delta + \sum_{j=\delta}^{\infty} a_j \zeta_{\delta-j}$ belongs to $\mathcal{H}_X(0)$ while $\sum_{j=0}^{\delta-1} a_j \zeta_{\delta-j}$ is orthogonal to $\mathcal{H}_X(0)$. ∎

The next lemma shows that in principle the moving average coefficients $\{a_j : j \geq 0\}$ of X_t can be found in terms of its autocorrelation function $R(t)$. In what follows we use the convention $a_j = 0$, for $j < 0$.

Lemma 4.5 *If X_t has the one sided moving representation $X_t = \sum_{j=0}^{\infty} a_j \zeta_{t-j}$ in terms of its innovation sequence, then*

$$R(t) = \sigma^2 \sum_{j=0}^{\infty} a_{j+t} \overline{a_j}, \quad t \in \mathbb{Z},$$

or in matrix form

$$\mathbf{R} = \sigma^2 \mathbf{\Gamma} \mathbf{\Gamma}^*. \tag{4.28}$$

Here for each $i, j \in \mathbb{Z}$ the ijth entry of the matrices $\mathbf{R}$ and $\mathbf{\Gamma}$ are defined by $R^{ij} = R(j-i)$ and $\Gamma^{ij} = a_{j-i}$, respectively.

Thus the moving average coefficients and hence the predictor coefficients $\{a_j\}$ of a regular sequence can be obtained from triangular or Cholesky factorization (4.28) of its covariance matrix $\mathbf{R}$. See Sections 4.2.5.3 and 8.5.3.3 for further discussion, including proofs, of Cholesky factorizations of positive definite and nonnegative definite matrices.

4.2.4 Spectral Theory and Prediction

Here we discuss how, using the Kolmogorov isomorphism (4.13), one can translate some prediction problems to spectral domain problems. However, we will omit the development of the optimal predictor in the spectral domain. Although interesting, this would take us a bit too far from the most practical issues of prediction, namely, finite prediction. A complete coverage of infinite past prediction can be found in Doob [49], Wiener and Masani [224, 225], and

Masani [152, 153]. Our presentation of infinite prediction theory is the same as developed in the Wiener and Masani references.

Recall the spectral representations (4.5) and (4.6):

$$R(t) = \int_0^{2\pi} e^{it\lambda} F(d\lambda) = \int_0^{2\pi} e^{it\lambda} dF(\lambda),$$

where the spectral distribution function $F(\lambda)$ of the stationary sequence X_t was chosen to be nondecreasing and left continuous on $[0, 2\pi)$ with $F(0) = 0$. By well known properties of such functions, (e.g., see [195, Chapter 5, Theorem 2]) F has a derivative $F'(\lambda)$ (in the usual sense), for a.e. λ, which is nonnegative valued and belongs to $L^1[0, 2\pi)$, and we have

$$F(\lambda) - F(0) \geq \int_0^{\lambda} F'(\theta) d\theta, \text{ for all } \lambda \in [0, 2\pi).$$

If the equality holds for all $\lambda \in [0, 2\pi)$, we say $F(\lambda)$ is absolutely continuous and denote its derivative by $f(\lambda)$ and call it the *spectral density* of the stationary sequence X_t. If this is the case, then

$$R(t) = \langle X_t, X_0 \rangle = E\{X_t \overline{X_0}\} = \int_0^{2\pi} e^{it\lambda} f(\lambda) d\lambda,$$

which means the spectral measure $F(d\lambda)$ of X_t is also absolutely continuous (a.c.) with respect to the Lebesgue measure and its Radon–Nikodym derivative is $f(\lambda)$.

The next theorem gives sufficient conditions for a process to have a moving average representation.

Theorem 4.4

(a) *A stationary process X_t has a moving average representation*

$$X_t = \sum_{k=-\infty}^{\infty} b_k \varepsilon_{t-k} \quad \textit{with} \quad \sum_{k=-\infty}^{\infty} |b_k|^2 < \infty \textit{ and } \langle \varepsilon_t, \varepsilon_s \rangle = \delta_{s-t} \sigma^2$$

if and only if its spectral measure is absolutely continuous. If this is the case, then

$$f(\lambda) = \frac{\sigma^2}{2\pi} |\varphi(e^{-i\lambda})|^2, \textit{ with } \varphi(e^{-i\lambda}) = \sum_{k=-\infty}^{\infty} b_k e^{-ik\lambda}.$$

(b) *If the moving average is one sided, that is, $b_k = 0$ for every $k < 0$, then*

$$\varphi = \sum_{k=0}^{\infty} b_k e^{-ik\lambda}$$

and either $\varphi_+(z) = \sum_{k=0}^{\infty} b_k z^k$ *is identically zero or* $\ln f \in L^1[0, 2\pi)$ *and*

$$\ln \frac{\sigma^2 |b_0|^2}{2\pi} \leq \frac{1}{2\pi} \int_0^{2\pi} \ln f(\lambda) d\lambda.$$

Proof. Suppose X_t has the moving average in (a); then one can easily conclude that

$$R(t) = \langle X_t, X_0 \rangle = \Big\langle \sum_{k=-\infty}^{\infty} b_{t+k}\varepsilon_{-k}, \sum_{k=-\infty}^{\infty} b_k \varepsilon_{-k} \Big\rangle = \sigma^2 \sum_{k=-\infty}^{\infty} b_{t+k}\overline{b_k}.$$

On the other hand, since $\sum_{k=-\infty}^{\infty} |b_k|^2 < \infty$ the function φ is in $L^2[0, 2\pi)$ and we can write

$$\begin{aligned}
\frac{\sigma^2}{2\pi} \int_0^{2\pi} e^{it\lambda} |\varphi(e^{-i\lambda})|^2 d\lambda &= \frac{\sigma^2}{2\pi} \int_0^{2\pi} e^{it\lambda} \varphi(e^{-i\lambda}) \overline{\varphi(e^{-i\lambda})} d\lambda \\
&= \frac{\sigma^2}{2\pi} \int_0^{2\pi} e^{it\lambda} \Big(\sum_{k=-\infty}^{\infty} b_k e^{-ik\lambda} \Big) \Big(\sum_{k=-\infty}^{\infty} \overline{b}_k e^{ik\lambda} \Big) \\
&= \frac{\sigma^2}{2\pi} \int_0^{2\pi} \Big(\sum_{k=-\infty}^{\infty} b_k e^{i(t-k)\lambda} \Big) \Big(\sum_{k=-\infty}^{\infty} \overline{b}_k e^{ik\lambda} \Big) \\
&= \frac{\sigma^2}{2\pi} \int_0^{2\pi} \Big(\sum_{k=-\infty}^{\infty} b_{t+k} e^{-ik\lambda} \Big) \Big(\sum_{k=-\infty}^{\infty} \overline{b}_k e^{ik\lambda} \Big) \\
&= \sigma^2 \sum_{k=-\infty}^{\infty} b_{t+k} \overline{b_k}.
\end{aligned}$$

Therefore we arrive at

$$R(t) = \frac{\sigma^2}{2\pi} \int_0^{2\pi} e^{it\lambda} |\varphi(e^{-i\lambda})|^2 d\lambda,$$

which, by virtue of uniqueness of the Fourier transform, implies that X_t has a spectral density f, which is of the form $f(\lambda) = \frac{\sigma^2}{2\pi} |\varphi(e^{-i\lambda})|^2$.

Conversely, suppose X_t has an absolutely continuous spectral measure. We know that its spectral density $f(\lambda)$ is nonnegative for a.e. λ and it belongs to $L^1[0, 2\pi)$. Therefore it can be factored as $f(\lambda) = |\varphi(\lambda)|^2$. Clearly, $\varphi(\lambda)$ belongs to $L^2[0, 2\pi)$ and has a Fourier series $\varphi(\lambda) = \sum_{k=-\infty}^{\infty} b_k e^{-ik\lambda}$ with

$\sum_{k=-\infty}^{\infty} |b_k|^2 < \infty$. Now letting $\zeta(\Delta) = \int_\Delta \varphi^{-1}(\lambda)\xi(d\lambda)$ we can write

$$\begin{aligned} X_t &= \int_0^{2\pi} e^{it\lambda}\xi(d\lambda) = \int_0^{2\pi} e^{it\lambda}\varphi(\lambda)\varphi^{-1}(\lambda)\xi(d\lambda) \\ &= \int_0^{2\pi} e^{it\lambda}\varphi(\lambda)\zeta(d\lambda) = \int_0^{2\pi} e^{it\lambda}\Big(\sum_{k=-\infty}^{\infty} b_k e^{-ik\lambda}\Big)\zeta(d\lambda) \\ &= \sum_{k=-\infty}^{\infty} b_k \int_0^{2\pi} e^{i(t-k)\lambda}\zeta(d\lambda) = \sum_{k=-\infty}^{\infty} b_k \varepsilon_{t-k} \end{aligned}$$

where $\varepsilon_t = \int_0^{2\pi} e^{it\lambda}\zeta(d\lambda)$ is clearly a white noise.

For (b), since $\varphi \in L^2[0, 2\pi)$ it follows that $\varphi_+(z)$ belongs to the Hardy class H_2 and hence either $\varphi(e^{i\lambda}) \in L^2$ vanishes identically or $\ln \varphi(e^{i\lambda})$ is integrable on $[0, 2\pi)$ and

$$\ln|\varphi_+(0)| \leq \frac{1}{2\pi}\int_0^{2\pi} \ln|\varphi(e^{-i\lambda})|d\lambda.$$

Multiplying both sides of this by 2 and substituting b_0 for $\varphi_+(0)$,

$$\ln|b_0|^2 \leq \frac{1}{2\pi}\int_0^{2\pi} \ln|\varphi(e^{-i\lambda})|^2 d\lambda.$$

Now substituting for $|\varphi(e^{-i\lambda})|^2$ from $f(\lambda) = \frac{\sigma^2}{2\pi}|\varphi(e^{-i\lambda})|^2$, we get

$$\ln|b_0|^2 \leq \frac{1}{2\pi}\int_0^{2\pi} \ln\frac{2\pi f(\lambda)}{\sigma^2}d\lambda,$$

which leads us to

$$\ln|b_0|^2 \leq \frac{1}{2\pi}\int_0^{2\pi} [\ln f(\lambda) + \ln(2\pi) - \ln(\sigma^2)]d\lambda.$$

Transporting the last term on the right hand side to the left and combining the resulting two terms on the right hand side, we get

$$\ln\frac{\sigma^2|b_0|^2}{2\pi} \leq \frac{1}{2\pi}\int_0^{2\pi} \ln f(\lambda)d\lambda. \qquad \blacksquare$$

Combining this theorem with the Wold Decomposition Theorem we get the following important result.

Theorem 4.5 *Suppose X_t is a stationary sequence with spectral distribution function F and spectral density f. Let F_Y and F_Z denote the spectral distribution functions of singular and regular components Y_t and Z_t in its Wold decomposition (see Theorem 4.3); then*

(a) $F = F_Y + F_Z$;

(b) F_Z *is absolutely continuous and its spectral density is of the form*

$$f_Z(\lambda) = \frac{\sigma^2}{2\pi}|\varphi(e^{-i\lambda})|^2$$

for some $\varphi(e^{-i\lambda}) = 1 + \sum_{k=1}^{\infty} a_k e^{-ik\lambda}$ *with* $a_k = \langle X_0, \zeta_k\rangle$;

(c) $\ln f_Z \in L^1[0, 2\pi)$ *and*

$$\ln \frac{\sigma^2}{2\pi} \leq \frac{1}{2\pi}\int_0^{2\pi} \ln F_Z(\lambda) d\lambda.$$

Proof. Since $\{Y_t\} \perp \{Z_t\}$, we have

$$\langle X_t, X_0\rangle = \langle Y_t + Z_t, Y_0 + Z_0\rangle = \langle Y_t, Y_0\rangle + \langle Z_t, Z_0\rangle,$$

which implies

$$\int_0^{2\pi} e^{it\lambda} dF_\lambda = \int_0^{2\pi} e^{it\lambda} d(F_Y)_\lambda + \int_0^{2\pi} e^{it\lambda} d(F_Z)_\lambda.$$

This in turn implies $F = F_Y + F_Z + \textit{constant}$. However, the constant must be zero because all three spectral distribution functions F, F_Y, and F_Z vanish at 0. This proves (a). Parts (b) and (c) follow from last theorem because by the Wold Decomposition Theorem 4.3,

$$Z_t = \sum_{k=0}^{\infty} a_k \zeta_{t-k}, \quad \langle \zeta_t, \zeta_s\rangle = \delta_{s-t}\sigma^2, \quad a_0 = 1 \text{ and } \sum_{k=0}^{\infty} |a_k|^2 < \infty. \quad ∎$$

Now we can prove the following two useful theorems.

Theorem 4.6 *The stationary sequence* X_t *is nondeterministic if and only if it has a spectral distribution function* F *such that* $\ln F' \in L^1[0, 2\pi)$. *In this case, the prediction error* σ^2 *is given by*

$$\sigma^2 = 2\pi \exp\Big(\frac{1}{2\pi}\int_0^{2\pi} \ln F'(\lambda) d\lambda\Big).$$

Proof. Let X_t be a nondeterministic stationary sequence with Wold decomposition $X_t = Y_t + Z_t$. Denoting the spectral distribution function of Z_t by F_Z, it follows from part (c) of Theorem 4.5 that $\ln F_Z' \in L^1[0, 2\pi)$ and

$$\ln \frac{\sigma^2}{2\pi} \leq \frac{1}{2\pi}\int_0^{2\pi} \ln F_Z'(\lambda) d\lambda \leq \frac{1}{2\pi}\int_0^{2\pi} \ln F'(\lambda) d\lambda. \tag{4.29}$$

Since the sequence X_t is nondeterministic, σ^2 is positive and hence the last integral cannot become $-\infty$. But neither can it be $+\infty$ because, by Jensen's inequality, it is dominated by $\ln \frac{1}{2\pi}\int_0^{2\pi} F'(\lambda)d\lambda$, which is finite since F' belongs to $L^1[0,2\pi)$ and has nonnegative values. Therefore $\ln F' \in L^1[0,2\pi)$.

Next, suppose that X_t is such that $\ln F' \in L^1[0,2\pi)$ and consider the innovation $\zeta_0 = X_0 - (X_0|\mathcal{H}_X(-1))$; since the projection $(X_0|\mathcal{H}_X(-1)) \in \mathcal{H}_X(-1)$, we can write

$$\zeta_0 = \lim_{t\to 0} u_t$$

for some u_t of the form $u_t = X_0 - \sum_{k=1}^{t} a_k^t X_{-k}$. By the Kolmogorov isomorphism we can write

$$\begin{aligned}
\| u_t \|^2 &= \int_0^{2\pi} | 1 - \sum_{k=1}^{t} a_k^t e^{it\lambda} |^2 \, dF(\lambda) \\
&\geq \int_0^{2\pi} | 1 - \sum_{k=1}^{t} a_k^t e^{it\lambda} |^2 \, F'(\lambda)d\lambda.
\end{aligned}$$

Dividing both sides by 2π and then taking the natural logarithm of both sides, we can write

$$\begin{aligned}
\ln \frac{\| u_t \|^2}{2\pi} &\geq \ln\Big(\frac{1}{2\pi}\int_0^{2\pi} \Big|1 - \sum_{k=1}^{t} a_k^t e^{it\lambda}\Big|^2 F'(\lambda)d\lambda\Big) \\
&\geq \frac{1}{2\pi}\int_0^{2\pi} \ln\Big(\Big|1 - \sum_{k=1}^{t} a_k^t e^{it\lambda}\Big|^2 F'(\lambda)\Big)d\lambda \\
&\geq \frac{1}{2\pi}\int_0^{2\pi} \ln\Big|1 - \sum_{k=1}^{t} a_k^t e^{it\lambda}\Big|^2 + \frac{1}{2\pi}\int_0^{2\pi} \ln F'(\lambda)d\lambda \\
&\geq \ln(1) + \frac{1}{2\pi}\int_0^{2\pi} \ln F'(\lambda)d\lambda \\
&\geq \frac{1}{2\pi}\int_0^{2\pi} \ln F'(\lambda)d\lambda.
\end{aligned}$$

The second inequality above follows from the Jensen inequality (see [195, page 110]) while the fourth one follows from Jensen's formula (see [3, page 206]). Now taking the limit of both sides of the inequality

$$\ln \frac{\| u_t \|^2}{2\pi} \geq \frac{1}{2\pi}\int_0^{2\pi} \ln F'(\lambda)d\lambda,$$

obtained above, we get

$$\ln \frac{\sigma^2}{2\pi} = \ln \frac{\| \zeta_0 \|^2}{2\pi} \geq \int_0^{2\pi} \ln F'(\lambda) d\lambda.$$

Since by assumption the integral is finite, we conclude that $\sigma > 0$, that is, X_t is nondeterministic. The error formula is a result of this last inequality and (4.29). ∎

The following theorem reveals the close correspondence that exists between the spectral measures (or distribution functions) associated with the components of the Wold decomposition of X_t, on one hand, and the Lebesgue decomposition of F_X on the other hand. For the Lebesgue decomposition of a measure and its properties one can refer to Halmos [86, Section 32] or Royden [195, Chapter 11].

Theorem 4.7 (Concordance) *The spectral distribution functions of Wold decomposition components Z_t and Y_t of a nondeterministic sequence X_t are given by*

$$F_Z = F^{ac} \text{ and } F_Y = F^s,$$

where F^{ac}, F^s are the absolutely continuous and singular part of $F = F_X$, respectively.

Proof. Since we have already seen that the spectral distribution function of X_t is given by $F = F_Z + F_Y$ and that F_Z is absolutely continuous, it suffices to show that $F'_Y = 0$ a.e. Now from Theorem 4.5(a) we can write

$$F'(\lambda) = F'_Z(\lambda) + F'_Y(\lambda) = \sigma^2 |\varphi(e^{-i\lambda})|^2 + F'_Y(\lambda), \text{ with } \varphi \in L^2[0, 2\pi).$$

Taking the natural logarithm of both sides and then integrating results in

$$\begin{aligned}
\int_0^{2\pi} \ln F'(\lambda) d\lambda &= \int_0^{2\pi} \ln \left[F'_Z(\lambda)(1 + \frac{F'_Y(\lambda)}{\sigma^2 |\varphi(e^{-i\lambda})|^2}) \right] d\lambda \\
&= \int_0^{2\pi} \ln F'_Z(\lambda) d\lambda + \int_0^{2\pi} \ln \left[1 + \frac{F'_Y(\lambda)}{\sigma^2 |\varphi(e^{-i\lambda})|^2} \right] d\lambda \\
&\geq 2\pi \ln \frac{\sigma^2}{2\pi} + \int_0^{2\pi} \ln \left[1 + \frac{F'_Y(\lambda)}{\sigma^2 |\varphi(e^{-i\lambda})|^2} \right] d\lambda.
\end{aligned}$$

The last inequality follows from Theorem 4.5(c) and the fact that X_t is assumed to be non deterministic. But by Theorem 4.6 the integral on the left-hand side of the last inequality is $2\pi \ln(\sigma^2/2\pi)$. Hence the integral on the right-hand side of the last inequality is nonpositive but its integrand is clearly nonnegative. Therefore

$$\ln \left[1 + \frac{F'_Y(\lambda)}{\sigma^2 |\varphi(e^{-i\lambda})|^2} \right] = 0, \quad \text{for a.e. } \lambda$$

and hence

$$\frac{F'_Y(\lambda)}{|\varphi(e^{-i\lambda})|^2} = 0, \qquad \text{for} \quad \text{a.e.} \quad \lambda.$$

Therefore the numerator must vanish. That is, $F'_Y(\lambda) = 0$, for a.e. λ. ∎

The following theorem, whose proof now is straightforward (and hence omitted), gives a spectral criterion for regularity.

Theorem 4.8 *A stationary sequence X_t is regular if and only if it has an absolutely continuous spectral distribution function F with a spectral density f such that $\ln f \in L^1[0, 2\pi)$. If this is the case, then*

$$\sigma^2 = 2\pi \exp\Big(\frac{1}{2\pi}\int_0^{2\pi} \ln f(\lambda)d\lambda\Big).$$

4.2.5 Finite Past Prediction

The problem we address here is that of predicting $X(t+\delta)$, for a stationary process X_t, based on the n observations $\{X_{t-n+1}, X_{t-n+2}, \dots, X_t\}$. The best linear predictor here, of course, is the orthogonal projection of $X_{t+\delta}$ onto $\mathcal{M}(t; n) = \text{sp}\{X_s : t-n < s \le t\}$ and we will thus denote it by

$$\widehat{X}_{t+\delta,n} = (X_{t+\delta}|\mathcal{M}(t;n)). \tag{4.30}$$

Throughout this section we consider only real random sequences and only two values of δ, namely, $\delta = 1$ and $\delta = -n$. Beginning with $\delta = 1$ we seek the coefficients in the forward prediction

$$\widehat{X}_{t+1,n} = \sum_{j=1}^{n} \alpha^t_{nj} X_{t-j+1}. \tag{4.31}$$

The normal equations arising from the properties of projection are

$$\langle X_{t+1} - \widehat{X}_{t+1,n}, X_s\rangle = 0, \quad t-n+1 \le s \le t,$$

or

$$\sum_{j=1}^{n} \alpha^t_{nj} R(t-j+1-s) = R(t+1-s), \quad t-n+1 \le s \le t. \tag{4.32}$$

These equations can be expressed in matrix form as

$$\begin{bmatrix} R(1) \\ R(2) \\ \vdots \\ R(n) \end{bmatrix} = \begin{bmatrix} R(0) & \cdots & R(n-1) \\ R(1) & \cdots & R(n-2) \\ \vdots & \vdots & \vdots \\ R(n-1) & \cdots & R(0) \end{bmatrix} \begin{bmatrix} \alpha^t_{n1} \\ \alpha^t_{n2} \\ \vdots \\ \alpha^t_{nn} \end{bmatrix}. \tag{4.33}$$

It is clear that solution of this equation, namely, $\{\alpha_{nj}^t : j = 1, 2, \ldots, n\}$ is independent of t—a fact that could be argued from the stationarity of X_t as well. So from now on we suppress t and express the last equation in compact form

$$\mathbf{R}_n^1 = \mathbf{R}_n \boldsymbol{\alpha}_n, \tag{4.34}$$

where $\mathbf{R}_n^1$ denotes the $n \times 1$ matrix on the left-hand side of the above equation.

For any $\boldsymbol{\alpha}_n = [\alpha_{n1}\ \alpha_{n2}\ \ldots \alpha_{nn}]'$ that solves (4.33), the prediction error

$$\varepsilon_{t+1,n} = X_{t+1} - \widehat{X}_{t+1,n} \tag{4.35}$$

has variance, again independent of t, given by

$$\begin{aligned}(\sigma_n^b)^2 &= \| X_{t+1} - \widehat{X}_{t+1,n} \|^2 = \langle X_{t+1} - \widehat{X}_{t+1,n}, X_{t+1} \rangle \\ &= R(0) - \sum_{j=1}^{n} \alpha_{nj} R(j) = R(0) - (\mathbf{R}_n^1)' \boldsymbol{\alpha}_n. \end{aligned} \tag{4.36}$$

Now let us examine the case $\delta = -n$ (backward prediction), where we predict X_{t-n} based on $\mathcal{M}(t; n) = \text{sp}\{X_s : t - n < s \leq t\}$. The coefficients β_{nj} in the best linear estimator

$$\widehat{X}_{t-n,n} = \sum_{j=1}^{n} \beta_{nj} X_{t-j+1} \tag{4.37}$$

are determined by the matrix equation

$$\begin{bmatrix} R(n) \\ \vdots \\ R(2) \\ R(1) \end{bmatrix} = \begin{bmatrix} R(0) & \cdots & R(n-1) \\ R(1) & \cdots & R(n-2) \\ \vdots & \vdots & \vdots \\ R(n-1) & \cdots & R(0) \end{bmatrix} \begin{bmatrix} \beta_{n1} \\ \beta_{n2} \\ \vdots \\ \beta_{nn} \end{bmatrix}, \tag{4.38}$$

which in the compact format becomes

$$\mathbf{R}_1^n = \mathbf{R}_n \boldsymbol{\beta}_n. \tag{4.39}$$

$\mathbf{R}_1^n$ is clearly the flip of $\mathbf{R}_n^1$, that is, $\mathbf{R}_1^n(j) = \mathbf{R}_n^1(n-j+1)$. The same is true about the column matrices $\boldsymbol{\beta}_n$ and $\boldsymbol{\alpha}_n$. That is, for any $k = 1, 2, \ldots, n$,

$$\beta_{nk} = \alpha_{n(n-k+1)} \quad \text{and} \quad (\mathbf{R}_n^1)_k = (\mathbf{R}_1^n)_{n-k+1}. \tag{4.40}$$

The prediction error at time $t - n$, namely,

$$\varepsilon_{t-n,n} = X_{t-n} - \widehat{X}_{t-n,n}, \tag{4.41}$$

has variance

$$\begin{aligned}(\sigma_n^f)^2 &= \|X_{t-n} - (X_{t-n} | \mathcal{M}(t; n))\|^2 = \langle X_{t-n} - \widehat{X}_{t-n,n}, X_{t-n} \rangle \\ &= R(0) - \sum_{j=1}^{n} \beta_{nj} R(-n+j-1) = R(0) - (\mathbf{R}_1^n)' \boldsymbol{\beta}_n,\end{aligned}$$

which, by virtue of (4.40), is identical to $(\sigma_n^f)^2 = R(0) - (\mathbf{R}_n^1)'\boldsymbol{\alpha}_n$ (see (4.36)). Thus $\sigma_n^f = \sigma_n^b$ and so from now on we write both of them as σ_n. The prediction coefficients $\boldsymbol{\alpha}_n$ and $\boldsymbol{\beta}_n$ are unique if and only if $\mathbf{R}_n$ is invertible, and this is true if and only if any n consecutive members $\{X_s, X_{s+1}, ..., X_{s+n}\}$ of the process are LI [56]. The normal equations producing the solutions $\boldsymbol{\alpha}_n$ and $\boldsymbol{\beta}_n$ are sometimes called the Yule–Walker equations.

Some important properties of $\widehat{X}_{t+1,n}$ and $\widehat{X}_{t-n,n}$ and their errors are now given in a manner suggested by the presentation in [183, Chapter 7]. Here we use σ^2 to denote the prediction error of predicting X_{t+1} on its infinite past. That is, $\sigma^2 = \|X_{t+1} - \widehat{X}_{t+1}\|^2$, with $\widehat{X}_{t+1} = (X_{t+1}|\mathcal{H}_X(t))$.

Proposition 4.5 *If X_t is stationary and $\widehat{X}_{t+1,n}$ and $\widehat{X}_{t-n,n}$ are the best linear predictors of X_{t+1} and X_{t-n} based on n observations $\{X_{t-n+1}, X_{t-n}, \ldots, X_t\}$, then*

(a) *σ_n is nonincreasing and bounded below by σ; that is, $\sigma_1 \geq \sigma_2, \ldots, \sigma$;*

(b) *$\sigma_n \to \sigma$, as $n \to \infty$;*

(c) $\lim_{n\to\infty} \widehat{X}_{t+1,n} = \widehat{X}_{t+1}$*;*

(d) *if $\sigma_m = 0$ for some m, then $\sigma_n = 0$ for all $n \geq m$;*

(e) *$\mathbf{R}_{n+1}$ is invertible for some n if and only if $\sigma_n > 0$ for that n. In that case $\mathbf{R}_n$ is also invertible and*

$$\sigma_n^2 = R(0) - (\mathbf{R}_n^1)'\mathbf{R}_n^{-1}\mathbf{R}_n^1 = \frac{|\mathbf{R}_{n+1}|}{|\mathbf{R}_n|}; \tag{4.42}$$

(f) *if $\sigma_n > 0$, then* $\operatorname{rank} \mathbf{R}_{n+1} = \operatorname{rank}(\mathbf{R}_n) + 1$*;*

(g) *if X_t is nondeterministic, then $\sigma > 0$ and $|\mathbf{R}_n| \neq 0$ for all $n \geq 1$. If this is the case, we have further that*

$$\sigma = \exp\left(\lim \frac{1}{n} \ln \sqrt{|\mathbf{R}_n|}\right) > 0. \tag{4.43}$$

Proof. For (a), σ_n is bounded and nonincreasing because of the top line of (4.36). For (b), this follows from $\lim_{n\to\infty} \mathcal{M}(t;n) = \operatorname{sp}\{X_s : s \leq t\}$ in conjunction with the fact that the predictor $\widehat{X}_{t+1}$ that achieves error σ^2 can be approximated arbitrarily closely by elements of $\operatorname{sp}\{X_s : s \leq t\}$. For (c), since $(\widehat{X}_{t+1} - X_{t+1}) \perp (\widehat{X}_{t+1,n} - \widehat{X}_{t+1})$, we can write

$$\begin{aligned} \sigma_n^2 = \| \widehat{X}_{t+1,n} - X_{t+1} \|^2 &= \| \widehat{X}_{t+1,n} - \widehat{X}_{t+1} + \widehat{X}_{t+1} - X_{t+1} \|^2 \\ &= \| \widehat{X}_{t+1,n} - \widehat{X}_{t+1} \|^2 + \sigma^2, \end{aligned} \tag{4.44}$$

and hence

$$\| \widehat{X}_{t+1,n} - \widehat{X}_{t+1} \|^2 = \sigma_n^2 - \sigma^2.$$

This in conjunction of part(b) gives

$$\lim_{n\to\infty} \| \widehat{X}_{t+1,t;n} - \widehat{X}_{t+1} \| = 0$$

which completes the proof. Part (d) is an immediate consequence of part (a). For part (e), $\sigma_n^2 > 0$ means that $X_{t+1} \notin \mathcal{M}(t;n)$ which, in turn, means that $\{X_{t-n+1}, X_{t-n}, \ldots, X_t, X_{t+1}\}$ is LI. The left equality in (4.42) follows from (4.39) and (4.36). To prove the right equality, we apply the result of Problem 4.14 at the end of this chapter to the partitioned matrix

$$\mathbf{R}_{n+1} = \begin{bmatrix} R(0) & (\mathbf{R}_n^1)' \\ \mathbf{R}_n^1 & \mathbf{R}_n \end{bmatrix}$$

and obtain

$$|\mathbf{R}_{n+1}| = [R(0) - (\mathbf{R}_n^1)'\mathbf{R}_n^{-1}\mathbf{R}_n^1]|\mathbf{R}_n| = \sigma_n^2|\mathbf{R}_n|.$$

For (f), if $\sigma_n > 0$ then both $\mathbf{R}_{n+1}$ and $\mathbf{R}_n$ are invertible and have respective ranks of $n+1$ and n. For (g), since X_t is nondeterministic, $\sigma > 0$ and, consequently, for every positive integer n, $\sigma_n > 0$ and $\mathbf{R}_n$ is invertible. Therefore we can write

$$\ln \sigma = \lim_{n\to\infty} \frac{1}{n} \sum_{k=1}^{n} \ln \sigma_n = \lim_{n\to\infty} \frac{1}{2n} \sum_{k=1}^{n} \ln \frac{|\mathbf{R}_{k+1}|}{|\mathbf{R}_k|} = \lim_{n\to\infty} \ln \sqrt{|\mathbf{R}_n|},$$

which implies (4.43). ∎

4.2.5.1 Partial Autocorrelations For a general second order random sequence X_t and any $n \geq 0$, its nth partial autocorrelation at time t is defined to be

$$\begin{aligned} \pi(t, n+1) &= \operatorname{Corr}\left(X_{t+1} - \widehat{X}_{t+1,n},\ X_{t-n} - \widehat{X}_{t-n,n}\right) \\ &= \frac{\langle \varepsilon_{t+1,n},\ \varepsilon_{t-n,n}\rangle}{\sigma_n^f \sigma_n^b} = \frac{\langle \varepsilon_{t+1,n}, \varepsilon_{t-n,n}\rangle}{\sigma_n^2}, \end{aligned} \tag{4.45}$$

which gives the immediate interpretation that $\pi(t, n+1)$ is the correlation of the prediction errors $\varepsilon_{t+1,n}$ with $\varepsilon_{t-n,n}$. Another interpretation is that $\pi(t, n+1)$ is the correlation between X_{t+1} and X_{t-n} when the effects on the variables $\{X_{t-n+1}, X_{t-n}, \ldots, X_t\}$ are removed. Note that when $n = 0$, we obtain $\pi(t,1) = \operatorname{Corr}(X_{t+1}, X_t)$, since there are none in between.

For stationary processes we expect, for each n, $\pi(t, n+1)$ to be constant with respect to t. The following result shows that this is in fact true.

Lemma 4.6 *If X_t is stationary, then each $\pi(t, n+1)$ is independent of t and hence from now on will be denoted by $\pi(n+1)$.*

Proof. By the remarks preceding Proposition 4.5, it is clear that the denominator of (4.45), defining the partial autocorrelation, is independent of t. So we only need to check time independence of its numerator and that is clear from

$$\begin{aligned} \langle \varepsilon_{t+1,n}, \varepsilon_{t-n,n} \rangle &= \left\langle X_{t+1} - \sum_{j=1}^{n} \alpha_{nj} X_{t-j+1},\ X_{t-n} - \sum_{j=1}^{n} \beta_{nj} X_{t-j+1} \right\rangle \\ &= R(n+1) - \sum_{j=1}^{n} \alpha_{nj} R(n-j+1) - \sum_{j=1}^{n} \beta_{nj} R(j) \\ &\quad + \sum_{j=1}^{n} \sum_{k=1}^{n} \alpha_{nj} \beta_{nk} R(k-j). \end{aligned} \tag{4.46}$$

Again we note that the vectors $\boldsymbol{\alpha}_n$ and $\boldsymbol{\beta}_n$ need not be unique solutions to the forward and backward Yule–Walker equations because as long as they are solutions, they represent the projections. The expression (4.46) for $\pi(n+1)$ can be shortened, since for each k in the last line of that equation,

$$\sum_{j=1}^{n} \alpha_{nj} R(k-j) = R(k),$$

thus causing the cancellation of the last two items and producing

$$\pi(n+1) = \frac{R(n+1) - (\mathbf{R}_1^n)' \boldsymbol{\alpha}_n}{\sigma_n^2}. \tag{4.47}$$

Due to the "flip" relationships between $\boldsymbol{\alpha}_n$ and $\boldsymbol{\beta}_n$ as well as between $\mathbf{R}_1^n$ and $\mathbf{R}_n^1$, equation (4.47) can be written as

$$\pi(n+1) = \frac{R(n+1) - (\mathbf{R}_n^1)' \boldsymbol{\beta}_n}{\sigma_n^2}. \tag{4.48}$$

4.2.5.2 Durbin–Levinson Algorithm The idea of the Durbin–Levinson algorithm is to find a computationally economical way to compute $\boldsymbol{\alpha}_{n+1}$ given the vector of predictor coefficients $\boldsymbol{\alpha}_n$ (a solution of (4.31)). To do this, write the matrix equation (4.33) with $n+1$ replacing n, as

$$\begin{bmatrix} \mathbf{R}_n^1 \\ R(n+1) \end{bmatrix} = \begin{bmatrix} \mathbf{R}_n & \mathbf{R}_1^n \\ (\mathbf{R}_1^n)' & R(0) \end{bmatrix} \begin{bmatrix} \boldsymbol{\alpha}_u \\ \alpha_l \end{bmatrix}. \tag{4.49}$$

We seek the vector of coefficients $\boldsymbol{\alpha}'_{n+1} = [\boldsymbol{\alpha}'_u \alpha_l]'$. Writing the two equations separately produces

$$\mathbf{R}_n^1 = \mathbf{R}_n\boldsymbol{\alpha}_u + \mathbf{R}_1^n\alpha_l \quad \text{and} \quad R(n+1) = (\mathbf{R}_1^n)'\boldsymbol{\alpha}_u + \alpha_l R(0). \tag{4.50}$$

Since $\mathbf{R}_n^1 = \mathbf{R}_n\boldsymbol{\alpha}_n$ it is natural to try $\boldsymbol{\alpha}_u = \boldsymbol{\alpha}_n + \mathbf{w}$, which transforms the preceding equations into

$$\mathbf{0} = \mathbf{R}_n\mathbf{w} + \mathbf{R}_1^n\alpha_l \quad \text{and} \quad R(n+1) = (\mathbf{R}_1^n)'(\boldsymbol{\alpha}_n + \mathbf{w}) + \alpha_l R(0), \tag{4.51}$$

respectively. But the top line is solved by $\mathbf{w} = -\alpha_l\boldsymbol{\beta}_n$. Substituting this expression for $\mathbf{w}$ in the bottom line and using equations (4.36) and (4.47), we get

$$\alpha_l = \frac{R(n+1) - (\mathbf{R}_1^n)'\boldsymbol{\alpha}_n}{R(0) - (\mathbf{R}_1^n)'\boldsymbol{\beta}_n} = \pi(n+1).$$

Note that α_l is the last coordinate of the vector $\boldsymbol{\alpha}_{n+1}$ of regression coefficients as described in [28, Section 3.4]. In other words, $\alpha_l = \alpha_{(n+1)(n+1)}$. Hence the last equation gives

$$\pi(n+1) = \alpha_{(n+1)(n+1)} = \beta_{(n+1)1}.$$

Given $\boldsymbol{\alpha}_n$ and $\boldsymbol{\beta}_n$, if $\pi_n \neq 0$ (meaning both X_{t+1} and X_{t-n} are LI of $\mathcal{M}(t, n-1)$), we determine α_l from the preceding and then $\boldsymbol{\alpha}_u = \boldsymbol{\alpha}_n - \alpha_l\boldsymbol{\beta}_n$. If $\pi_n = 0$ then (4.49) is solved by $\alpha_l = 0$ and $\boldsymbol{\alpha}_u = \boldsymbol{\alpha}_n$, which makes perfect sense because X_{t-n} does not add any new information. For the backward coefficients, we solve for $\boldsymbol{\beta}_{n+1}$ (predicting to time $t-n$ based on a sample of size $n+1$ into the future) in terms of $\boldsymbol{\beta}_n$. Beginning with (4.38) we obtain

$$\begin{bmatrix} R(n+1) \\ \mathbf{R}_1^n \end{bmatrix} = \begin{bmatrix} R(0) & (\mathbf{R}_n^1)' \\ \mathbf{R}_n^1 & \mathbf{R}_n \end{bmatrix} \begin{bmatrix} \beta_u \\ \boldsymbol{\beta}_l \end{bmatrix}, \tag{4.52}$$

which leads, as above, to

$$\beta_u = \frac{R(n+1) - (\mathbf{R}_n^1)'\boldsymbol{\beta}_n}{R(0) - (\mathbf{R}_n^1)'\boldsymbol{\alpha}_n} = \pi(n+1),$$

with $\boldsymbol{\beta}_{n+1} = [\beta_u\ \boldsymbol{\beta}'_l]'$, where $\boldsymbol{\beta}_l = \boldsymbol{\beta}_n - \beta_u\,\boldsymbol{\alpha}_n$. Given that we wish to compute the coefficients up through some $n = n_0$, we begin with $n = 1$ and directly obtain

$$\beta_{11} = \alpha_{11} = \frac{R(1)}{R(0)}.$$

What we need for the second step, namely, computing coordinates of $\boldsymbol{\alpha}_2$ and $\boldsymbol{\beta}_2$, are $\boldsymbol{\alpha}_1 = [\alpha_{11}]$ and $\boldsymbol{\beta}_1 = [\beta_{11}]$, which we already have. The process continues recursively up to $n = n_0$.

4.2.5.3 Cholesky Decomposition and Innovation Algorithm A variation of the Durbin–Levinson idea is the *innovation algorithm*. This algorithm is useful for recursively computing finite past prediction coefficients without the need for an explicit matrix inversion as in (4.33) or (4.34). Additionally, it does not depend on the stationarity of X_t, a feature we will utilize in Chapter 8, where the issue of deficient rank covariance matrices is also treated.

We will now show how the innovation algorithm is essentially connected to the Cholesky decomposition (or factorization) of a positive definite and therefore invertible covariance matrix $\mathbf{R}$. The case of a rank deficient $\mathbf{R}$ will be treated in Chapter 8.

Proposition 4.6 (Cholesky Decomposition) *If the $n \times n$ matrix $\mathbf{R}$ is positive definite, then there exists a lower triangular matrix $\boldsymbol{\Theta}$ for which*

$$\mathbf{R} = \boldsymbol{\Theta}\,\boldsymbol{\Theta}'. \tag{4.53}$$

This factor $\boldsymbol{\Theta}$ is unique if we demand its diagonal elements to be positive.

Proof. Recall [56] that $\mathbf{R}$ is positive definite if and only if there exist n LI random variables $\{X_1, X_2, \ldots, X_n\}$ of finite variance such that $\mathbf{R} = \text{Cov}\,(\mathbf{X}, \mathbf{X})$, where $\mathbf{X} = [X_1, X_2, \ldots, X_n]'$. Here we sketch a proof based on the Gram–Schmidt orthogonalization procedure applied to the vectors $\{X_1, X_2, \ldots, X_n\}$.

To start the Gram–Schmidt procedure, set $Y_1 = X_1$ and then $\eta_1 = Y_1/\|Y_1\|$. Since the set $\{X_1, X_2\}$ is LI, the vector $Y_2 = X_2 - P_{\mathcal{M}_1}X_2$ is not null; then set $\eta_2 = Y_2/\|Y_2\|$. It is clear that $\{\eta_1, \eta_2\}$ are orthonormal and that $X_1 = \theta_{11}\eta_1$ and $X_2 = \theta_{21}\eta_1 + \theta_{22}\eta_2$. Assuming the orthonormal set $\{\eta_1, \eta_2, \ldots, \eta_k\}$ has been determined from $\{X_1, X_2, \ldots, X_k\}$, the linear independence of $\{X_1, X_2, \ldots, X_k\}$ implies $Y_{k+1} = X_{k+1} - P_{\mathcal{M}_k}X_{k+1}$ is not null; thus we set $\eta_{k+1} = Y_{k+1}/\|Y_{k+1}\|$ and we can write

$$\begin{bmatrix} X_1 \\ X_2 \\ \vdots \\ X_{k+1} \end{bmatrix} = \begin{bmatrix} \theta_{11} & 0 & 0\cdots & \\ \theta_{21} & \theta_{22} & 0 & \cdots \\ \vdots & \vdots & \vdots & \\ \theta_{k+1,1} & \theta_{k+1,2} & \cdots & \theta_{k+1,k+1} \end{bmatrix} \begin{bmatrix} \eta_1 \\ \eta_2 \\ \vdots \\ \eta_{k+1} \end{bmatrix}.$$

Thus $\mathbf{X}_n = \boldsymbol{\Theta}\boldsymbol{\eta}$. Since $\{X_1, X_2, \ldots, X_n\}$ is LI, the preceding holds for $k+1 = n$ and the set $\{\eta_1, \eta_2, \ldots, \eta_n\}$ is orthonormal. This leads finally to

$$\mathbf{R} = E\mathbf{X}_n\,\mathbf{X}_n' = \boldsymbol{\Theta}E\{\boldsymbol{\eta}\boldsymbol{\eta}'\}\boldsymbol{\Theta}',$$

which, since $E\{\boldsymbol{\eta}\boldsymbol{\eta}'\} = \mathbf{I}$, is (4.53) with $\boldsymbol{\Theta}$ lower triangular as required. ∎

Discussions of the Cholesky decomposition (or factorization) may be found in many references. For example, see [207], and for a matrix oriented proof, see Golub and Van Loan [79, Theorem 5.2-3].

The preceding sketch only shows the existence of $\mathbf{\Theta}$ appearing in the Cholesky decomposition. The connection to finite past prediction is readily made by recognizing that η_{k+1} is prediction error vector $X_{k+1} - P_{\mathcal{M}_k} X_{k+1}$ normalized to unit length, where we recognize that

$$\mathcal{M}_k = \text{sp}\{X_1, X_2, \ldots, X_k\} = \text{sp}\{\eta_1, \eta_2, \ldots, \eta_k\}, \quad 1 \le k \le n.$$

Since $X_{k+1} = \sum_{j=1}^{k+1} \theta_{k+1,j}\eta_j$ can be uniquely decomposed

$$X_{k+1} = P_{\mathcal{M}_k} X_{k+1} + [X_{k+1} - P_{\mathcal{M}_k} X_{k+1}]$$

and then noting $\sum_{j=1}^{k} \theta_{k+1,j}\eta_j \in \mathcal{M}_k$ and $\eta_{k+1} \perp \mathcal{M}_k$, we can easily make the identifications

$$\widehat{X}_{k+1} = P_{\mathcal{M}_k} X_{k+1} = \sum_{j=1}^{k} \theta_{k+1,j}\eta_j \tag{4.54}$$

and

$$X_{k+1} - P_{\mathcal{M}_k} X_{k+1} = \theta_{k+1,k+1}\eta_{k+1}. \tag{4.55}$$

So at row $k+1$, the first k terms form the least square predictor $P_{\mathcal{M}_k} X_{k+1}$ of X_{k+1} and the last term $\theta_{k+1,k+1}\eta_{k+1}$ is the prediction error whose norm is $\theta_{k+1,k+1}$. Thus we see that computing the Cholesky decomposition for $\mathbf{R}$ is the same problem as computing the coefficients for the predictor expressed in terms of the prediction errors (the finite past innovations).

Finally, we come to the innovation algorithm, which gives a method of recursively computing the $(k+1)$st row of $\mathbf{\Theta}$ given the coefficients of the first k rows.

Proposition 4.7 (The Innovation Algorithm) *If the $n \times n$ matrix $\mathbf{R}$ is positive definite then the lower triangular matrix $\mathbf{\Theta}$ in (4.53) can be computed recursively as follows. First set $\theta_{11} = [R(1,1)]^{1/2}$. The remainder of the coefficients $\theta_{k+1,j}$ are computed left to right beginning with $k = 1$ (row 2) as follows. For $j = 1, 2, \ldots, k$, set*

$$\theta_{k+1,j} = \frac{1}{\theta_{jj}}\Big[R(k+1,j) - \sum_{m=1}^{j-1} \theta_{jm}\theta_{k+1,m}\Big]. \tag{4.56}$$

For the diagonal term, set

$$\theta_{k+1,k+1} = \Big[R(k+1,k+1) - \sum_{j=1}^{k} \theta_{k+1,j}^2\Big]^{1/2}. \tag{4.57}$$

Subsequent rows ($k = 2, 3, \ldots, n-1$) are computed in increasing order.

Proof. Let $\mathbf{X} = [X_1, X_2, \ldots, X_n]'$ be a random vector whose components are LI and $\mathbf{R} = \text{Cov}(\mathbf{X}, \mathbf{X})$. Since every subset is also LI, its corresponding submatrix of $\mathbf{R}$ is positive definite. The process is started by setting $\widehat{X}_1 = \mathbf{0}$ so that $X_1 - \widehat{X}_1$ and hence $\|X_1 - \widehat{X}_1\| = \|X_1\|$ and $X_1 = \theta_{11}\eta_1$ leads us to $\theta_{11}^2 = R(1,1)$. Suppose now the first k rows of $\boldsymbol{\Theta}$ have been determined. By the Gram–Schmidt construction, we first have $\langle X_{k+1} - \widehat{X}_{k+1}, \eta_j \rangle = 0$ for $j = 1, 2, \ldots, k$. But this may be written

$$\langle X_{k+1}, \eta_j \rangle = \langle \widehat{X}_{k+1}, \eta_j \rangle = \theta_{k+1,j},$$

where the last equality follows from (4.54). Using $\eta_j = \theta_{jj}^{-1}(X_j - \widehat{X}_j)$ in the preceding,

$$\begin{aligned}
\theta_{k+1,j} &= \left\langle X_{k+1}, \frac{X_j - \widehat{X}_j}{\theta_{jj}} \right\rangle \\
&= \frac{1}{\theta_{jj}} \left[\langle X_{k+1} X_j \rangle - \langle X_{k+1}, \widehat{X}_j \rangle \right] \\
&= \frac{1}{\theta_{jj}} \left[R(k+1, j) - \sum_{m=1}^{j-1} \theta_{jm} \langle X_{k+1}, \eta_m \rangle \right] \\
&= \frac{1}{\theta_{jj}} \left[R(k+1, j) - \sum_{m=1}^{j-1} \theta_{jm} \theta_{k+1,m} \right],
\end{aligned}$$

giving (4.56). Note that for computing $\theta_{k+1,j}$, only θ's from previous rows and $\theta_{k+1,m}$ for $m < j$ are needed. The diagonal term in row $k+1$ is computed by

$$\begin{aligned}
\theta_{k+1,k+1} &= \|X_{k+1} - \widehat{X}_{k+1}\| \\
&= \left\| X_{k+1} - \sum_{j=1}^{k} \theta_{k+1,j} \eta_j \right\| \\
&= \left[R(k+1, k+1) - \sum_{j=1}^{k} \theta_{k+1,j}^2 \right]^{1/2},
\end{aligned}$$

giving (4.57). ∎

4.3 MULTIVARIATE SPECTRAL THEORY

Let us recall from Definition 1.5 that a second order q-variate sequence $\mathbf{X}_t = [X_t^1, X_t^2, \ldots, X_t^q]'$ is stationary if $m^j(t) \equiv m^j$ and $R^{jk}(s,t) = R^{jk}(s-t)$ for all $s, t \in \mathbb{Z}$ and all $j, k = 1, 2, \ldots, q$.

4.3.1 Unitary Shift

Multivariate sequences also have a unitary shift defined on their time domain. The time domain of an q-variate random sequence is defined as

$$\mathcal{H}_{\mathbf{X}} = \overline{\text{sp}}\{\sum_{j=1}^{k} a_j X_{t_j}^{p_j} \; : \; k \geq 1, \; a_j \in \mathbb{C}, \; 1 \leq p_j \leq q, \; t_j \in \mathbb{Z}\}.$$

Proposition 4.8 *A zero mean q-variate sequence $\mathbf{X}_t$ is stationary if and only if there exists a unitary operator U defined on $\mathcal{H}_{\mathbf{X}}$ such that*

$$X_{t+1}^j = U X_t^j, \tag{4.58}$$

for each $t \in \mathbb{Z}$ and $1 \leq j \leq q$.

Proof. Since unitary operators preserve inner products (4.58), we get

$$R^{jk}(s,t) = \langle X_s^j, X_t^k \rangle = \langle U X_s^j, U X_t^k \rangle = \langle X_{s+1}^j, X_{t+1}^k \rangle = R^{jk}(s+1, t+1)$$

and this implies $R^{jk}(s,t) = R^{jk}(s-t)$. To prove the converse we set

$$L_{\mathbf{X}} = \text{sp}\{X_t^p : t \in \mathbb{Z}, 1 \leq p \leq q\}$$

so that $\mathcal{H}_{\mathbf{X}} = \overline{L_{\mathbf{X}}}$ and for any $\mathbf{z} = \sum_{j=1}^k a_j X_{t_j}^{p_j}$ in $L^{\mathbf{X}}$ we define $U\mathbf{z} = \sum_{j=1}^k a_j X_{t_j+1}^{p_j}$. Using stationarity one can show, just as was done above in the univariate case, that U is well defined, linear, and preserves inner products as a map from $L_{\mathbf{X}}$ onto $L_{\mathbf{X}}$. Then we similarly extend U to a unitary map from $\mathcal{H}_{\mathbf{X}}$ to $\mathcal{H}_{\mathbf{X}}$. ∎

In other words, a multivariate stationary sequence has a single unitary operator U acting as the shift operator for all its components. This is a basic characteristic of multivariate stationary processes. Sometimes we will express $X_{t+1}^j = U X_t^j$ for $t \in \mathbb{Z}$ and $j = 1, 2, \ldots, q$ in a brief form by $\mathbf{X}_{t+1} = U\mathbf{X}_t, \; t \in \mathbb{Z}$.

The issue of a nonzero mean is similar to the univariate case. Specifically,

$$m_j = \int X_t^j(\omega) P(d\omega) = \langle X_t^j, 1 \rangle.$$

Thus a constant mean says that each component sequence X_t^j has the fixed projection m_j onto 1. Again it is not necessarily true that $1 \in \mathcal{H}_{\mathbf{X}}$ but if so, it still remains true that $U1 = 1$. See problems 4.1 and 4.2.

4.3.2 Spectral Representation

The spectral representation for multivariate stationary sequences follows from an application of

$$U^t = \int_0^{2\pi} e^{it\lambda} dE(\lambda)$$

(Spectral Theorem for Unitary Operators (Theorem 3.7)) to the unitary operator U that gives $\mathbf{X}_{t+1} = U\mathbf{X}_t$ in Proposition 4.8.

Denoting E as the corresponding spectral measure, we can define a column vector valued random measure by

$$\boldsymbol{\xi}(d\lambda) = [\xi^j(d\lambda)]_{j=1}^q = [E(d\lambda)X_0^j]_{j=1}^q$$

and write, for $j = 1, 2, \ldots, q$,

$$X_t^j = \Big(\int_0^{2\pi} e^{it\lambda} E(d\lambda)\Big) X_0^j = \int_0^{2\pi} e^{it\lambda} E(d\lambda) X_0^j = \int_0^{2\pi} e^{it\lambda} \xi^j(d\lambda).$$

The countable additivity of $\boldsymbol{\xi}$ and orthogonality of its increments,

$$\langle \xi^i(\Delta), \xi^j(\Delta')\rangle = 0, \text{ whenever } \Delta \cap \Delta' = \emptyset$$

for any $i, j = 1, 2, ..., q$, follow from the properties of the spectral measure E. These remarks in conjunction with Theorem 4.2 applied component-wise yields the following.

Theorem 4.9 *If $\mathbf{X}_t$ is a q-variate stationary sequence, then*

(a) *there exists an q-variate vector measure $\boldsymbol{\xi}$, called its random spectral measure or simply its random measure, such that*

$$\mathbf{X}_t = \int_0^{2\pi} e^{it\lambda} \boldsymbol{\xi}(d\lambda);$$

(b) *the spectral measure $\mathbf{F}$ of $\mathbf{X}_t$ defined by*

$$\mathbf{F}(\Delta) = \Big[\langle \boldsymbol{\xi}^i(\Delta), \boldsymbol{\xi}^j(\Delta)\rangle\Big]_{i,j=1}^q, \text{ for any Borel subset } \Delta \text{ of } [0, 2\pi)$$

is a nonnegative definite matrix valued measure;

(c) *the covariance has the spectral representation*

$$\mathbf{R}(\tau) = \int_0^{2\pi} e^{i\tau\lambda} \mathbf{F}(d\lambda) = \int_0^{2\pi} e^{i\tau\lambda} d\mathbf{F}_\lambda,$$

where its matrix valued distribution $\mathbf{F}_\lambda$ *of* $\mathbf{X}_t$ *is related to its spectral measure* $\mathbf{F}(d\lambda)$ *just as in the univariate case.*

By the Lebesgue decomposition we can always write $\mathbf{F} = \mathbf{F}^{ac} + \mathbf{F}^{s}$. We use $\mathbf{F}'$ to denote the Radon–Nikodym derivative of its absolutely continuous part $\mathbf{F}^{ac}$ with respect to Lebesgue measure. If $\mathbf{F}$ is absolutely continuous (w.r.t. Lebesgue measure) then we denote $\mathbf{f} = \mathbf{F}'$ and call it the spectral density of $\mathbf{X}_t$.

4.3.3 Mean Ergodic Theorem

For multivariate stationary sequences, the mean ergodic theorem addresses mean-square convergence of

$$\mathbf{S}_N(\lambda) = \frac{1}{2N+1} \sum_{t=-N}^{N} \mathbf{X}_t e^{-i\lambda t} \tag{4.59}$$

at $\lambda = 0$. By considering Proposition 4.2 applied to each component of $\mathbf{X}_t$, we obtain the following.

Proposition 4.9 *If* $\mathbf{X}_t$ *is a stationary sequence, then*

$$\lim_{N\to\infty} \mathbf{S}_N(\lambda) = \boldsymbol{\xi}(\{\lambda\}). \tag{4.60}$$

Extending Definition 4.3, if $E\{\mathbf{X}_t\} = \mathbf{m}$, we will say that $\mathbf{X}_t$ is mean ergodic if

$$\lim_{N\to\infty} \mathbf{S}_N(0) = \mathbf{m} \tag{4.61}$$

component-wise, in the mean-square sense. Then, as in the univariate case, $\mathbf{S}_N(0) \to \mathbf{m}$ if and only if the atom of $\boldsymbol{\xi}(\cdot)$ at $\{0\}$ is the vector of constants $\mathbf{m}$, or in terms of $\mathbf{F}$, $F_{jj}(\{0\}) = |m_j|^2$.

4.3.4 Spectral Domain

In this section we introduce the spectral domain of a multivariate stationary sequence as a Hilbert space of functions. Then the Kolmogorov isomorphism between time and spectral domains of such a sequence is presented and its importance in prediction theory is discussed. We show that to any random variable $Y \in \mathcal{H}_X$ there corresponds a vector function

$$\boldsymbol{\varphi}(\lambda) = [\varphi^j(\lambda)]_{j=1}^q = [\varphi^1(\lambda), \varphi^2(\lambda), ..., \varphi^q(\lambda)]^*$$

in the spectral domain such that

$$Y = \int_0^{2\pi} \boldsymbol{\varphi}^*(\lambda)\boldsymbol{\xi}(d\lambda) = \int_0^{2\pi} \sum_{j=1}^q \varphi^j(\lambda)\xi^j(d\lambda).$$

To be more precise, let $\mathbf{F} = [F^{ij}]_{i,j=1}^{q}$ be the spectral measure of $\mathbf{X}_t$ as defined in part (c) of Theorem 4.9 and let μ be any complex valued measure with respect to which all diagonal measures F^{jj} of $\mathbf{F}$ are absolutely continuous. For example, one can take $\mu = \sum_{j=1}^{q} F^{jj}$. In view of the nonnegative definiteness of $\mathbf{F}(\Delta)$ for any Borel set Δ, the vanishing of all diagonal elements $F^{jj}(\Delta)$ implies that $F^{ij}(\Delta) = 0$ for any $i, j = 1, 2, ..., q$. Hence each F^{ij} is absolutely continuous with respect to this measure μ. Define the spectral density $\mathbf{f}_\mu$ by

$$f_\mu^{ij}(\lambda) = \frac{dF^{ij}}{d\mu}(\lambda); \quad \text{for any } i, j = 1, 2, ..., q.$$

Since the matrix $\mathbf{F}(\Delta)$ is nonnegative definite for each Borel set Δ, the matrix function

$$\mathbf{f}_\mu(\lambda) = [f_\mu^{ij}(\lambda)]$$

will be positive definite for almost every λ. We define the space $L^2(\mathbf{f}_\mu)$ to be the set of all vector valued functions $\boldsymbol{\varphi}(\lambda) = [\varphi^j(\lambda)]_{j=1}^{q}$ for which

$$[\boldsymbol{\varphi}^*(\lambda)\mathbf{f}_\mu(\lambda)\boldsymbol{\varphi}(\lambda)] = \sum_{i,j=1}^{q} \overline{\varphi^i(\lambda)}\varphi^j(\lambda) f_\mu^{ij}(\lambda) \in L^1(\mu).$$

If both vector functions $\boldsymbol{\varphi}$ and $\boldsymbol{\psi}$ belong to $L^2(\mathbf{f}_\mu)$ then it is easy to see that the function $[\boldsymbol{\varphi}^*\mathbf{f}_\mu\boldsymbol{\psi}] \in L^1(\mu)$. In fact, since $\mathbf{f}$ is nonnegative definite, we have

$$|\boldsymbol{\varphi}^*(\lambda)\mathbf{f}_\mu(\lambda)\boldsymbol{\psi}(\lambda)| \leq [\boldsymbol{\varphi}^*(\lambda)\mathbf{f}_\mu(\lambda)\boldsymbol{\varphi}(\lambda)]^{1/2}[\boldsymbol{\psi}^*(\lambda)\mathbf{f}_\mu(\lambda)\boldsymbol{\psi}(\lambda)]^{1/2}$$

for almost every λ w.r.t μ. Hence by the Cauchy–Schwarz inequality $\boldsymbol{\varphi}^*\mathbf{f}_\mu\boldsymbol{\psi}$ belongs to $L^1(\mu)$. We define an inner product on $L^2(\mathbf{f}_\mu)$ by

$$\langle \boldsymbol{\varphi}, \boldsymbol{\psi} \rangle_\mu = \int [\boldsymbol{\varphi}^*(\lambda)\mathbf{f}_\mu(\lambda)\boldsymbol{\psi}(\lambda)]\mu(d\lambda).$$

One can easily check that neither the spectral domain $L^2(\mathbf{f}_\mu)$ nor its inner product depends on the choice of the measure μ. We denote this (μ-invariant) common space by $L^2(\mathbf{F})$ and this common inner product by $\langle \boldsymbol{\varphi}, \boldsymbol{\psi} \rangle$: that is,

$$\langle \boldsymbol{\varphi}, \boldsymbol{\psi} \rangle = \int \boldsymbol{\varphi}^*\mathbf{F}(d\lambda)\boldsymbol{\psi} = \int \boldsymbol{\varphi}^*(\lambda)\mathbf{f}_\mu(\lambda)\boldsymbol{\psi}(\lambda)\mu(d\lambda).$$

From now on we set the auxiliary measure to $\mu = \sum_1^q F^{jj}$ and suppress all μ indices. If we identify two functions $\boldsymbol{\varphi}$ and $\boldsymbol{\psi}$ in $L^2(\mathbf{F})$ when

$$\int [\boldsymbol{\varphi} - \boldsymbol{\psi}]^*\mathbf{F}(d\lambda)[\boldsymbol{\varphi} - \boldsymbol{\psi}] = 0,$$

then $L^2(\mathbf{F})$ becomes an inner product space. The following lemma shows that this space, which is called *the spectral domain* of $\mathbf{X}_t$, is complete and hence a Hilbert space.

Lemma 4.7 *The inner product space $L^2(\mathbf{F})$ with norm $\|\varphi\| = \sqrt{\langle\varphi, \varphi\rangle}$ is complete.*

Proof. It is easy to see that the spectral density $\mathbf{f}$ is different from zero almost everywhere w.r.t. μ. In fact, if $\mathbf{f} = [f^{ij}] = [F^{ij}(d\lambda)/\mu(d\lambda)]$ is zero on a set Δ then $F^{jj}(\Delta) = \int_\Delta f^{jj}(\lambda)\mu(d\lambda) = 0, j = 1, ..., q$. Hence $\mu(\Delta) = \sum_1^q F^{jj}(\Delta) = 0$. For each $\lambda \in [0, 2\pi)$ let $m(\lambda)$ denote the smallest nonzero eigenvalue of the matrix $\mathbf{f}(\lambda)$; further denote $S(\lambda)$ as the subspace of all q-dimensional row vectors $\varphi = [\varphi^j]$ satisfying the condition $\varphi^*(\lambda)\mathbf{f}(\lambda) = 0$ and $R(\lambda)$ the orthogonal complement of $S(\lambda)$ in $\mathbb{C}^q$. It follows readily from the non negativeness of the matrix $\mathbf{f}(\lambda)$ for almost every λ that $\|\varphi\| = 0$ if and only if $\varphi(\lambda) \in S(\lambda)$ for almost every λ w.r.t. μ. It is easily seen that for any row vector function $\varphi \in L^2(\mathbf{F})$ one can find another such function $\widetilde{\varphi} \in L^2(\mathbf{F})$ such that $\|\varphi - \widetilde{\varphi}\| = 0$ and $\widetilde{\varphi} \in R(\lambda)$ for almost every λ w.r.t. μ, and for these λ's

$$\varphi^*(\lambda)\mathbf{f}(\lambda)\varphi(\lambda) \geq m(\lambda)\sum_{j=1}^{q}|\widetilde{\varphi}^j(\lambda)|^2.$$

Now let $\{\varphi_p\}$ be a Cauchy sequence in $L^2(\mathbf{F})$ and $\{\widetilde{\varphi}_p\}$ be its corresponding sequence as we just defined. We may assume without loss of any generality that $\varphi_p = \widetilde{\varphi}_p$ for any postive integer p. Given any $\epsilon > 0$, since $\{\varphi_p\}$ is Cauchy, there exists a positive integer N such that

$$\|\varphi_p - \varphi_{p'}\| < \sqrt{\epsilon}, \quad \text{whenever } p, p' > N.$$

This means

$$\int m(\lambda)\sum_{j=1}^{q}|\varphi_p^j(\lambda) - \varphi_{p'}^j(\lambda)|^2\mu(d\lambda) < \epsilon, \quad \text{whenever } p, p' > N,$$

which in turn implies that, for each $j = 1, 2, .., q$

$$\int |\varphi_p^j(\lambda)\sqrt{m(\lambda)} - \varphi_{p'}^j(\lambda)\sqrt{m(\lambda)}|^2\mu(d\lambda) < \epsilon, \text{ whenever } p, p' > N.$$

So for each $j = 1, 2, ..., q$ the sequence $\{\varphi_p^j(\lambda)\sqrt{m(\lambda)}\}$ is Cauchy in the Hilbert space $L^2(\mu)$. Hence for each $j = 1, 2, ..., q$ there is a function $\phi^j \in L^2(\mu)$ such that

$$\lim_{p\to\infty} \varphi_p^j\sqrt{m(\cdot)} = \phi^j, \quad \text{in the } L^2(\mu) \text{ sense.}$$

Hence for each $j = 1, 2, ..., q$ there is a subsequence of φ_p^j, which we again denote by φ_p^j and some sets Λ^j with $\mu(\Lambda^j) = 0$ such that

$$\lim_{p\to\infty} \varphi_p^j(\lambda)\sqrt{m(\lambda)} = \phi^j(\lambda), \qquad \text{for every } \lambda \in \Lambda^j.$$

Taking

$$\varphi^j(\lambda) = \frac{\phi^j(\lambda)}{\sqrt{m(\lambda)}}$$

and noting that $m(\lambda)$ is positive for almost every λ with respect to μ, we see that for each $j = 1, 2, ..., q$

$$\lim_{p\to\infty} \varphi_p^j(\lambda) = \varphi^j(\lambda), \quad \text{for almost all } \lambda \in \Lambda^j.$$

Letting $\Lambda = \cup_j \Lambda^j$, clearly $\mu(\Lambda) = 0$ and for every $\lambda \in \Lambda$, and hence almost every λ, we have

$$\lim_{p\to\infty} \boldsymbol{\varphi}_p^*(\lambda)\mathbf{f}(\lambda)\boldsymbol{\varphi}_p(\lambda) = \boldsymbol{\varphi}^*(\lambda)\mathbf{f}(\lambda)\boldsymbol{\varphi}(\lambda).$$

As the sequence $\{\boldsymbol{\varphi}_p\}$ is Cauchy and hence bounded, there exists a constant M such that

$$\|\boldsymbol{\varphi}_p\|^2 = \int \boldsymbol{\varphi}_p^*(\lambda)\mathbf{f}(\lambda)\boldsymbol{\varphi}_p(\lambda)\mu(d\lambda) \le M, \qquad \text{for all } p \in \mathbb{Z}.$$

By Fatou's lemma,

$$\int \boldsymbol{\varphi}^*(\lambda)\mathbf{f}(\lambda)\boldsymbol{\varphi}(\lambda) \le M,$$

which shows that the row vector function $\boldsymbol{\varphi}$ belongs to $L^2(\mathbf{F})$. Now for any $\epsilon > 0$ there exists a positive N such that

$$\|\boldsymbol{\varphi}_p^* - \boldsymbol{\varphi}_{p'}^*\| < \epsilon, \qquad \text{whenever } p,\ p' > N.$$

Fixing $p > N$ and letting $p' \to \infty$ along those values of p' for which $\boldsymbol{\varphi}_{p'}(\lambda) \to \boldsymbol{\varphi}(\lambda)$ almost everywhere w.r.t. μ, and using Fatou's lemma again, we obtain

$$\|\boldsymbol{\varphi}_p^* - \boldsymbol{\varphi}^*\| < \epsilon, \text{ for any } p > N. \qquad \blacksquare$$

The Kolmogorov Isomorphism. Now we can establish a multivariate extension of the Kolmogorov isomorphism, which can transfer some prediction problems from the time domain to the spectral domain, where Fourier analysis can be used to solve the problem there and then transfer the result back to the time domain. Let $\varphi = [\varphi^j]_{j=1}^q$ be a row vector such that

$$\int |\varphi^j(\lambda)|^2 f^{jj}(\lambda)(d\lambda) < \infty, \;\; j = 1, 2, \ldots, q. \tag{4.62}$$

For each j the integral $\int \varphi^j \xi^j d(\lambda)$ exists and is a random variable in $\mathcal{H}_{\mathbf{X}}$, the time domain. We correspond to this vector function $\boldsymbol{\varphi} \in L^2(\mathbf{F})$ the random variable Y defined by

$$Y = \sum_{j=1}^{q} \int \varphi^j(\lambda)\xi^j(d\lambda),$$

which is clearly in $\mathcal{H}_{\mathbf{X}}$. This correspondence is an isometric mapping because if

$$Z = \sum_{j=1}^{q} \int \psi^j(\lambda)\xi^j(d\lambda)$$

is another such random variable then one can easily check that

$$\langle Y, Z\rangle = \int \boldsymbol{\varphi}^*(\lambda)d\mathbf{F}(\lambda)\boldsymbol{\psi}(\lambda) = \langle \boldsymbol{\varphi}, \boldsymbol{\psi}\rangle.$$

This correspondence can be extended by linearity to the set of all finite linear combination of $\boldsymbol{\varphi}$'s satisfying (4.62) and then continuity to their span closure, which is $L^2(\mathbf{F})$. Using the standard arguments one can show that this extension remains to be an isometry. Now since the range of this mapping contains all random variables X_t^j (because $X_t^j = \int \varphi_t^j(\lambda)\xi(d\lambda)$ with $\varphi_t^j = [e^{it\lambda}\delta_{j-k}]_{k=1}^q$), this mapping is an isometry from the spectral domain $L^2(\mathbf{F})$ onto the time domain $\mathcal{H}_X$. So any element Y in the time domain of a multivariate stationary sequence has a spectral representation

$$Y = \int \boldsymbol{\varphi}^*(\lambda)\xi(d\lambda)$$

for some vector function $\boldsymbol{\varphi} \in L^2(\mathbf{F})$. Note that any q-variate vector random variable $\mathbf{Y} = [Y^k]_{k=1}^q$ can also be represented as $\mathbf{Y} = \int \boldsymbol{\varphi}(\lambda)\boldsymbol{\xi}(d\lambda)$, where $\boldsymbol{\varphi}$ is an $q \times q$ matrix valued function. In particular, we can write $\mathbf{X}_t = \int e^{it\lambda}\boldsymbol{\xi}(d\lambda)$.

It should be remarked that if T is a linear operator on the time domain, then the equation

$$Y' = TY = T\Big[\int \boldsymbol{\varphi}(\lambda)\boldsymbol{\xi}(d\lambda)\Big] = \int \boldsymbol{\varphi}'(\lambda)\boldsymbol{\xi}(d\lambda) = \int [T'\boldsymbol{\varphi}](\lambda)\xi(d\lambda)$$

defines a linear operator T' on the spectral domain, namely, $T'\boldsymbol{\varphi} = \boldsymbol{\varphi}'$. Conversely, to any linear operator T' in the spectral domain there corresponds a linear operator T in the time domain. In particular, one can see that multiplication by $e^{i\lambda}$ is the operator in the spectral domain corresponding to the unitary shift operator U we defined earlier in the time domain.

4.4 MULTIVARIATE PREDICTION THEORY

In this section we study the prediction of multivariate stationary random sequences. The presentation is based mostly on the Wiener and Masani cite-WandM1,WandM2, Masani [153] and Rozanov [201] treatments.

4.4.1 Infinite Past, Regularity and Singularity

We start by generalizing some more definitions to the multivariate case.

Definition 4.7 *Let $\mathbf{X}_t = [X_t^j]_{j=1}^q$ be a second order random sequence. We define its past up to and including time t and remote past to be*

$$\mathcal{H}_{\mathbf{X}}(t) = \overline{\text{sp}}\{X_s^j : s \le t,\ 1 \le j \le q\} \tag{4.63}$$

and

$$\mathcal{H}_{\mathbf{X}}(-\infty) = \bigcap_{t \in \mathbb{Z}} \mathcal{H}_{\mathbf{X}}(t), \tag{4.64}$$

respectively.

Definition 4.8 *A q-variate stationary random sequence $\mathbf{X}_t$ is called purely nondeterministic (or regular) if*

$$\mathcal{H}_{\mathbf{X}}(-\infty) = \{\mathbf{0}\}$$

and is called deterministic (or singular) if

$$\mathcal{H}_{\mathbf{X}}(-\infty) = \mathcal{H},$$

or equivalently if $\mathcal{H}_{\mathbf{X}}(s) = \mathcal{H}_{\mathbf{X}}(t)$, for all $s \in \mathbb{Z}$.

Definition 4.9 *For any q-variate vector $\mathbf{X} = [X^j]_{j=1}^q$, with all its q components belonging to some Hilbert space $\mathcal{H}$, and for any subspace $\mathcal{M}$ of this Hilbert space, we define $(\mathbf{X}|\mathcal{M})$ to be*

$$(\mathbf{X}|\mathcal{M}) = [(X^j|\mathcal{M})]_{j=1}^q.$$

We refer to this as the projection of $\mathbf{X}$ on $\mathcal{M}$ and hence sometimes denote it by $P_{\mathcal{M}}(\mathbf{X})$. By the predictor $\widetilde{\mathbf{X}}_{t+1}$ of $\mathbf{X}_t$ based on $\mathcal{H}_{\mathbf{X}}(t)$ here we mean

$$\widetilde{\mathbf{X}}_{t+1} = P_{\mathcal{H}_{\mathbf{X}}(t)}\mathbf{X}_{t+1} = (\mathbf{X}_{t+1}|H_{\mathbf{X}}(t)).$$

For more on this see Problem 4.13.

4.4.2 Wold Decomposition

Again, as in the univariate case, we characterize the relationship between the propagating unitary operator U defined in (4.58) and the time domain subspaces. The proofs follow exactly as in the univariate case (see Lemma 4.1).

Lemma 4.8 *If $\mathbf{X}_t$ is m-variate stationary with unitary shift U, then*

(a) $\mathcal{H}_{\mathbf{X}}(t+1) = U\mathcal{H}_{\mathbf{X}}(t)$;

(b) $\mathcal{H}_{\mathbf{X}} = U\mathcal{H}_{\mathbf{X}}$;

(c) $\mathcal{H}_{\mathbf{X}}(-\infty) = U\mathcal{H}_{\mathbf{X}}(-\infty)$.

Proposition 4.10 (Wold Decomposition) *Any multivariate stationary sequence $\mathbf{X}_t$ can be uniquely decomposed as the sum of two orthogonal stationary sequences $\mathbf{Y}_t$ and $\mathbf{Z}_t$,*

$$\mathbf{X}_t = \mathbf{Y}_t + \mathbf{Z}_t \tag{4.65}$$

such that

(a) $\mathcal{H}_{\mathbf{Z}}(t) = \mathcal{H}_{\mathbf{Y}}(t) \oplus \mathcal{H}_{\mathbf{Z}}(t)$;

(b) $\mathbf{Y}_t$ *is deterministic and* $\mathbf{Z}_t$ *purely nondeterministic;*

(c) $U_Y = U|_{\mathcal{H}_{\mathbf{X}}(-\infty)}$ *and* $U_Z = U|_{\mathcal{H}_{\mathbf{X}}(-\infty)^\perp}$;

(d) $\mathbf{F}_X = \mathbf{F}_Y + \mathbf{F}_Z$.

Proof. For each $t \in \mathbb{Z}$ set

$$\mathbf{Y}_t = (\mathbf{X}_t|\mathcal{H}_{\mathbf{X}}(-\infty)) \text{ and } \mathbf{Z}_t = \mathbf{X}_t - \mathbf{Y}_t.$$

Since for each s, $\mathbf{Y}_s \in \mathcal{H}_{\mathbf{X}}(-\infty)$ and for each t, $\mathbf{Z}_t = \mathbf{X}_t - \mathbf{Y}_t = \mathbf{X}_t - (\mathbf{X}_t|\mathcal{H}_{\mathbf{X}}(-\infty)) \perp \mathcal{H}_{\mathbf{X}}(-\infty)$, the components of $\mathbf{Y}_s$ are orthogonal to the components of $\mathbf{Z}_t$ and this implies $\mathcal{H}_{\mathbf{Y}} \perp \mathcal{H}_{\mathbf{Z}}$. For each $t \in \mathbb{Z}, \mathbf{Y}_t \in \mathcal{H}_{\mathbf{X}}(-\infty) \subseteq \mathcal{H}_{\mathbf{X}}(t)$ and hence $\mathbf{Z}_t = \mathbf{X}_t - \mathbf{Y}_t \in \mathcal{H}_{\mathbf{X}}(t)$. Therefore $\mathcal{H}_{\mathbf{Y}}(t) \oplus \mathcal{H}_Z(t) \subseteq \mathcal{H}_{\mathbf{X}}(t)$. In order to complete the proof of (a) it suffices to show that $\mathcal{H}_{\mathbf{X}}(t) \subseteq \mathcal{H}_{\mathbf{Y}}(t) \oplus \mathcal{H}_{\mathbf{Z}}(t)$. Pick $\mathbf{w} \in \mathcal{H}_{\mathbf{X}}(t)$; then each component w^j of $\mathbf{w}$ satisfies $w^j = \lim w_n^j$, where each $w_n^j \in \mathcal{M}_{\mathbf{X}}(t)$ and hence $w_n^j = u_n^j + v_n^j$, $u_n^j \in L_{\mathbf{Y}}(t)$, $v_n^j \in L_{\mathbf{Z}}(t)$. But since w_n^j must be Cauchy and $\|u_n^j - u_m^j\| \leq \|w_n^j - w_m^j\|$, and similarly for v_n^j, then $u_n^j \to u^j \in \mathcal{H}_Y(t)$, $v_n^j \to v^j \in \mathcal{H}_{\mathbf{Z}}(t)$. Taking the limit of $w_n^j = u_n^j + v_n^j$, we get $w^j = u^j + v^j$, which shows $\mathcal{H}_{\mathbf{X}}(t) \subset \mathcal{H}_{\mathbf{Y}}(t) \oplus \mathcal{H}_{\mathbf{Z}}(t)$.

For (b), we first show that $\mathcal{H}_{\mathbf{X}}(-\infty) = \mathcal{H}_{\mathbf{Y}}(t)$ for all $t \in \mathbb{Z}$, which implies $\mathbf{Y}_t$ is deterministic. Since it is clear that $\mathcal{H}_{\mathbf{Y}}(t) \subset \mathcal{H}_{\mathbf{X}}(-\infty)$ for each t, we show that $\mathcal{H}_{\mathbf{Y}}(t) \subset \mathcal{H}_{\mathbf{X}}(-\infty)$. For contradiction, suppose for fixed t there is a nonzero $\mathbf{u} \in \mathcal{H}_{\mathbf{X}}(-\infty) \ominus \mathcal{H}_{\mathbf{Y}}(t)$; then for each $s \leq t$ we have $\mathbf{u} \perp \mathbf{Y}_s$ and $\mathbf{u} \perp Z_s$ so that $\mathbf{u} \perp \mathbf{X}_s = \mathbf{Y}_s + \mathbf{Z}_s$ (component-wise). Hence $\mathbf{u} \perp \mathcal{H}_{\mathbf{X}}(t)$, and therefore $\mathbf{u} \perp \mathcal{H}_{\mathbf{X}}(-\infty)$ component-wise, implying $\mathbf{u} = \mathbf{0}$, which is a contradiction.

For (c), using Lemma 4.1, for any $t \in \mathbb{Z}$, we can write

$$\mathbf{Y}_{t+1} = (\mathbf{X}_{t+1}|H_{\mathbf{X}}(\infty)) = (U\mathbf{X}_t|H_{\mathbf{X}}(\infty)) = U(\mathbf{X}_t|H_{\mathbf{X}}(-\infty)),$$

from which one can conclude that $\mathbf{Y}_t$ is stationary. This also shows that U and $U_{\mathbf{Y}}$ act the same on $H_{\mathbf{Y}}$, which together with $H_{\mathbf{X}}(-\infty) = H_{\mathbf{Y}}$, proved above, give $U_{\mathbf{Y}} = U|_{H_{\mathbf{X}}(-\infty)}$. The statements about $\mathbf{Z}_t$ follow easily. Part (d) is easy to see. ∎

4.4.3 Innovations and Rank

Here we discuss notions of rank and innovation for multivariate stationary processes.

Definition 4.10 *The innovation space of the m-variate sequence $\mathbf{X}_t$ at time t is defined to be*

$$\mathcal{I}_{\mathbf{X}}(t) = \mathcal{H}_{\mathbf{X}}(t) \ominus \mathcal{H}_{\mathbf{X}}(t-1) = \{X \in \mathcal{H}_{\mathbf{X}}(t) : X \perp \mathcal{H}_{\mathbf{X}}(t-1)\}, \qquad (4.66)$$

which can also be identified through

$$\mathcal{H}_{\mathbf{X}}(t) = \mathcal{I}_{\mathbf{X}}(t) \oplus \mathcal{H}_{\mathbf{X}}(t-1).$$

Now we give some basic facts generalizing those for univariate case given in Lemma 4.2 .

Lemma 4.9 *If $\mathbf{X}_t$ is an q-variate stationary sequence with Wold decomposition $\mathbf{X}_t = \mathbf{Y}_t + \mathbf{Z}_t$, then*

(a) $\mathcal{I}_{\mathbf{X}}(t+1) = U\mathcal{I}_{\mathbf{X}}(t)$;

(b) $\mathcal{I}_{\mathbf{X}}(t) \perp \mathcal{H}_{\mathbf{X}}(-\infty), \quad \mathcal{I}_{\mathbf{X}}(t) = \mathcal{I}_{\mathbf{Z}}(t)$;

(c) $\dim \mathcal{I}_{\mathbf{X}}(t+1) = \dim \mathcal{I}_{\mathbf{X}}(t) \leq m$;

(d) $\mathbf{X}_t$ *singular* $\Rightarrow \dim \mathcal{I}_{\mathbf{X}}(t) \equiv 0$.

Proof. For (a), we follow the same reasoning as in the univariate case, namely,

$$\begin{aligned} \mathcal{I}_{\mathbf{X}}(t+1) &= \mathcal{H}_{\mathbf{X}}(t+1) \ominus \mathcal{H}_{\mathbf{X}}(t) \\ &= U\mathcal{H}_{\mathbf{X}}(t) \ominus U\mathcal{H}_{\mathbf{X}}(t-1) \\ &= U[\mathcal{H}_{\mathbf{X}}(t) \ominus \mathcal{H}_{\mathbf{X}}(t-1)], \end{aligned}$$

where the last line follows from the fact that U is unitary.

For (b), let $\mathbf{u} \in \mathcal{I}_{\mathbf{X}}(t)$ and $\mathbf{v} \in \mathcal{H}_{\mathbf{X}}(-\infty)$. Thus $\mathbf{u} \in \mathcal{H}_{\mathbf{X}}(t)$ and $\mathbf{u} \perp \mathcal{H}_{\mathbf{X}}(t-1)$ but $\mathbf{v} \in \mathcal{H}_{\mathbf{X}}(s)$ for all s and in particular for $s = t-1$ so componentwise $\langle u^j, v^k \rangle = 0$. To see the second part of (b) write

$$\begin{aligned}
\mathcal{I}_{\mathbf{X}}(t) = \mathcal{H}_{\mathbf{X}}(t) \ominus \mathcal{H}_{\mathbf{X}}(t-1) &= [\mathcal{H}_{\mathbf{Y}}(t) \oplus \mathcal{H}_{\mathbf{Z}}(t)] \ominus [\mathcal{H}_{\mathbf{Y}}(t-1) \oplus \mathcal{H}_{\mathbf{Z}}(t-1)] \\
&= [\mathcal{H}_{\mathbf{Y}}(t) \ominus \mathcal{H}_{\mathbf{Y}}(t-1)] \oplus [\mathcal{H}_{\mathbf{Z}}(t) \ominus \mathcal{H}_{\mathbf{Z}}(t-1)] \\
&= [\mathcal{H}_{\mathbf{Z}}(t) \ominus \mathcal{H}_{\mathbf{Z}}(t-1)] = \mathcal{I}_{\mathbf{Z}}(t),
\end{aligned}$$

because for all s, t we have $\mathcal{H}_{\mathbf{Z}}(s) \perp \mathcal{H}_{\mathbf{Y}}(t)$ and $\mathcal{H}_{\mathbf{Y}}(t) \ominus \mathcal{H}_{\mathbf{Y}}(t-1) = \{0\}$.

For (c), suppose $\dim \mathcal{I}_{\mathbf{X}}(t) = q'$, for some $q' \leq q$ and let $\{\eta_1, \eta_2, \ldots, \eta_{q'}\}$ be a basis for $\mathcal{I}_{\mathbf{X}}(t)$. Since $\mathcal{I}_{\mathbf{X}}(t+1) = U\mathcal{I}_{\mathbf{X}}(t)$ where U is unitary, then we only need to observe that $\{\eta'_j = U\eta_j : j = 1, 2, \ldots, q'\}$ will be a basis for $\mathcal{I}_{\mathbf{X}}(t+1)$; thus $\dim \mathcal{I}_{\mathbf{X}}(t+1) = q'$.

For (d), if $\mathbf{X}_t$ is singular, then $\mathbf{X}_t \in \mathcal{H}_{\mathbf{X}}(t-1)$, for all t, and hence $\mathcal{I}_{\mathbf{X}}(t) = \mathcal{H}_{\mathbf{X}}(t) \ominus \mathcal{H}_{\mathbf{X}}(t-1) = \{0\}$ for all t, meaning $\dim \mathcal{I}_{\mathbf{X}}(t) \equiv 0$. ∎

Now we further examine the case when $\mathbf{X}_t$ is not singular, so the regular component $\mathbf{Z}_t$ of the Wold decomposition (4.65) has at least one component of positive length: that is, $\|Z_t^j\| > 0$ for some $j \geq 1$ and each $t \in \mathbb{Z}$. The innovation vector defined here by

$$\begin{aligned}
\boldsymbol{\zeta}_t &= [\mathbf{X}_t - P_{\mathcal{H}_{\mathbf{X}}(t-1)}\mathbf{X}_t] \in \mathcal{I}_{\mathbf{X}}(t) \\
&= [\mathbf{Z}_t - P_{\mathcal{H}_{\mathbf{Z}}(t-1)}\mathbf{Z}_t] \in \mathcal{I}_{\mathbf{Z}}(t)
\end{aligned} \tag{4.67}$$

cannot be null. For if $\boldsymbol{\zeta}_t = \mathbf{0}$, then $\mathbf{Z}_t \in \mathcal{H}_{\mathbf{Z}}(t-1)$, which together with $\|Z_t^j\| > 0$ for some j contradicts that $\mathbf{Z}_t$ is regular. It is clear from (4.67) that $\text{sp}\{\boldsymbol{\zeta}_t\} \subset \mathcal{I}_{\mathbf{X}}(t)$. In fact we have $\mathcal{I}_{\mathbf{X}}(t) = \text{sp}\{\boldsymbol{\zeta}_t\}$. To see this, first write

$$\mathbf{X}_t = \mathbf{X}_t - P_{\mathcal{H}_{\mathbf{X}}(t-1)}\mathbf{X}_t) + P_{\mathcal{H}_{\mathbf{X}}(t-1)}\mathbf{X}_t = \boldsymbol{\zeta}_t + P_{\mathcal{H}_{\mathbf{X}}(t-1)}\mathbf{X}_t,$$

where $\boldsymbol{\zeta}_t \perp \mathcal{H}_{\mathbf{X}}(t-1)$. Then suppose $\mathbf{Y} \in \mathcal{H}_{\mathbf{X}}(t)$ and $\mathbf{Y} \perp \zeta_t^j, j = 1, 2, \ldots, q$. Then $\mathbf{Y} \in \mathcal{H}_{\mathbf{X}}(t-1)$ and so $\mathbf{Y} \perp \mathcal{I}_{\mathbf{X}}(t)$, yielding $\mathcal{I}_{\mathbf{X}}(t) \subset \text{sp}\{\boldsymbol{\zeta}_t\}$. Thus we have proved part (a) of the following lemma.

Lemma 4.10 *If $\boldsymbol{\zeta}_t$ is the innovation process of an q-variate stationary sequence $\mathbf{X}_t$, then*

(a) $\mathcal{I}_{\mathbf{X}}(t) = \text{sp}\{\boldsymbol{\zeta}_t\}, t \in \mathbb{Z}$;

(b) *$\boldsymbol{\zeta}_t$ is stationary and has the same shift as $\mathbf{X}_t$;*

(c) *$\mathbf{X}_t$ and $\boldsymbol{\zeta}_t$ are jointly stationary, that is, $\langle \mathbf{X}_t, \boldsymbol{\zeta}_s \rangle = \langle \mathbf{X}_{t+1}, \boldsymbol{\zeta}_{s+1} \rangle$, $s, t \in \mathbb{Z}$;*

(d) $\mathbf{\Sigma} = \text{Cov}\,(\boldsymbol{\zeta}_t, \boldsymbol{\zeta}_t)$ *is independent of* t *and we have*

$$[\langle \zeta_s^j, \zeta_t^k \rangle]_{j,k=1}^q = \mathbf{\Sigma}\ \delta_{s-t}, s, t \in \mathbb{Z};$$

(e) *any future innovation of* $\mathbf{X}_t$ *is orthogonal to the past of the sequence* $\mathbf{X}_t$. *In fact, for any positive integer* k *and any integer* t *we have*

$$\langle \boldsymbol{\zeta}_t, \mathbf{X}_{t-k} \rangle = \mathbf{0}, \ \textit{and}\ \langle \boldsymbol{\zeta}_t, \mathbf{X}_t \rangle = \mathbf{\Sigma}.$$

Proof. For (b), since the unitary shift operator U commutes with the projection onto $\mathcal{H}_{\mathbf{X}}(-\infty)$, then for every t

$$\begin{aligned}\boldsymbol{\zeta}_t &= \mathbf{X}_t - P_{\mathcal{H}_{\mathbf{X}}(-\infty)}\mathbf{X}_t = U\mathbf{X}_{t-1} - P_{\mathcal{H}_{\mathbf{X}}(-\infty)}U\mathbf{X}_{t-1} \\ &= U[\mathbf{X}_{t-1} - P_{\mathcal{H}_{\mathbf{X}}(-\infty)}\mathbf{X}_{t-1}] = U\boldsymbol{\zeta}_{t-1},\end{aligned} \tag{4.68}$$

showing that U is the shift for $\boldsymbol{\zeta}_t$. Hence $\boldsymbol{\zeta}_t$ is stationary.

For (c), it suffices to note because of (b) that $\boldsymbol{\zeta}_t$ and $\mathbf{X}_t$ have the same unitary shift.

For (d), note that $\text{Cov}\,(\boldsymbol{\zeta}_t, \boldsymbol{\zeta}_t)$ is independent of t due to (c), hence proving the equation when $s = t$. Now if $s < t$, then $\boldsymbol{\zeta}_s \in \mathcal{H}_{\mathbf{X}}(s)$ and $\boldsymbol{\zeta}_t = \mathbf{X}_t - \widetilde{\mathbf{X}}_t \perp \mathcal{H}_{\mathbf{X}}(s)$. Therefore $\boldsymbol{\zeta}_s \perp \boldsymbol{\zeta}_t$, which means $[\langle \zeta_s^j, \zeta_t^k \rangle]_{j,k=1}^q = \mathbf{0}$.

For (e), for any integer $k \geq 1$, $\mathbf{X}_{t-k} \in \mathcal{H}_X(t-1)$ and $\boldsymbol{\zeta}_t \perp \mathcal{H}_{\mathbf{X}}(t-1)$. Therefore we have $\langle \boldsymbol{\zeta}_t, \mathbf{X}_{t-k} \rangle = \mathbf{0}$. Using the top line of (4.67) to express $\mathbf{X}_t$ it is clear that

$$\langle \boldsymbol{\zeta}_t, \mathbf{X}_t \rangle = \langle \boldsymbol{\zeta}_t, \boldsymbol{\zeta}_t + P_{\mathcal{H}_{\mathbf{X}}(t-1)}\mathbf{X}_t \rangle = \langle \boldsymbol{\zeta}_t, \boldsymbol{\zeta}_t \rangle = \mathbf{\Sigma}. \quad \blacksquare$$

Next we turn to the notion of rank of a multivariate stationary sequence $\mathbf{X}_t$. In the time domain, where we work with a process and its innovation, there are two types of rank that we consider: its *process rank* denoted by $p_{\mathbf{X}}$, which is the dimension of $\text{sp}\{X_t^j : 1 \leq j \leq q\}$ and its *innovation rank*, denoted by $r_{\mathbf{X}}$, which is the dimension of $\text{sp}\{\zeta_t^j : 1 \leq j \leq q\}$. These, which because of stationarity are independent of t, turn out to be equal to $\text{rank}\,\mathbf{R}(0)$ and $\text{rank}\,\mathbf{\Sigma}$, respectively. Working with the spectral domain of $\mathbf{X}_t$, we say $\mathbf{X}_t$ has *spectral rank* $s_{\mathbf{X}}$ if it possesses a spectral density matrix $\mathbf{f}(\lambda)$ having rank $s_{\mathbf{X}}$ for a.e. λ.

The innovation rank is the most informative one because it describes the number of new (i.e., LI) random variables entering $\mathcal{H}_{\mathbf{X}}(t)$ at each time step and hence it has a direct bearing on prediction and on the complexity of the sequence. From now on the rank of $\mathbf{X}_t$ will mean its innovation rank. Here $\dim \mathcal{I}_{\mathbf{X}}(t)$ stands for the dimension of the set $\{\zeta_t^j : 1 \leq j \leq q\}$ and by the dimension of any finite subset A of a vector space we mean dimension of span of A, which turns out to be the maximum number of LI vectors in A.

From the preceding remarks, any deterministic multivariate sequence $\mathbf{X}_t$ has rank zero. But when $\mathbf{X}_t$ is nonsingular, it must have a nontrivial regular part having $\dim \mathcal{I}_{\mathbf{Z}}(0) = \dim \mathcal{I}_{\mathbf{X}}(0) = r > 0$. Since $\mathcal{I}_{\mathbf{X}}(t) = \text{sp}\{\zeta_0^j : 1 \le j \le q\}$ and the number of LI vectors in $\{\zeta_0^j : 1 \le j \le q\}$ is given by rank $\text{Cov}(\boldsymbol{\zeta}_0, \boldsymbol{\zeta}_0) =$ rank $\boldsymbol{\Sigma}$, we have the following corollary.

Corollary 4.10.1 *The rank of a q-variate stationary sequence $\mathbf{X}_t$ is*

$$r_{\mathbf{X}} = \text{rank}\, \boldsymbol{\Sigma}. \tag{4.69}$$

Lemma 4.11 *If $\mathbf{X}_t$ is multivariate sequence, then $r_{\mathbf{X}} \le p_{\mathbf{X}}$.*

For any $\alpha_1, \alpha_2, \ldots, \alpha_q$ we can write

$$\begin{aligned}
\sum \alpha_j X_0^j = 0 \quad &\Rightarrow \quad \left(\sum \alpha_j X_0^j | \mathcal{M}_X(-1)\right) = 0 \\
&\Rightarrow \quad \sum \alpha_j X_0^j - \sum \alpha_j \left(X_0^j | \mathcal{M}_X(-1)\right) = 0 \\
&\Rightarrow \quad \sum \alpha_j \left(X_0^j - (X_0^j | \mathcal{M}_X(-1))\right) = 0 \\
&\Rightarrow \quad \sum \alpha_j \zeta_0^j = 0,
\end{aligned}$$

which shows that for any positive integer $k \le q$

$$\{\mathbf{X}_0^{t_j} : 1 \le j \le k\} \text{ is LI whenever } \{\zeta_0^{t_j} : 1 \le j \le k\} \text{ is so.}$$

Using this fact one can prove the following lemma.

Lemma 4.12 *If $\mathbf{A}$ is a $q \times q$ and $\mathbf{B}$ is a $q \times r$ matrix such that $\mathbf{A} = \mathbf{B}\mathbf{B}^*$, then $\mathbf{A}$ and $\mathbf{B}$ have the same rank.*

Proof. It suffices to show that $\mathbf{A}$ and $\mathbf{B}$ have the same row rank and this is the immediate consequence of the following claim. If A_j and B_j denote the jth row of $\mathbf{A}$ and $\mathbf{B}$, respectively, then

$$\sum_{j=1}^{q} \alpha_j A_j = 0 \Leftrightarrow \sum_{j=1}^{q} \alpha_j B_j = 0$$

for any scalars $\alpha_1, \alpha_2, \ldots, \alpha_q$.

To verify this claim, we first note that the jth row of $\mathbf{A}$ is

$$A_j = [\langle B_j, B_1^* \rangle, \langle B_j, B_2^* \rangle, \ldots, \langle B_j, B_q^* \rangle]$$

and hence

$$\begin{aligned}\sum_{j=1}^{q}\alpha_j A_j &= \Big[\sum_{j=1}^{q}\alpha_j\langle B_j, B_1^*\rangle, \sum_{j=1}^{q}\alpha_j\langle B_j, B_2^*\rangle, \ldots, \sum_{j=1}^{q}\alpha_j\langle B_j, B_q^*\rangle\Big] \\ &= \Big[\Big\langle\sum_{j=1}^{q}\alpha_j B_j, B_1^*\Big\rangle, \Big\langle\sum_{j=1}^{q}\alpha_j B_j, B_2^*\Big\rangle, \ldots, \Big\langle\sum_{j=1}^{q}\alpha_j B_j, B_q^*\Big\rangle\Big]. \quad (4.70)\end{aligned}$$

Now if $\sum_{j=1}^{q}\alpha_j B_j = 0$ then it is clear from (4.70) that each entry of the row vector $\sum_{j=1}^{m}\alpha_j A_j$ is zero, and hence $\sum_{j=1}^{q}\alpha_j A_j$ itself is zero. On the other hand, if we assume $\sum_{j=1}^{q}\alpha_j A_j$ is zero, then all entries must be zero, which according to (4.70) gives

$$\Big\langle\sum_{j=1}^{q}\alpha_j B_j, B_i^*\Big\rangle = 0, \quad i = 1, 2, \ldots, q.$$

But that means $\sum_{j=1}^{q}\alpha_j B_j$, which is in the row space of $\boldsymbol{\varphi}$, is also orthogonal to the row space. Therefore $\sum_{j=1}^{q}\alpha_j B_j = 0$ must be zero. ∎

Lemma 4.13 *Any nonnegative definite $q \times q$ matrix $\mathbf{f} = [f^{kj}]$ of rank r can be factored as*

$$\mathbf{f} = \mathbf{\Phi}\mathbf{\Phi}^*, \tag{4.71}$$

where $\mathbf{\Phi} = [\phi^{kj}]$ is a $q \times r$ rectangular matrix of rank r.

Proof. Let $\mathbf{G}$ be a $q \times q$ square root of the matrix $\mathbf{f}$ and $\mathbf{V}$ be a $q \times q$ unitary matrix that diagonalizes $\mathbf{G}$ as

$$\mathbf{D} = \mathbf{V}\mathbf{G}\mathbf{V}^* = [d^{jj}],$$

where $\mathbf{D}$ is diagonal with $d^{jj} > 0$ for all $j = 1, 2, \ldots, r$. Now let $\widetilde{\mathbf{D}}$ be the rectangular matrix obtained from $\mathbf{D}$ by omitting the last $m - r$ columns. It is easy to see that the matrix $\mathbf{\Phi} = \mathbf{V}^*\widetilde{\mathbf{D}}$ serves as the desired factor. The statement about the rank of $\mathbf{\Phi}$ is clear from the last lemma. See also our discussion on the Cholesky factorization in Chapter 8.

Lemma 4.14 *If $\mathbf{\Phi} = [\phi^{kj}]$ is an $q \times r$ rectangular matrix of rank r then there exists an $r \times q$ matrix $\mathbf{\Psi} = [\psi^{ik}]$ such that*

$$\mathbf{\Phi}\mathbf{\Psi} = \mathbf{I}_r, \tag{4.72}$$

with $\mathbf{I}_r$ the $r \times r$ identity matrix.

Proof. For each $i = 1, 2, \ldots, q$ the system of r linear equations

$$\sum_{k=1}^{q} \psi^{ik}\phi^{kj} = \delta_{i-j}, \ \ j = 1, 2, \ldots, r$$

with q unknowns ψ^{ik}, $k = 1, 2, \ldots, q$ has a solution because the rank of its coefficient matrix $\boldsymbol{\varphi}$ is q. If $\psi^{i1}, \psi^{i2}, \ldots, \psi^{iq}$ satisfy this system, then the matrix $\boldsymbol{\Psi} = [\psi^{ik}]$ is clearly the required factor. ∎

Now consider an q-variate stationary sequence $\mathbf{X}_t = [X_t^j]$ of rank r with spectral density $\mathbf{f}(\lambda) = [f^{kj}(\lambda)]$ and spectral measure $\boldsymbol{\xi}(d\lambda)$. By Lemma 4.13 we have the factorization $\mathbf{f}(\lambda) = \boldsymbol{\Phi}(\lambda)\boldsymbol{\Phi}^*(\lambda)$, where the factor $\boldsymbol{\Phi}(\lambda)$ is an $q \times r$ matrix valued function, each entry of which is in $L^2[0, 2\pi)$. So we can write

$$\phi^{jk}(\lambda) = \sum_{n=-\infty}^{\infty} a_n^{jk} e^{-i\lambda n}, \qquad \sum_{n=-\infty}^{\infty} |a_n^{jk}|^2 < \infty, 1 \leq j \leq q; 1 \leq k \leq r.$$

Thus there exists an $r \times q$ matrix valued function $\boldsymbol{\Psi} = [\psi^{ik}]$ that satisfies (4.71) and (4.72) and consequently

$$\boldsymbol{\Psi}(\lambda)\mathbf{f}(\lambda)\boldsymbol{\Psi}^*(\lambda) = \mathbf{I}_r, \ \text{for a.e. } \lambda.$$

Define $\Lambda^j(\Delta)$, for any $j = 1, 2, \ldots, r$ and any Borel subset of $[0, 2\pi)$, by

$$\Lambda^j(\Delta) = \int_{\Delta} \psi^j(\lambda)\boldsymbol{\xi}(d\lambda),$$

where ψ^j is the jth row of the matrix $\boldsymbol{\Psi}$. By this choice of $\boldsymbol{\Psi}$ the random measures $\Lambda^j, j = 1, 2, \ldots, r$ are mutually uncorrelated, namely, $E\{\Lambda^j(\Delta)\overline{\Lambda^{j'}(\Delta')}\} = 0, \ j \neq j'$. Moreover

$$E|\Lambda^j(d\lambda)|^2 = d\lambda, \ \ j = 1, 2, \ldots, r.$$

Consider the r-variate stationary sequence

$$\boldsymbol{\zeta}_t = \int_0^{2\pi} e^{it\lambda}\boldsymbol{\Lambda}(d\lambda).$$

The above-mentioned properties of measures Λ^j imply that the sequence $\boldsymbol{\zeta}_t = [\zeta_t^j]$ is uncorrelated. In fact one can see that the set $\{\zeta_t^j : t \in \mathbb{Z}, 1 \leq j \leq r\}$ is an orthonormal set. That is, for any $i, j = 1, 2, \ldots, q$ we have

$$E\zeta_t^i\overline{\zeta_s^j} = \begin{cases} 1 & \text{if } i = j \text{ and } t = s \\ 0 & \text{otherwise} \end{cases}.$$

Now one can easily check that $(\mathbf{\Phi\Psi}-\mathbf{I}_q)\mathbf{f}(\mathbf{\Phi\Psi}-\mathbf{I}_q)^* = \mathbf{0}$ almost everywhere. Therefore $\mathbf{\Phi\Psi}$ and $\mathbf{I}_q$ are identical in $L^2(F)$ and hence we can write

$$\mathbf{X}_t = \int_0^{2\pi} e^{it\lambda}\boldsymbol{\xi}(d\lambda) = \int_0^{2\pi} e^{it\lambda}\mathbf{I}_m\boldsymbol{\xi}(d\lambda) = \int_0^{2\pi} e^{it\lambda}\mathbf{\Phi}(\lambda)\mathbf{\Psi}(\lambda)\boldsymbol{\xi}(d\lambda),$$

which gives

$$\mathbf{X}_t = \int_0^{2\pi} e^{it\lambda}\mathbf{\Phi}(\lambda)\mathbf{\Lambda}(d\lambda). \tag{4.73}$$

Substituting the above Fourier expansion of φ in (4.73), we get the moving average representation

$$\mathbf{X}_t = \sum_{n=-\infty}^{\infty} \mathbf{A}_n\boldsymbol{\zeta}_{t-n},$$

with $\mathbf{A}_n = [A_n^{kj}]$ being some $q\times r$ matrix and $\boldsymbol{\zeta}_n = \int_0^{2\pi} e^{in\lambda}\mathbf{\Lambda}(d\lambda)$ being an orthonormal sequence. This establishes the following theorem.

Theorem 4.10 *If a q-variate stationary sequence $\mathbf{X}_t$ has spectral rank r then it has a moving average representation*

$$\mathbf{X}_t = \sum_{n=-\infty}^{\infty} \mathbf{A}_n\boldsymbol{\zeta}_{t-n}, \tag{4.74}$$

where (1) the $q\times r$ coefficients $\mathbf{A_n}$ have square summable entries: $\sum_n |A_n^{kj}|^2 < \infty$ for $k = 1,2,\ldots,q$, $j = 1,2,\ldots,r$, and (2) the components $\{\zeta_t^j : t\in\mathbb{Z},\ 1\le j\le r\}$ of the r-variate sequence $\boldsymbol{\zeta}_t$ form an orthonormal basis for $\mathcal{H}_X$.

Proof. Based on the remarks preceding the statement of theorem we only need to show the last statement about the basicity of $\boldsymbol{\zeta}_t$. Let $\mathbf{\Phi} = [\phi^{kj}]$ and $\mathbf{\Lambda} = [\Lambda^j]$ be the Fourier transforms of $\mathbf{A}_n = [A_n^{kj}]$ and $\boldsymbol{\zeta}_n = [\zeta_n^j]$, respectively; for example, $\phi^{kj}(\lambda) = \sum_{n=-\infty}^{\infty} a_n^{kj}e^{-i\lambda n}$. The subspace $\mathcal{H}_\mathbf{Z}$ generated by the values of the sequence ζ_n^j coincides with the space $\mathcal{H}_\mathbf{\Lambda}$ of elements of the form

$$\mathbf{u} = \int \mathbf{\Psi}(\lambda)\mathbf{\Lambda}(\lambda)d\lambda \text{ with } \int \|\mathbf{\Psi}(\lambda)\mathbf{\Lambda}(\lambda)\|^2 d\lambda < \infty.$$

From the representation (4.74) it is clear that $\mathcal{H}_\mathbf{X} \subseteq \mathcal{H}_{\boldsymbol{\zeta}}$. On the other hand, since ζ_n^j is assumed to form a basis for $\mathcal{H}_\mathbf{X}$ we have $\mathcal{H}_\mathbf{\Lambda} \subset \mathcal{H}_\mathbf{X}$. So $\mathcal{H}_\mathbf{X} = \mathcal{H}_{\boldsymbol{\zeta}}$, which by our choice of $\mathbf{\Lambda}$ implies $\mathcal{H}_\mathbf{\Lambda} \subseteq \mathcal{H}_\mathbf{X}$. We only need to show that $\mathbf{X}_t$ has rank r. That is, we must show that its spectral density, which turns out to be $\mathbf{f}(\lambda) = \mathbf{\Phi}(\lambda)\mathbf{\Phi}^*(\lambda)$, has rank r for a.e. λ. Suppose not: that is, suppose that on a set of positive measure, the rank of the matrix $\mathbf{f}(\lambda)$ is less than r; then there exists a vector function $\mathbf{\Psi}(\lambda) = [\psi^j(\lambda)]$ such that

$\int \|\mathbf{\Psi}(\lambda)\|^2 d\lambda \neq 0$ and $\mathbf{\Phi}(\lambda)\mathbf{\Psi}^*(\lambda) = \mathbf{0}$, for a.e. λ. This would mean that the element $\mathbf{u} = \int \mathbf{\Psi}(\lambda)\mathbf{\Lambda}(\lambda)d\lambda$ is in $\mathcal{H}_{\mathbf{\Lambda}}$ and at the same time orthogonal to $\mathcal{H}_X$, because

$$\langle X_t^k, \mathbf{u}\rangle = \int e^{it\lambda} \sum_{j=0}^{r} \Phi^{kj}(\lambda)\overline{\psi^j(\lambda)}d\lambda = 0, \text{ for all } k \text{ and } t$$

which is a contradiction. ∎

The moving average representation claimed in this theorem is not unique. In fact, every such factorization of the form (4.71) of the spectral density $\mathbf{f}$ gives rise to one such moving average representation. The moving average representation in the last result is two sided. However, one sided moving averages are more useful in studying prediction of regular stationary processes and other issues studied in the next subsection.

4.4.4 Regular Processes

We start with defining the multivariate versions of *uncorrelated* and *white noise* sequences. A multivariate sequence $\boldsymbol{\xi}_t = [\xi_t^j]_{j=1}^m$ is called *uncorrelated* if for any $t \neq s$,

$$\text{Cov}\,(\xi_t^j, \xi_s^k) = 0, \text{ for every } j, k = 1, 2, \ldots, q.$$

Uncorrelated multivariate sequences are not necessarily stationary. For a simple univariate example start with an orthonormal sequence Y_t and set $X_t = Y_t$ for even and $X_t = 0.5Y_t$ for odd $t's$. When an uncorrelated multivariate sequence $\boldsymbol{\varepsilon}_t$ is stationary it is called a *white noise* , and in this case we have $\text{Cov}\,(\boldsymbol{\varepsilon}_t, \boldsymbol{\varepsilon}_s) = \mathbf{\Sigma}\,\delta_{t-s}$. We call a white noise $\boldsymbol{\varepsilon}_t$ for which $\text{Cov}\,(\boldsymbol{\varepsilon}_t, \boldsymbol{\varepsilon}_s) = \mathbf{I}\,\delta_{t-s}$ a normalized white noise. As we see in the next lemma, any white noise can be expressed in terms of a normalized white noise.

Lemma 4.15 *Every white noise $\boldsymbol{\xi}_t$ of rank $r \leq q$ can be written as $\boldsymbol{\xi}_t = \mathbf{A}\boldsymbol{\varepsilon}_t$, where $\mathbf{A}$ is an $q \times r$ matrix and $\boldsymbol{\varepsilon}_t$ is an r-variate normalized white noise. In this case we also have* $\text{Cov}\,(\boldsymbol{\xi}_t, \boldsymbol{\xi}_t) = \mathbf{A}\mathbf{A}'$.

Proof. Let $\{\varepsilon_0^1, \varepsilon_0^2, \ldots, \varepsilon_0^r\}$ be an orthogonal basis for $\text{sp}\{\xi_0^j : 1 \leq j \leq q\}$ and $\boldsymbol{\varepsilon}_0 = [\varepsilon_0^1\ \varepsilon_0^2\ \ldots\ \varepsilon_0^r]'$. Now let $\mathbf{A}$ be the $m \times r$ matrix that expresses the components of $\boldsymbol{\xi}_0$ in terms of the orthonormal basis $\{\varepsilon_0^1, \varepsilon_0^2, \ldots, \varepsilon_0^r\}$; that is, $\boldsymbol{\xi}_0 = \mathbf{A}\boldsymbol{\varepsilon}_0$. Let U be the unitary shift for $\boldsymbol{\xi}_t$ and set $\boldsymbol{\varepsilon}_t = U^t\boldsymbol{\varepsilon}_0 = [U^t\varepsilon_0^1, U^t\varepsilon_0^2, \ldots, U^t\varepsilon_0^r]'$, for each integer t. It is easy to see that $\boldsymbol{\varepsilon}_t$ is a normalized white noise for which

$$\boldsymbol{\xi}_t = U^t\boldsymbol{\xi}_0 = U^t\mathbf{A}\boldsymbol{\varepsilon}_0 = \mathbf{A}U^t\boldsymbol{\varepsilon}_0 = \mathbf{A}\boldsymbol{\varepsilon}_t.$$

The statement about the covariance follows from $\boldsymbol{\xi}_t = \mathbf{A}\boldsymbol{\varepsilon}_t$ and the easily checked fact that $\text{Cov}\,(\boldsymbol{\varepsilon}_t, \boldsymbol{\varepsilon}_t) = \mathbf{I}_r$. ∎

The preceding lemma enables us to show any uncorrelated multivariate stationary sequence is regular. We start with the following lemma.

Lemma 4.16 *Every multivariate normalized white noise $\boldsymbol{\varepsilon}_t = [\varepsilon_t^j]$ is regular.*

Proof. The assumptions on sequence $\boldsymbol{\varepsilon}_t$ ensure that $\{\varepsilon_t^j : t \in \mathbb{Z}, 1 \leq j \leq m\}$ forms an orthonormal basis for $\mathcal{H}_{\boldsymbol{\varepsilon}}$. So for any fixed vector $\mathbf{u} \in \mathcal{H}_{\boldsymbol{\varepsilon}}(-\infty) \subset \mathcal{H}_{\boldsymbol{\varepsilon}}$ we can write

$$\mathbf{u} = \sum_{t=-\infty}^{\infty} \sum_{j=1}^{q} a_t^j \varepsilon_t^j, \quad \text{with } a_t^j = \langle \mathbf{u}, \varepsilon_t^j \rangle.$$

On the other hand, the expression

$$\mathbf{u} \in \mathcal{H}_{\boldsymbol{\varepsilon}}(-\infty) \subset \mathcal{H}_{\boldsymbol{\varepsilon}}(t-1) \subset \mathcal{H}_{\boldsymbol{\varepsilon}}(t), \ t \in \mathbb{Z},$$

in conjunction with fact that $\varepsilon_t^j \perp \mathcal{H}_{\boldsymbol{\varepsilon}}(t-1)$, implies that

$$a_t^j = \langle \mathbf{u}, \varepsilon_t^j \rangle = 0$$

for each $t \in \mathbb{Z}, 1 \leq j \leq q$. This, in turn, implies that $\mathbf{u} = \mathbf{0}$. Therefore we get $\mathcal{H}_{\boldsymbol{\varepsilon}}(-\infty) = \mathbf{0}$, which completes the proof. ∎

Proposition 4.11 *Every uncorrelated multivariate stationary sequence $\boldsymbol{\xi}_t$ is regular.*

Proof. By Lemma 4.15 we can write $\boldsymbol{\xi}_t = \mathbf{A}\boldsymbol{\varepsilon}_t$, for some $q \times r$ matrix $\mathbf{A}$ and some r-variate normalized white noise $\boldsymbol{\varepsilon}_t$. This implies that $\mathcal{H}_{\boldsymbol{\xi}}(t) \subset \mathcal{H}_{\boldsymbol{\varepsilon}}(t)$ for every integer t, and hence

$$\mathcal{H}_{\boldsymbol{\xi}}(-\infty) \subset \mathcal{H}_{\boldsymbol{\varepsilon}}(-\infty).$$

But $\mathcal{H}_{\boldsymbol{\varepsilon}}(-\infty) = \{\mathbf{0}\}$ by Lemma 4.16 and so $\mathcal{H}_{\boldsymbol{\xi}}(-\infty) = \{\mathbf{0}\}$. This complete the proof. ∎

Next we characterize all multivariate regular stationary sequences.

Proposition 4.12 (Moving Average Representation)

(a) *Any q-variate one sided moving average of an r-variate normalized white noise* ε_t*:*

$$\mathbf{X}_t = \sum_{s\geq 0} \mathbf{A}_s \varepsilon_{t-s} \tag{4.75}$$

with

$$\sum_{s\geq 0} |a_s^{jk}|^2 < \infty, \ 1 \leq j \leq q; \ 1 \leq k \leq r, \tag{4.76}$$

is stationary and regular.

(b) *Any q-variate regular stationary process* $\mathbf{X}_t$ *of rank* $r \leq q$ *has a one sided moving average representation*

$$\mathbf{X}_t = \sum_{s\geq 0} \mathbf{A}_s \varepsilon_{t-s}, \tag{4.77}$$

where ε_t *is an r-variate normalized white noise with matrices* $\mathbf{A}_t = [a_t^{jk}]$ *satisfying (4.76).*

Proof. For (a), suppose $\mathbf{X}_t$ is given by (4.75) with $\{a_t^{jk}\}$ and ε_t having the stated properties. First note that orthonormality of all the components $\{\varepsilon_t^j : t \in \mathbb{Z}, 1 \leq j \leq r\}$ of ε_t and the square summability (4.76) of $\{a_t^{jk}\}$ ensure that X_t^j is well defined for every integer t and any $j = 1, 2, \ldots, q$. It is obvious from (4.75) that $\mathcal{H}_X(t) \subset \mathcal{H}_\varepsilon(t)$ for every integer t and hence

$$\mathcal{H}_X(-\infty) \subset \mathcal{H}_\varepsilon(-\infty).$$

This, and the regularity of ε_t imply that $\mathbf{X}_t$ is regular. Now consider the unitary operator defined on $\mathcal{H}_\varepsilon$ via

$$U\varepsilon_t^j = \varepsilon_{t+1}^j, \ \ t \in \mathbb{Z}; \ \ j = 1, 2, ..., r.$$

The linearity and continuity of U implies that

$$U\mathbf{X}_t = U\Big(\sum_{j\geq 0} \mathbf{A}_j \varepsilon_{t-j}\Big) = \sum_{j\geq 0} \mathbf{A}_j U\varepsilon_{t-j} = \sum_{j\geq 0} \mathbf{A}_j \varepsilon_{t+1-j} = \mathbf{X}_{t+1},$$

showing that U serves as a unitary shift for $\mathbf{X}_t$, and hence $\mathbf{X}_t$ is stationary .

For (b), conversely suppose $\mathbf{X}_t$ is a regular q-variate stationary sequence of rank $r \leq q$. Since the innovation spaces $\mathcal{I}_\mathbf{X}(t)$ appearing in (4.66) are of dimension r and $\mathcal{I}_\mathbf{X}(t) \perp \mathcal{I}_\mathbf{X}(s)$ for $t \neq s$, we may express any vector $\mathbf{u} \in \mathcal{H}_\mathbf{X}(t)$ as

$$\mathbf{u} = \sum_{s\geq 0} \sum_{j=1}^{r} \langle \mathbf{u}, \varepsilon_{t-s}^j \rangle \varepsilon_{t-s}^j, \text{ with } \sum_{s\geq 0} |\langle \mathbf{u}, \varepsilon_{t-s}^j \rangle|^2 < \infty$$

where $\{\varepsilon_0^j : 1 \le j \le r\}$ is any fixed orthonormal basis for $\mathcal{I}_{\mathbf{X}}(0)$ and

$$\varepsilon_t^j = U^t \varepsilon_0^j, \quad t \in \mathbb{Z}, 1 \le j \le r.$$

In particular, taking $\mathbf{u} = X_t^k$, we get

$$X_t^k = \sum_{s\ge 0}\sum_{j=1}^{r} \langle X_t^k, \varepsilon_{t-s}^j\rangle \varepsilon_{t-s}^j \quad \text{with} \quad \sum_{s\ge 0} |\langle X_t^k, \varepsilon_{t-s}^j\rangle|^2 < \infty.$$

Thus we have shown that each component of $\mathbf{X}_t$ and hence $\mathbf{X}_t$ has the representation (4.77). It remains to show that each coefficient $\langle X_t^k, \varepsilon_{t-s}^j\rangle$ is independent of t. However this is an immediate consequence of uniqueness of such expansions together with

$$X_{t+1}^k = \sum_{s\ge 0} \langle X_{t+1}^k, \varepsilon_{t+1-s}^j\rangle \varepsilon_{t+1-s}$$

and

$$X_{t+1}^k = UX_t^k = \sum_{s\ge 0} \langle X_t^k, \varepsilon_{t-s}^j\rangle U[\varepsilon_{t-j}] = \sum_{s\ge 0} \langle X_t^k, \varepsilon_{t-s}^j\rangle \varepsilon_{t+1-j}.$$

Here U may be brought inside the sum due to convergence of the partial sums and continuity of U. The last equality follows from the fact that $\varepsilon_{t+1}^j = U\varepsilon_t^j$, a conclusion that may be drawn from item (a) of Lemma 4.3. Taking $a_s^{jk} = \langle X_t^k, \varepsilon_{t-s}^j\rangle$ one arrives at (4.76). ∎

4.4.5 Infinite Past Prediction

As in the univariate case in Section 4.2, we can evaluate the prediction error of $\widetilde{\mathbf{X}}_\delta$, $\delta \ge 1$, by expressing it in terms of the innovations of $\mathbf{X}_t$.

Proposition 4.13 *If $\mathbf{X}_t$ is a regular q-variate stationary sequence with one sided moving average (4.76), in terms of its innovation $\boldsymbol{\zeta}_t$, then its δ-step ahead predictor $\widetilde{\mathbf{X}}_\delta = (\mathbf{X}_\delta | \mathcal{H}_{\mathbf{X}}(0))$ based on its past is given by*

$$\widetilde{\mathbf{X}}_\delta = \sum_{s=\delta}^{\infty} \mathbf{A}_s \boldsymbol{\zeta}_{\delta-s}, \quad \delta \ge 1. \tag{4.78}$$

Its prediction error is given by

$$\mathbf{X}_\delta - \widetilde{\mathbf{X}}_\delta = \sum_{s=0}^{\delta-1} \mathbf{A}_s \boldsymbol{\zeta}_{\delta-s} \tag{4.79}$$

and has variance

$$\operatorname{Var}(\mathbf{X}_\delta - \widetilde{\mathbf{X}}_\delta) = \sum_{s=0}^{\delta-1} \mathbf{A}_s \mathbf{\Sigma} \mathbf{A}_s'. \tag{4.80}$$

Proof. It is clear that $\sum_{s=\delta}^{\infty} \mathbf{A}_s \boldsymbol{\zeta}_{\delta-s}$ belongs to $\mathcal{H}_{\mathbf{X}}(0)$ and $\sum_{s=0}^{\delta-1} \mathbf{A}_s \boldsymbol{\zeta}_{\delta-s}$ is orthogonal to $\mathcal{H}_{\mathbf{X}}(0)$. These in conjunction with uniqueness of projection gives

$$\widetilde{\mathbf{X}}_\delta = (\mathbf{X}_\delta | \mathcal{H}_{\mathbf{X}}(0)) = \sum_{s=\delta}^{\infty} \mathbf{A}_s \boldsymbol{\zeta}_{\delta-s}.$$

The error formulas are now immediate. ∎

Corollary 4.13.1 *Let $\mathbf{X}_t$ be an q-variate stationary sequence with innovation process $\boldsymbol{\zeta}_t$ and let $\mathbf{X}_t = \mathbf{Y}_t + \mathbf{Z}_t$ be its Wold decomposition (4.65) of Proposition 4.10 with its regular component $\mathbf{Z}_t$ having one sided moving average (4.76), namely, $\mathbf{Z}_t = \sum_{s\geq 0} \mathbf{A}_s \boldsymbol{\zeta}_{t-s}$. Then the δ-step ahead predictor of $\mathbf{X}_\delta$ based on its past $\mathbf{X}_0, \mathbf{X}_{-1}, \ldots$ is given by*

$$\widetilde{\mathbf{X}}_\delta = \sum_{s=\delta}^{\infty} \mathbf{A}_s \boldsymbol{\zeta}_{\delta-s} + \mathbf{Y}_\delta \quad \delta \geq 1. \tag{4.81}$$

In this case (4.79) and (4.80) remain true.

Proof. From the Wold Decomposition Theorem we see that $\mathcal{H}_{\mathbf{X}}(0) = \mathcal{H}_{\mathbf{X}}(-\infty) \oplus \mathcal{H}_{\mathbf{Z}}(0)$, which implies $\mathbf{Y}_\delta + \sum_{s=\delta}^{\infty} \mathbf{A}_s \boldsymbol{\zeta}_{\delta-s}$ belongs to $\mathcal{H}_{\mathbf{X}}(0)$. On the other hand, obviously $\sum_{s=0}^{\delta-1} \mathbf{A}_s \boldsymbol{\zeta}_{\delta-s}$ is orthogonal to $\mathcal{H}_{\mathbf{X}}(0)$, which completes the proof of the first part. The error formulas are again immediate. ∎

In principle, the moving average matrix coefficients $\mathbf{A}_s$ can be found in terms of the matrix autocorrelation function $\mathbf{R}(t)$. One can prove the following lemma as in its univariate version.

Corollary 4.13.2 *If $\mathbf{X}_t$ is a regular stationary sequence with one sided moving average representation $\mathbf{X}_t = \sum_{k=0}^{\infty} \mathbf{A}_k \boldsymbol{\zeta}_{t-k}$ in terms of its innovation process $\boldsymbol{\zeta}_t$ (4.76), then*

$$\mathbf{R}(t) = \sum_{s=0}^{\infty} \mathbf{A}_{s+t} \overline{\mathbf{A}_s}, \ t \in \mathbb{Z}.$$

Here we use the convention of taking $\mathbf{A}_s = \mathbf{0}$ for each negative s.

As in the univariate case one can write the last equation as

$$\tilde{\mathbf{R}} = \tilde{\mathbf{\Gamma}}\Sigma\tilde{\mathbf{\Gamma}}^*,$$

where for any $j, k = 0, 1, \ldots,$ the entries of the matrices $\tilde{\mathbf{R}}$ and $\tilde{\mathbf{\Gamma}}$ are defined by $\tilde{\mathbf{R}}^{kj} = \mathbf{R}(j-k)$ and $\tilde{\mathbf{R}}^{kj} = \mathbf{A}_{j-k}$, respectively. This again suggests that the moving average coefficients for a regular sequence and its predictor's coefficients $\{\mathbf{A}_k : k = 0, 1, \ldots\}$ can be obtained from a matricial Cholesky factorization of the covariance matrix $\tilde{\mathbf{R}}$. For further discussion see Propositions 4.6 and 8.9.

4.4.6 Spectral Theory and Rank

The notion of rank of multivariate stationary sequences can be described nicely in spectral terms. This is another reason that the innovation rank is considered a more natural notion than the rank of $\mathbf{R}(0)$. The following is due to Rozanov [199] who credits a 1941 note of Zasuhin [229].

Theorem 4.11 (Spectral Characterization of Regularity) *A q-variate stationary sequence $\mathbf{X}_t$ is regular if and only if all entries F^{jk} of its spectral measure $\mathbf{F}$ are absolutely continuous w.r.t. Lebesgue measure and its spectral density $\mathbf{f}(\lambda)$ can be factored as*

$$\mathbf{f}(\lambda) = \mathbf{\Phi}(\lambda)\mathbf{\Phi}^*(\lambda) \tag{4.82}$$

with factor $\mathbf{\Phi}(\lambda)$ being a $q \times r$ matrix valued function (for some $r \le q$) with entries in $L^2[0, 2\pi]$ such that

$$\phi^{jk}(\lambda) = \sum_{n=0}^{\infty} a_n^{jk} e^{-i\lambda n}, \quad \sum_{n=0}^{\infty} |a_n^{jk}|^2 < \infty, \; j = 1, 2, \ldots, q, \quad k = 1, 2, \ldots, r.$$

Proof. Suppose $\mathbf{X}_t$ is a regular q-variate stationary sequence. By part (b) of Proposition 4.11 it has one sided moving average representation (4.76)

$$\mathbf{X}_t = \sum_{s \ge 0} \mathbf{A}_s \boldsymbol{\zeta}_{t-s}$$

with properties specified there. The r-variate normalized white noise sequence $\boldsymbol{\zeta}_t$ has a spectral representation

$$\boldsymbol{\zeta}_t = \int_0^{2\pi} e^{it\lambda} \boldsymbol{\eta}(d\lambda),$$

where the r-variate random measure $\boldsymbol{\eta}(d\lambda)$ has orthogonal components and uniform spectral density,

$$\langle \eta^j(A), \eta^k(B) \rangle = \frac{\text{Leb}(A \cap B)}{2\pi} \delta_{j-k}.$$

Substituting for $\boldsymbol{\zeta}_t$ in (4.76) we get

$$\mathbf{X}_t = \int_0^{2\pi} \Big[\sum_{s=0}^{\infty} \mathbf{A}_s e^{-is\lambda}\Big] e^{it\lambda} \boldsymbol{\eta}(d\lambda) = \int_0^{2\pi} \boldsymbol{\Phi}(\lambda) e^{it\lambda} \boldsymbol{\eta}(d\lambda), \tag{4.83}$$

which gives

$$\mathbf{R}_X(\tau) = \int_0^{2\pi} \boldsymbol{\Phi}(\lambda)\boldsymbol{\Phi}^*(\lambda) e^{i\tau\lambda} d\lambda.$$

From this, by virtue of uniqueness of Fourier transforms, we arrive at $\mathbf{f}(\lambda) = \boldsymbol{\Phi}(\lambda)\boldsymbol{\Phi}^*(\lambda)$, which is the desired factorization. Conversely, suppose $\mathbf{X}_t$ is a q-variate stationary sequence having an absolutely continuous spectral measure with spectral density $\mathbf{f}(\lambda)$ satisfying (4.82). Consider an infinite dimensional Hilbert space with an orthonormal basis specially labeled as $\{\zeta_t^j : t \in \mathbb{Z}, 1 \leq j \leq r\}$. Now let $\mathbf{Y}_t$ be the q-variate sequence defined by

$$\mathbf{Y}_t = \sum_{s \geq 0} \mathbf{A}_s \boldsymbol{\zeta}_{t-s}.$$

It is clear that $\mathbf{Y}_t$ is regular and has autocovariance function

$$\mathbf{R}_Y(\tau) = \int_0^{2\pi} \boldsymbol{\Phi}(\lambda)\boldsymbol{\Phi}^*(\lambda) e^{i\tau\lambda} d\lambda.$$

This and (4.4.6) show that $\mathbf{X}_t$ and $\mathbf{Y}_t$ have the same correlation structure, and hence $\mathbf{X}_t$ is regular as well. ∎

Proposition 4.14 *The various ranks of a regular q-variate stationary sequence $\mathbf{X}_t$ satisfy the following inequalities:*

$$s = r \leq p.$$

Proof. The claim $r \leq p$ was proved in Lemma 4.11. In order to prove $r = s$ we first show $s \leq r$. To begin, the q-variate regular stationary process $\mathbf{X}_t$ of rank $r \leq q$ has a one sided moving average representation (4.76)

$$\mathbf{X}_t = \sum_{s \geq 0} \mathbf{A}_s \boldsymbol{\zeta}_{t-s},$$

where $\boldsymbol{\zeta}_t$ is an r-variate white noise. This forces its spectral density to factor as (4.82)

$$\mathbf{f}(\lambda) = \boldsymbol{\Phi}(\lambda)\boldsymbol{\Phi}^*(\lambda) \tag{4.84}$$

with $\boldsymbol{\Phi}(\lambda)$ being an $q \times r$ matrix valued function. Now, on one hand, since $\boldsymbol{\Phi}(\lambda)$ is $q \times r$ its rank can never exceed r. On the other hand, its rank, by Lemma 4.12, is a.e. equal to s. That is, $s \leq \text{rank}\, \boldsymbol{\Phi}(\lambda) \leq r$.

Since $\mathbf{f}(\lambda)$ has rank s, then for a.e. λ we have $\mathbf{f}(\lambda) = \mathbf{\Phi}(\lambda)\mathbf{\Phi}^*(\lambda)$, where $\mathbf{\Phi}(\lambda)$ is $q \times s$. Let $\boldsymbol{\eta}$ be a s-dimensional vector of orthogonally scattered random measures and set $\mathbf{Y}_t = \int_0^{2\pi} \mathbf{\Phi}(\lambda)e^{it\lambda}\boldsymbol{\eta}(d\lambda)$. Then $\mathbf{R}_Y(\tau) = \mathbf{R}_X(\tau)$, so $\mathbf{X}_t$ and $\mathbf{Y}_t$ are essentially the same. But then $\mathbf{\Phi}(\lambda)$ is $L^2[0, 2\pi)$ because $\mathbf{f}(\lambda)$ is $L^1[0, 2\pi)$. So $\mathbf{\Phi}(\lambda)$ has an L^2 Fourier series, $\mathbf{\Phi}(\lambda) = \sum_{j=0}^{\infty} \mathbf{A}'_j e^{ij\lambda}$, leading to

$$\begin{aligned} \mathbf{Y}_t &= \sum_{j=0}^{\infty} \mathbf{A}'_j \int_0^{2\pi} e^{i\lambda(t-j)}\boldsymbol{\eta}(d\lambda) \\ &= \sum_{j=0}^{\infty} \mathbf{A}'_j \boldsymbol{\varepsilon}'_{t-j}, \end{aligned}$$

where $\boldsymbol{\varepsilon}'_j$ is an s-vector of mutually orthogonal white sequences. But this means that the innovation space of $\mathbf{Y}_t$ is of dimension at most s, meaning $r \leq s$. ∎

4.4.7 Spectral Theory and Prediction

If $\mathbf{X}_t$ is a q-variate regular sequence of rank r, then by part (b) of Proposition 4.12 it has a one sided moving average representation (with convergence in the mean-square sense, component-wise)

$$\mathbf{X}_t = \sum_{k=0}^{\infty} \mathbf{A}_k \boldsymbol{\zeta}_{t-k},$$

in terms of a sequence of $r \times m$ matrices $\mathbf{A}_k$ and an r-variate white noise $\boldsymbol{\zeta}_t$ with $\mathcal{I}(t) = \text{sp}\{\boldsymbol{\zeta}_t\}$. Hence, as discussed earlier, one can write the predictor of $\mathbf{X}_\delta$ based on its past $..., \mathbf{X}_{-1}, \mathbf{X}_0$ as

$$\widetilde{\mathbf{X}}_\delta = \sum_{k=\delta}^{\infty} \mathbf{A}_k \boldsymbol{\zeta}_{t-k}. \tag{4.85}$$

But this representation of the predictor is in terms of innovations $\boldsymbol{\zeta}_t$, which cannot usually be observed. In order to resolve this problem note that for each integer t, the innovation $\boldsymbol{\zeta}_t \in \mathcal{H}_X(t)$ and thus is a limit of some finite linear combinations of $\mathbf{X}_t$, namely,

$$\boldsymbol{\zeta}_t = \lim_n \sum_{k=0}^{n} \mathbf{C}_k(n)\mathbf{X}_{t-k},$$

where $\mathbf{C}_k(n)$ is independent of t is due to the stationarity of $\mathbf{X}_t$. Under suitable conditions on $\mathbf{f}$, (cf. [152,156,159,225]) one can show that the limiting process can be achieved by a series,

$$\boldsymbol{\zeta}_t = \sum_{k=0}^{\infty} \mathbf{C}_k \mathbf{X}_{t-k},$$

with convergence in mean-square sense. Substituting for $\boldsymbol{\zeta}_t$ from last equation into (4.85) we obtain

$$\widetilde{\mathbf{X}}_\delta = \sum_{k=0}^{\infty} \mathbf{D}_{\delta k} \mathbf{X}_{t-k}, \quad \text{with } \mathbf{D}_{\delta k} = \sum_{n=0}^{k} \mathbf{A}_{\delta+n} \mathbf{C}_{k-n}.$$

This expresses the predictor in terms of the past of the process itself, which is observable. Algorithms for determining the coefficients $\mathbf{A}_k, \mathbf{C}_k$, and subsequently $\mathbf{D}_k$ from the spectral density, which generalizes our presentation in the univariate case, is available in the literature. For a full account of this one can refer to [152, 224, 225].

Lemma 4.17 *Let $\mathbf{X}_t$ be a regular full rank q-variate stationary sequence with spectral distribution function $\mathbf{F}$ and spectral density $\mathbf{f}$. Suppose $\boldsymbol{\zeta}_t$ and $\boldsymbol{\Phi}$ denote its q-variate innovation and generating function, respectively, then*

(a) *$e^{it\lambda}\boldsymbol{\Phi}^{-1}$ is in the spectral domain $L^2(\mathbf{F})$ of $\mathbf{X}_t$ and corresponds to $\boldsymbol{\zeta}_t$ in its time domain $\mathcal{H}_{\mathbf{X}}$;*

(b) *for any $\boldsymbol{\Psi} \in L^2(\mathbf{F})$, $\boldsymbol{\Psi\Phi} \in L^2[0, 2\pi)$;*

(c) *for any $\boldsymbol{\Psi} \in L^2(\mathbf{F})$, if $\mathbf{A}_k$ is the k-th Fourier coefficient of $\boldsymbol{\Psi\Phi}$, then*

$$\lim_{n\to\infty} \Big(\sum_{-n}^{n} \mathbf{A}_k e^{ik\lambda}\Big)\boldsymbol{\Phi}^{-1} = \boldsymbol{\Psi}, \text{ in the } L^2(\mathbf{F}) \text{sense.}$$

Theorem 4.12 *Suppose the spectral measure F of a random sequence $\mathbf{X}_t$ is absolutely continuous w.r.t. Lebesgue measure and its spectral density $\mathbf{f}$ satisfies the boundedness condition :*

$$\nu \mathbf{I} < \mathbf{f} < \mu \mathbf{I}, \quad \nu > 0. \tag{4.86}$$

Then for any $\delta > 1$,

$$\widetilde{\mathbf{X}}_\delta = \sum_{k=0}^{\infty} \mathbf{D}_{\delta k} \mathbf{X}_{-k}, \quad \textit{with } \mathbf{D}_{\delta k} = \sum_{n=0}^{k} \mathbf{A}_{\delta+n} \mathbf{C}_{k-n}.$$

The function $\mathbf{\Phi} = \sum_{k=0}^{\infty} \mathbf{A}_k e^{-ik\lambda}$ *is called the spectral generating function of* $\mathbf{X}_t$ *and* $\mathbf{\Phi}^{-1} = \sum_{k=0}^{\infty} \mathbf{C}_k e^{-ik\lambda}$ *is its inverse. In this case the prediction error matrix of lag* δ *is given by*

$$\sum_{n=1}^{\delta-1} \mathbf{A}_n \mathbf{R}(0) \mathbf{A}_n^*,$$

where $\mathbf{R}(0) = \mathrm{Cov}\,(\mathbf{X}_0, \mathbf{X}_0)$.

Note that when the boundedness condition (4.86) is assumed, the corresponding processes turn out to be of full rank. Therefore this last theorem holds only for full rank processes. For some generalization of this result to nonfull rank case, see [156] and [159].

4.4.8 Finite Past Prediction

The problem addressed here is that of predicting a member, say $\mathbf{X}(t+\delta)$, of a q-variate stationary sequence $\mathbf{X}_t$ based on on a finite number of observations $\{\mathbf{X}_{t-n+1}, \mathbf{X}_{t-n}, \ldots, \mathbf{X}_t\}$ in its past. We take the best linear predictor to mean the component-wise orthogonal projection of $\mathbf{X}_{t+\delta}$ onto $\mathcal{M}_{\mathbf{X}}(t;n) = \mathrm{sp}\{X_s^j : t-n < s \leq t, 1 \leq j \leq q\}$, namely,

$$\widehat{\mathbf{X}}_{t+\delta,n} = (\mathbf{X}_{t+\delta} | \mathcal{M}_{\mathbf{X}}(t;n)). \tag{4.87}$$

As in the univariate case we are especially interested in $\delta = 1$ and $\delta = -n$. We will treat only the case of $\delta = 1$ and assume the process is real. Completion of some details, being similar to the univariate case, are suggested as a problem. Assuming $\delta = 1$, we seek the coefficients of the following linear expression:

$$\widehat{\mathbf{X}}_{t+1,n} = \sum_{j=1}^{n} \mathbf{A}_{nj} \mathbf{X}_{t-j+1}. \tag{4.88}$$

The multivariate normal equations arising from the properties of projection can be written in terms of all the components of $\mathbf{X}_{t+1} - \widehat{\mathbf{X}}_{t+1,n}$ and $\mathbf{X}_s$, but this can be more conveniently expressed as

$$\langle \mathbf{X}_{t+1} - \widehat{\mathbf{X}}_{t+1,n}, \mathbf{X}_s \rangle = \mathbf{0}, \quad t-n+1 \leq s \leq t,$$

where the Gramian $\langle \mathbf{X}, \mathbf{Y} \rangle$ of any two q-vectors $\mathbf{X} = [X^1, X^2, \ldots, X^q]'$ and $\mathbf{Y} = [Y^1, Y^2, \ldots, Y^q]'$ is the $q \times q$ matrix whose (i,j)th entry is $\langle X^i, Y^j \rangle$.

These normal equations can be expressed in terms of the autocorrelation function

$$\sum_{j=1}^{n} \mathbf{A}_{nj} \mathbf{R}(t-j+1-s) = \mathbf{R}(t+1-s), \quad t-n+1 \leq s \leq t, \tag{4.89}$$

or in matrix form

$$\begin{bmatrix} \mathbf{R}(1) \\ \mathbf{R}(2) \\ \vdots \\ \mathbf{R}(n) \end{bmatrix} = \begin{bmatrix} \mathbf{R}(0) & \cdots & \mathbf{R}(-n+1) \\ \mathbf{R}(1) & \cdots & \mathbf{R}(-n+2) \\ \vdots & \vdots & \vdots \\ \mathbf{R}(n-1) & \cdots & \mathbf{R}(0) \end{bmatrix} \begin{bmatrix} \mathbf{A}_{n1} \\ \mathbf{A}_{n2} \\ \vdots \\ \mathbf{A}_{nn} \end{bmatrix}. \tag{4.90}$$

Finally, the normal equations can be expressed as

$$\tilde{\mathbf{R}}_n^{\mathbf{1}} = \tilde{\mathbf{R}}_n \tilde{\mathbf{A}}_n. \tag{4.91}$$

As we see, these matrices $\tilde{\mathbf{R}}_n^{\mathbf{1}}, \tilde{\mathbf{R}}_n$ and hence $\tilde{\mathbf{A}}_n$ do not depend on t, which is of course due to the stationarity of $\mathbf{X}_t$. For any $\tilde{\mathbf{A}}_n = [\mathbf{A}_{n1}\mathbf{A}_{n2}\ldots\mathbf{A}_{nn}]'$ that solves (4.90), the prediction error

$$\zeta_{t+1,n} = \mathbf{X}_{t+1} - \widehat{\mathbf{X}}_{t+1,n} \tag{4.92}$$

has covariance

$$\begin{aligned} \mathbf{\Sigma}_n(t+1) &= \text{Var}\,(\mathbf{X}_{t+1} - \widehat{\mathbf{X}}_{t+1,n}) \\ &= \langle \mathbf{X}_{t+1} - \widehat{\mathbf{X}}_{t+1,n},\ \mathbf{X}_{t+1} \rangle \\ &= \mathbf{R}(0) - \sum_{j=1}^{n} \mathbf{A}_{nj}\mathbf{R}(j) \\ &= \mathbf{R}(0) - (\tilde{\mathbf{R}}_n^1)'\tilde{\mathbf{A}}_n. \end{aligned} \tag{4.93}$$

The second line follows from the fact that $\langle(\mathbf{X}_{t+1} - \widehat{\mathbf{X}}_{t+1,n},\ \widehat{\mathbf{X}}_{t+1,n}\rangle = \mathbf{0}$. The last line shows that $\mathbf{\Sigma}_n(t+1)$ is independent of t and so from now on we denote it by $\mathbf{\Sigma}_n$.

The corresponding relationship for predicting $\mathbf{X}_{t-n}$ based on $\mathcal{M}(t;n) = \text{sp}\{X_s^j : t-n < s \le t, 1 \le j \le q\}$ follows exactly as in the univariate case (see discussion following (4.36) leading up to (4.42)).

In the following proposition we extend to the multivariate case some of the results we proved earlier for the univariate case. Here we use the following notation: $\mathbf{\Sigma} = [\sigma^{ij}] = \text{Var}\,(\mathbf{X}_{t+1} - \widehat{\mathbf{X}}_{t+1})$ and $\mathbf{\Sigma}_n = \text{Var}\,(\mathbf{X}_{t+1} - \widehat{\mathbf{X}}_{t+1,n})$. Recall that for two $n \times n$ matrices $\mathbf{A}$ and $\mathbf{B}$ we write $\mathbf{A} \ge \mathbf{B}$ if $\mathbf{A} - \mathbf{B}$ is nonnegative definite.

Proposition 4.15 *If $\mathbf{X}_t$ is an q-variate stationary sequence, then*

(a) *for any two positive integers n and k with $k > n$,*

$$\mathbf{\Sigma}_n \ge \mathbf{\Sigma}_k \ge \mathbf{\Sigma};$$

(b) *for every $j = 1, 2, \ldots, q$, the sequence σ_n^{jj} is bounded and nonincreasing;*

(c) *whenever* $\sigma_n^{jj} = 0$ *for some* j *and some* n, *then* $\sigma_m^{jj} = 0$ *for any* $m \geq n$;

(d) $\sigma_n^{jj} \to \sigma^{jj}$, *for any* $j = 1, 2, \ldots, q$;

(e) $\lim_{n\to\infty} \mathbf{\Sigma}_n = \mathbf{\Sigma}$, *entry-wise;*

(f) $\lim_{n\to\infty} \widehat{\mathbf{X}}_{t+1,n} = \widehat{\mathbf{X}}_{t+1}$;

(g) *rank of* $\mathbf{\Sigma}_n$ *is nonincreasing and nullity of* $\mathbf{\Sigma}_n$ *is nondecreasing;*

(h) $\lim_{n\to\infty} |\mathbf{\Sigma}_n| = |\mathbf{\Sigma}|$;

(i) *if* $\tilde{\mathbf{R}}_n$ *is invertible then*

$$|\mathbf{\Sigma}_n| \quad = \quad |\mathbf{R}(0) - (\tilde{\mathbf{R}}_n^1)'\tilde{\mathbf{R}}_n^{-1}\tilde{\mathbf{R}}_n^1| = \frac{|\tilde{\mathbf{R}}_{n+1}|}{|\tilde{\mathbf{R}}_n|};$$

(k) *whenever* $\mathbf{X}_t$ *is full rank and nondeterministic, then* $|\tilde{\mathbf{R}}_n| \neq 0$ *for all* $n \geq 1$, *and we have*

$$|\mathbf{\Sigma}| = \exp\left(\lim \frac{1}{n} \ln |\tilde{\mathbf{R}}_n|\right) > 0.$$

Proof. For (a), it is not hard to see that for any $n' \geq n$

$$\begin{aligned}
\mathbf{\Sigma}_n &= \operatorname{Var}(\mathbf{X}_{t+1} - \widehat{\mathbf{X}}_{t+1,n}) \\
&= \operatorname{Var}(\mathbf{X}_{t+1} - \widehat{\mathbf{X}}_{t+1,n'} + \widehat{\mathbf{X}}_{t+1,n'} - \widehat{\mathbf{X}}_{t+1,n}) \\
&= \operatorname{Var}(\mathbf{X}_{t+1} - \widehat{\mathbf{X}}_{t+1,n'}) + \operatorname{Var}(\widehat{\mathbf{X}}_{t+1,n'} - \widehat{\mathbf{X}}_{t+1,n}) \\
&= \mathbf{\Sigma}_{n'} + \operatorname{Var}(\widehat{\mathbf{X}}_{t+1,n'} - \widehat{\mathbf{X}}_{t+1,n}).
\end{aligned}$$

So we have $\mathbf{\Sigma}_n - \mathbf{\Sigma}_{n'} = \operatorname{Var}(\widehat{\mathbf{X}}_{t+1,n'} - \widehat{\mathbf{X}}_{t+1,n})$. But the term on the right-hand side, being a covariance, is nonnegative, thus proving the first inequality in (a). The second one is similar. Part (b) is an immediate consequence of part (a), noting that diagonal elements of a nonnegative definite matrix are nonnegative. Part (c) is immediate from part (b). The limit argument in (d) is actually univariate and can be argued as we did in part (b) of Proposition 4.5. For (e), it suffices to show that

$$\lim_{n\to\infty} (\sigma_n^{jk} - \sigma^{jk}) = 0, \; j, k = 1, 2, \ldots, q.$$

But since $\mathbf{\Sigma}_n - \mathbf{\Sigma}$ is nonnegative and hence its diagonal entries dominate the rest, it suffices to show

$$\lim_{n\to\infty} (\sigma_n^{jj} - \sigma^{jj}) = 0$$

and this was proved in part (d). For (f), we can write for any $j = 1, 2, \ldots, q$

$$\begin{aligned} \sigma_n^{jj} = \|\widehat{X}^j_{t+1,n} - X^j_{t+1}\|^2 &= \|\widehat{X}^j_{t+1,n} - \widehat{X}^j_{t+1} + \widehat{X}^j_{t+1} - X^j_{t+1}\|^2, \\ &= \|\widehat{X}^j_{t+1,n} - \widehat{X}^j_{t+1}\|^2 + \sigma^{jj} \end{aligned} \tag{4.94}$$

which implies

$$\|\widehat{X}^j_{t+1,n} - \widehat{X}^j_{t+1}\|^2 = \sigma_n^{jj} - \sigma^{jj}.$$

By virtue of the convergence statement of part (d), this completes the proof of (f). For (g), since it is well known that null $\mathbf{\Sigma}_n$ + rank $\mathbf{\Sigma}_n = q$, it is enough to verify the statement only for rank. For this note that for each positive integer n, rank $\mathbf{\Sigma_n} = \dim \mathcal{I}_n$ is the maximum number of LI vectors in the generators $\{X^j_{t+1} - \widehat{X^j}_{t+1,n} : 1 \le j \le q\}$ of $\mathcal{I}_n$. So it suffices to show that for any two integers $n' \ge n$ the maximum number of LI vectors in the generators of $\mathcal{I}'_n$ does not exceed ($\le$) the maximum number of LI vectors in the generators of $\mathcal{I}_n$. The inequality is true if every LD set of generators from $\mathcal{I}_n$ is also LD in $\mathcal{I}'_n$. We prove the latter by considering the subsets of the generators in order, as in the Gram-Schmidt process. Indeed, suppose the $s(\le q)$ vectors $\{X^j_{t+1} - \widehat{X^j}_{t+1,n} : 1 \le j \le s\}$ are LD; then there exists scalars $\{\alpha_j : 1 \le j \le s\}$, not all zero, such that

$$0 = \sum_{j=1}^{s} \alpha_j (X^j_{t+1} - \widehat{X^j}_{t+1,n}) = \sum_{j=1}^{s} \alpha_j X^j_{t+1} - \sum_{j=1}^{s} \alpha_j \Big(X^j_{t+1} | \mathcal{M}(t,n) \Big).$$

Thus $\sum_{j=1}^{s} \alpha_j X^j_{t+1}$ is in $\mathcal{M}(t,n)$ and hence in $\mathcal{M}(t,n')$, which in turn implies

$$\sum_{j=1}^{s} \alpha_j (X^j_{t+1} - \widehat{X^j}_{t+1,n'}) = 0.$$

Therefore $\{X^j_{t+1} - \widehat{X^j}_{t+1,t;n} : 1 \le j \le s\}$ are LD, which completes the proof (see also problem 4.16). Part (h) follows from part (e) and the fact that the determinant of a matrix is an algebraic function of its entries and hence a continuous function of them. The first equality in part (i) follows from taking the determinant of both sides of (4.93). For the second equality, the partitioning

$$\tilde{\mathbf{R}}_{n+1} = \begin{bmatrix} \mathbf{R}(0) & (\tilde{\mathbf{R}}^1_n)' \\ \tilde{\mathbf{R}}^1_n & \tilde{\mathbf{R}}_n \end{bmatrix} \tag{4.95}$$

in conjunction with the fact that if $\mathbf{A}$ and $\mathbf{D}$ are invertible $n \times n$ and $m \times m$ matrices then

$$\begin{vmatrix} \mathbf{A} & \mathbf{B} \\ \mathbf{C} & \mathbf{D} \end{vmatrix} = |\mathbf{A}||\mathbf{D} - \mathbf{C}\mathbf{A}^{-1}\mathbf{D}|,$$

gives us

$$|\tilde{\mathbf{R}}_{n+1}| = \left|[\mathbf{R}(0) - (\tilde{\mathbf{R}}_n^1)'\tilde{\mathbf{R}}_n^{-1}\tilde{\mathbf{R}}_n^1]\right| |\tilde{\mathbf{R}}_n| = |\mathbf{\Sigma}_n||\tilde{\mathbf{R}}_n|$$

(see [183, Ch. 7, problem 14] and [148, Appendix A]). For (k), since $|\mathbf{\Sigma}_n| \to |\mathbf{\Sigma}|$ by part (h), we obtain

$$\ln |\mathbf{\Sigma}| = \lim_{n\to\infty} \frac{1}{n}\sum_{k=1}^{n} |\mathbf{\Sigma}_n| = \lim_{n\to\infty} \frac{1}{n}\sum_{k=1}^{n} \ln \frac{|\tilde{\mathbf{R}}_{k+1}|}{|\tilde{\mathbf{R}}_k|} = \lim_{n\to\infty} \ln |\tilde{\mathbf{R}}_n|. \qquad \blacksquare$$

PROBLEMS AND SUPPLEMENTS

4.1 If X_t is a stationary process with mean m and shift U, then 1 is an eigenvector of U with eigenvalue of 1. For a proof, let $\{\phi_j, j = 1, 2, \dots\}$ be a complete orthonormal set generated from $X_0, X_1, X_{-1}, X_2, X_{-2}, \dots$ obtained by the Gram–Schmidt procedure. Now note that each ϕ_k is a finite linear combination $\phi_k = \sum_{j=1}^{n_k} a_j^{(k)} X_{t_j}$ so that $\langle \phi_k, 1\rangle = \sum_{j=1}^{n_k} a_j^{(k)} \langle X_{t_j}, 1\rangle = \sum_{j=1}^{n_k} a_j^{(k)} m$; from this it follows that

$$\langle \phi_k, U1\rangle = \sum_{j=1}^{n_k} a_j^{(k)} \langle X_{t_j}, U1\rangle = \sum_{j=1}^{n_k} a_j^{(k)} \langle U^{-1}X_{t_j}, 1\rangle = \sum_{j=1}^{n_k} a_j^{(k)} m.$$

Thus the vectors 1 and $U1$ have the same Fourier coefficients with respect to the complete orthonormal sequence $\{\phi_j : j = 1, 2, \dots\}$.

4.2 Show that if $\mathbf{X}_t$ is q-variate stationary with mean $\mathbf{m}$ and $1 \in \mathcal{H}$ then 1 is an eigenvector of U with eigenvalue of 1. The sketch for the univariate case can be modified to give $\langle \phi_k, 1\rangle = \langle \phi_k, U1\rangle$ for arbitrary k and hence the proof. The main point to note is that $\langle X_t^p, 1\rangle = m_p = \langle U^{-1} X_j^p, 1\rangle$.

4.3 Show that if X_t is a stationary process with mean zero then

$$P_{\text{sp}\{1, X_1, X_2, \dots, X_n\}} X_{n+1} = P_{\text{sp}\{X_1, X_2, \dots, X_n\}} X_{n+1}.$$

4.4 Consider $X_t = \varepsilon_t - \alpha\varepsilon_{t-1}$, with ε_t a white noise. Show that when $|\alpha| < 1$ the least-square predictor of X_{t+1} in terms of its past $\mathcal{H}(t)$ is given by

$$\widehat{X}_{t+1} = -\sum_{j=1}^{\infty} \alpha^j X_{t+1-j}.$$

What happens if $\alpha = 1$?

4.5 Let X_t be a mean zero stationary process with autocovariance $R(\cdot)$. Show that the series $\sum_{k=1}^{n} \theta_k X_k$ converges in the mean-square sense if $\sum_{i=0}^{\infty}\sum_{j=0}^{\infty} \theta_i\theta_j R(i-j)$ is finite.

4.6 Show that $\sigma_\zeta^2 > 0$ if and only if X_t has a nontrivial regular part.

4.7 Show that item (c) of Lemma 4.1 is equivalent to saying that U commutes with the projection onto $\mathcal{H}(-\infty)$.

4.8 Determine the autocovariance function of the stationary process with spectral density $f(\lambda) = (\pi - |\lambda|)/\pi^2, \quad \lambda \in [-\pi, \pi)$.

4.9 Show that a stationary sequence X_t is regular if and only if $\lim_{n\to\infty}(X_t|\mathcal{H}_X(t-n)) = 0$. Here is a sketch of the "only if" part. If X_t is regular, it has a representation (4.21)

$$\begin{aligned} \| (X_t|\mathcal{H}_X(t-n)) \|^2 &\le \| (X_t|\mathcal{H}_\zeta(t-n)) \|^2 \\ &\le \| \sum_{k=n+1}^{\infty} b_k\zeta_{t-k} \|^2 \\ &\le \sigma^2 \sum_{k=n+1}^{\infty} |b_k|^2 \to 0, \end{aligned}$$

as $n \to \infty$, as claimed.

4.10 An alternative proof that the infinite moving average $\mathbf{X}_t$ in (4.76) is stationary is to take $t \ge s$ and show that $\mathrm{Cov}\,(\mathbf{X}_t, \mathbf{X}_s)$ depends only on $t - s$. Complete the proof by showing this component-wise: that is, by taking any $1 \le j \le q$ and any integers s and t with $t \ge s$ and showing $R^{jk}(t, s) = R^{jk}(t+1, s+1)$.

4.11 Give an example of a regular q-variate sequence for which rank $R(0) = q$ and rank $(\Gamma_\zeta) < q$. Now give an q-variate sequence that is singular but rank $R(0) = q$.

4.12 Complete the details appearing in (4.88) through Proposition 4.15 for the estimator $\widehat{\mathbf{X}}_{t+\delta,t;n} = (\mathbf{X}_{t+\delta}|\mathcal{M}_{\mathbf{X}}(t;n))$ when $\delta = -n$.

4.13 With the notation of Definition 4.9, show that $(\mathbf{X}|\mathcal{M})$ is the unique q-variate vector such that

$$|||\mathbf{X} - (\mathbf{X}|\mathcal{M})||| = \inf\{|||\mathbf{X} - \mathbf{Y}||| : \text{ components of } \mathbf{Y} \text{ are in } \mathcal{M}\},$$

with $|||\mathbf{X}|||$ representing the Euclidean norm $\mathrm{tr}\left(\mathbf{X}\mathbf{X}^*\right)$ of $\mathbf{X}$.

4.14 Show that for any $n \times n$ invertible matrix $\mathbf{A}$ and any $m \times m$ invertible matrix $\mathbf{D}$ we have

$$(\mathbf{A} + \mathbf{BDB}')^{-1} = \mathbf{A}^{-1} - \mathbf{A}^{-1}\mathbf{B}(\mathbf{B}'\mathbf{A}^{-1}\mathbf{B} + \mathbf{D}^{-1})^{-1}\mathbf{B}'\mathbf{A}^{-1}$$

and

$$\det\begin{bmatrix} \mathbf{A} & \mathbf{B} \\ \mathbf{C} & \mathbf{D} \end{bmatrix} = \det\mathbf{A}\det(\mathbf{D} - \mathbf{C}\mathbf{A}^{-1}\mathbf{B}).$$

4.15 Suppose X_t and Y_t are stationary processes satisfying $X_t - \alpha X_{t-1} = Z_t$ and $Y_t - \alpha Y_{t-1} = X_t + Z_t$, with X_t and Z_t being two uncorrelated white noises $WN(0, \sigma^2)$ and $|\alpha| < 1$. Find spectral density of Y_t.

4.16 If $\mathcal{M}$ and $\mathcal{N}$ are two subspaces of a vector space $\mathcal{X}$ such that for any k LI vectors in N one can find k LI vectors in $\mathcal{M}$, then $\dim M \geq \dim N$. Hint: The dimension of any vector space is the maximum number of LI vectors found in that vector space.

CHAPTER 5

HARMONIZABLE SEQUENCES

In various applications, both data and physical models suggest the absence of stationarity. This motivates the extension of the well developed theory of stationary processes to some classes of nonstationary ones. One such class is that of periodically correlated processes, which is the subject of this book. Another large and useful class is that of harmonizable processes. The concept of harmonizable processes was introduced by Loève (see [139] for a discussion) and was studied by Cramèr in his paper on nonstationary processes [35]. The idea there was to extend the harmonic integral representations (4.4) and (4.5) of Theorem 4.2, namely,

$$X_t = \int_0^{2\pi} e^{it\lambda} \xi(d\lambda)$$

and

$$R(s,t) = R(s-t) = \int_0^{2\pi} e^{i(s-t)\lambda} F(d\lambda)$$

Periodically Correlated Random Sequences:Spectral Theory and Practice. By H.L. Hurd and A.G. Miamee

beyond the class of stationary processes. In the stationary case, ξ is a Hilbert valued orthogonally scattered vector measure on the Borel subsets $\mathcal{B}$ of $[0, 2\pi)$ but in the case of harmonizable sequences, studied in this chapter, the vector measure ξ is not necessarily orthogonally scattered. Thus we briefly present integration with respect to such general measures (see [48, Chapter 1]).

5.1 VECTOR MEASURE INTEGRATION

In this section we review basic properties of vector measures and present integration of scalar functions with respect to such measures.

Definition 5.1 *Let $\mathcal{F}$ be a field of subsets of a set Ω. A set function ξ from $\mathcal{F}$ to a Hilbert space $\mathcal{H}$ is called a finitely additive vector measure, or simply a vector measure, if for any two disjoint sets E_1 and E_2 in $\mathcal{F}, \xi(E_1 \bigcup E_2) = \xi(E_1) + \xi(E_2)$ and it is called countably additive if*

$$\xi\Big(\bigcup_{n=1}^{\infty} E_n\Big) = \sum_{n=1}^{\infty} \xi(E_n)$$

for any pairwise disjoint sequence E_n of sets in $\mathcal{F}$ for which $\bigcup_{n=1}^{\infty} E_n$ is also in $\mathcal{F}$.

■ EXAMPLE 5.1 A finitely additive vector measure

Let $T : L^{\infty}[0, 1] \to \mathcal{H}$ be a linear transformation. For each Lebesgue measurable subset E of $[0, 1]$, we define $\zeta(E)$ to be $T(1_E)$. Then by linearity of T it is clear that ζ is a finitely additive vector measure that in general may fail to be countably additive.

■ EXAMPLE 5.2 A countably additive vector measure

Let $T : L^1[0, 1] \to \mathcal{H}$ be a continuous linear operator and consider the measure η defined by $\eta(E) = T(1_E)$, for each Lebesgue measurable subset E of $[0, 1]$. For each such set E we have $\|\eta(E)\| \leq \|T\|\lambda(E)$; here and throughout this chapter λ stands for Lebesgue measure. Consequently, for any sequence $\{E_n\}_{n=1}^{\infty}$ of disjoint Lebesgue measurable subsets of $[0, 1]$, we can write

$$\begin{aligned}
\lim_{m\to\infty} \Big\|\eta\Big(\bigcup_{n=1}^{\infty} E_n\Big) - \sum_{n=1}^{m} \eta\Big(E_n\Big)\Big\| &= \lim_{m\to\infty} \Big\|\eta\Big(\bigcup_{n=m+1}^{\infty} E_n\Big)\Big\| \\
&\leq \|T\| \lim_{m\to\infty} \lambda\Big(\bigcup_{n=m+1}^{\infty} E_n\Big) = 0,
\end{aligned}$$

which means η is countably additive.

In the study of vector measures and harmonizable processes, two notions of variation are used.

Definition 5.2 *If ξ is a vector measure on a field $\mathcal{F}$ of subsets of Ω, then its total variation (or simply variation) is the extended real valued set function $|\xi| : \mathcal{F} \to \mathbb{R}^+$ defined by*

$$|\xi|(B) = \sup \sum_{i=1}^{n} \|\xi(B_i)\|, \quad B \in \mathcal{F}$$

and its semivariation is the extended real valued set $\|\xi\| : \mathcal{F} \to \mathbb{R}^+$ defined by

$$\|\xi\|(B) = \sup \Big\| \sum_{i=1}^{n} \alpha_i \xi(B_i) \Big\|, \quad B \in \mathcal{F}$$

where the sup is over all finite $\mathcal{F}$-partitions $\{B_i\}$ of B and all finite collection of scalars α_i with $|\alpha_i| \leq 1$. The measure ξ is said to be of bounded variation (of bounded semivariation) if $|\xi|(\Omega) < \infty$ (or if $\|\xi\|(\Omega) < \infty$, respectively).

The countably additive vector measure ζ in Example 5.1 is of bounded semivariation because for any finite partition B_i of Ω and any scalars α_i with $|\alpha_i| \leq 1$, we have

$$\Big\| \sum_{i=1}^{n} \alpha_i \zeta(B_i) \Big\| = \Big\| \sum_{i=1}^{n} \alpha_i T(1_{B_i}) \Big\| \leq \Big\| T\Big(\sum_{i=1}^{n} \alpha_i 1_{B_i} \Big) \Big\| \leq \|T\| \Big\| \sum_{i=1}^{n} \alpha_i 1_{B_i} \Big\| \leq \|T\|.$$

However, ζ is not of finite variation. To see this consider the finite partition of $[0,1]$ consisting of the intervals $E_k = (2^{-k}, 2^{-k+1}]$, for $k = 1, 2, ..., n$ and $E_{n+1} = [0, 2^{-n})$, and note that for each positive integer n

$$\sum_{k=1}^{n+1} \|\eta(E_k)\| = n + 1.$$

This means $|\eta|([0,1])$ is infinite.

The vector measure η in Example 5.2 is of bounded variation. This is an immediate consequence of the fact that for any finite partition $\{B_i\}$ of Ω we have

$$\Big\| \sum_{i=1}^{n} \eta(B_i) \Big\| = \Big\| \sum_{i=1}^{n} T(1_{B_i}) \Big\| \leq \sum_{i=1}^{n} \|T(1_{B_i})\| \leq \|T\| \sum_{i=1}^{n} \lambda(1_{B_i}) \leq \|T\| \lambda(\Omega).$$

Proposition 5.1 *For any vector measure ξ on $\mathcal{F}$ and any set $B \in \mathcal{F}$,*

$$\|\xi\|(B) = \sup\{|x^* \xi|(B) : x^* \in X^*, \|x^*\| \leq 1\}.$$

Proof. If $\{B_1, B_2, ..., B_m\}$ is a finite $\mathcal{F}$ partition of B and $\alpha_1, \alpha_2, ..., \alpha_m$ are scalars with $|\alpha_n| \le 1, n = 1, 2, ..., m$, then

$$\begin{aligned}\Big\|\sum_{n=1}^{m} \xi(B_n)\Big\| &= \sup\Big\{\Big|x^*\Big(\sum_{n=1}^{m} \alpha_n \xi(B_n)\Big)\Big| : x^* \in X^*, \|x^*\| \le 1\Big\}\\ &\le \sup\Big\{\sum_{n=1}^{m} |x^*\xi(B_n)| : x^* \in X^*, \|x^*\| \le 1\Big\}\\ &\le \sup\Big\{|x^*\xi|(B) : x^* \in X^*, \|x^*\| \le 1\Big\}.\end{aligned}$$

This implies $\|\xi\|(B) \le \sup\{|x^*\xi|(B) : x^* \in X^*, \|x^*\| \le 1\}$. Now for any $x^* \in X^*$ with $\|x^*\| \le 1$ and any finite $\mathcal{F}$ partition $\{B_1, B_2, ..., B_m\}$ of B we arrive at the following inequalities:

$$\begin{aligned}\sum_{n=1}^{m} |x^*(\xi(B_n))| &= \sum_{n=1}^{m} \text{sgn}\,[x^*(\xi(B_n))] x^*(\xi(B_n))\\ &= \Big|x^*\Big(\sum_{n=1}^{m} \text{sgn}\,[x^*(\xi(B_n))]\xi(B_n)\Big)\Big|\\ &\le \Big\|\sum_{n=1}^{m} \text{sgn}\,[x^*(\xi(B_n))\xi(B_n)]\Big\|,\end{aligned}$$

which implies the needed reverse inequality. ∎

The next few propositions give some basic properties of variation and semi-variation.

Proposition 5.2 *For any vector measure ξ and any set $B \in \mathcal{F}$,*

$$\|\xi\|(B) \le |\xi|(B).$$

Proof. It is immediate from the definitions. ∎

Proposition 5.3 *The variation $|\xi|$ of any vector measure ξ on a field $\mathcal{F}$ is a monotone finitely additive set function.*

Proof. Let E and F be any two disjoint members of $\mathcal{F}$ and let $\{B_i : i = 1, 2, ..., n\}$ be a finite collection of disjoint sets in $\mathcal{F}$ with $B_i \subseteq E \bigcup F$. Let $E_i = E \cap B_i$, $F_i = F \cap B_i$, then

$$\sum_{i=1}^{n} \|\xi(B_i)\| \le \sum_{i=1}^{n} \|\xi(E_i)\| + \sum_{i=1}^{n} \|\xi(F_i)\| \le |\xi|(E) + |\xi|(F),$$

which means

$$|\xi|(E \cup F) \leq |\xi|(E) + |\xi|(F).$$

So if $|\xi|(E \cup F) = \infty$, the inequality in the last equation must be equality and we are done. Now if $|\xi|(E \cup F) < \infty$, then there are finitely many subsets $\{E_i : i = 1, 2, ..., n\}$ and $\{F_j : j = 1, 2, ..., n\}$ of members $\mathcal{F}$ with $E_i \subseteq E, F_j \subseteq F$ such that

$$|\xi|(E) < \sum_{i=1}^{n} \|\xi(E_i)\| + \epsilon \text{ and } |\xi|(F) < \sum_{j=1}^{m} \|\xi(F_j)\| + \epsilon.$$

This gives

$$|\xi|(E) + |\xi|(F) \leq \sum_{i=1}^{n} \|\xi(E_i)\| + \epsilon + \sum_{j=1}^{m} \|\xi(F_j)\| + \epsilon \leq |\xi|(E \cup F) + 2\epsilon,$$

which, since ϵ is an arbitrary positive real number, implies the reverse inequality

$$|\xi|(E) + |\xi|(F) \leq |\xi|(E \cup F). \quad \blacksquare$$

Proposition 5.4 *A vector measure of bounded variation is countably additive if and only if its variation is countably additive.*

Proof. For a proof one can refer to [48]. ∎

Proposition 5.5 *If ξ is a measure on a sigma field $\mathcal{F}$ with values in some Hilbert space $\mathcal{H}$, then*

(a) $\|\xi\|$ *is a monotone set function on* $\mathcal{F}$*;*

(b) $\sup\{\|\xi(C)\| : C \in \mathcal{F}, C \subseteq B\} \leq \|\xi\|(B)$, *for any* $B \in \mathcal{F}$*;*

(c) $\|\xi\|$ *is subadditive on* $\mathcal{F}$*: that is, for any sequence* $\{B_n\}$ *of sets in* $\mathcal{F}$

$$\|\xi\|(\cup_{n=1}^{\infty} B_n) \leq \sum_{n=1}^{\infty} \|\xi\|(B_n);$$

(d) $\|\xi\|(B) \leq 4 \sup\{\|\xi(C)\| : C \in \mathcal{F}, C \subseteq B\} < \infty, \quad B \in \mathcal{F}.$

Proof. Parts (a) and (b) are immediate consequences of the definition of semivariation $\|\xi\|$. For (c), without loss of generality we assume that sets E_n are disjoint. For any finite disjoint partition $\{F_1, F_2, ..., F_k\}$ of $E = \cup_{n=1}^{\infty} E_n$, and for each n, the collection $\{E_n \cap F_1, E_n \cap F_2, ..., E_n \cap F_k\}$ forms a finite disjoint partition for E_n. Thus if $|\alpha_j| \leq 1$, for $j = 1, 2, ..., k$

$$\left\|\sum_{j=1}^{k} \alpha_j \xi(F_j)\right\| = \left\|\sum_{j=1}^{k}\sum_{n=1}^{\infty} \alpha_j \xi(E_n \cap F_j)\right\| \leq \sum_{n=1}^{\infty} \|\xi\|(E_n),$$

which implies

$$\|\xi\|(E) \leq \sum_{n=1}^{\infty} \|\xi\|(E_n).$$

For (d) first assume that the Hilbert space $\mathcal{H}$ is real. In that case if $\pi = \{B_1, B_2, ..., B_m\}$ is a disjoint $\mathcal{F}$-partition for B, set

$$\begin{aligned}\pi^+ &= \{n : 1 \leq n \leq m, x^*\xi(B_n) \geq 0\}, \text{ and} \\ \pi^- &= \{n : 1 \leq n \leq m, x^*\xi(B_n) < 0\},\end{aligned}$$

then for any $x^* \in X^*$ with $\|x^*\| \leq 1$, we can write

$$\begin{aligned}\sum_{1\leq n\leq m} |x^*\xi(B_n)| &= \sum_{n\in\pi^+} x^*\xi(B_n) - \sum_{n\in\,\pi^-} x^*\xi(B_n) &(5.1)\\ &= x^*\Big[\sum_{n\in\pi^+} \xi(B_n)\Big] - x^*\Big[\sum_{n\in\pi^-} \xi(B_n)\Big] &(5.2)\\ &\leq 2\sup\{\|\xi(C)\| : C \in \mathcal{F}, C \subseteq B\}, &(5.3)\end{aligned}$$

as required. One can then check that for complex Hilbert spaces, by splitting $x^*\xi$ into its real and imaginary parts and applying estimate 5.3 to each, the estimate continues to hold with 4 instead of 2. ∎

In contrast to the complex valued measures, the variation of a countably additive vector measure ξ need not be finite. However, parts (b) and (d) of Proposition 5.5 imply the following.

Proposition 5.6 *A vector measure is bounded if and only if it is of bounded semivariation.*

This fact suggests using semivariation instead of variation in developing an integration theory with respect to vector measures. In fact this theory has been successfully developed and such an integral has been defined for a wide class of scalar valued functions (see [53, Section IV.10]). In the rest of this section, we develop this integral for the smaller class of bounded measurable functions, which is large enough for the scope of the present book.

Definition 5.3 (Integral of Simple Functions) *Let $\mathcal{F}$ be a field of subsets of Ω and $\xi : \mathcal{F} \to \mathcal{H}$ be a bounded vector measure. We define the integral of a simple function $f = \sum_{i=1}^{n} \alpha_i 1_{E_i}$, where α_i are nonzero scalars and E_i are pairwise disjoint members of $\mathcal{F}$, by*

$$\int_{\Omega} f d\xi = \sum_{i=1}^{n} \alpha_i \xi (E_i) .$$

It is easy to show that this integral, on the set of simple functions, is well defined. Furthermore, if we define the sup norm of f as $\|f\|_\infty = \sup\{f(\omega) : \omega \in \Omega\}$, then we have

$$\left\| \int_{\Omega} f d\xi \right\| = \left\| \sum_{i=1}^{n} \alpha_i \xi(E_i \right\| = \|f\|_\infty \left\| \sum_{i=1}^{n} \frac{\alpha_i}{\|f\|_\infty} \xi(E_i) \right\| \leq \|f\|_\infty \|\xi\|(\Omega).$$

The last equality is a consequence of the definition of variation once we notice that $\alpha_i / \|f\|_\infty \leq 1$ for $i = 1, 2, ..., n$. It is also easy to see that this integral is linear: that is, if $f = \sum_{i=1}^{n} \alpha_i 1_{E_i}$ and $g = \sum_{j=1}^{m} \beta_j 1_{F_j}$ are two simple functions then

$$\int_{\Omega} (cf + g) d\xi = c \int_{\Omega} f \, d\xi + \int_{\Omega} g \, d\xi.$$

This integral can be extended to all uniform limits of simple functions. In fact, let f be the uniform limit of a sequence $\{f_n\}$ of simple functions. Thus $\{f_n\}$ is Cauchy in sup norm. On the other hand, we can write

$$\left\| \int_{\Omega} f_n \, d\xi - \int_{\Omega} f_m d\,\xi \right\| = \left\| \int_{\Omega} (f_n - f_m) d\,\xi \right\| \leq \|f_n - f_m\|_\infty \|\xi\|(\Omega),$$

which shows that $\{\int_{\Omega} f_n d\,\xi\}$ is a Cauchy sequence in the Hilbert space $\mathcal{H}$ and thus has a limit. We define $\int f d\xi$ to be this limit: that is, we define

$$\int_{\Omega} f d\,\xi := \lim_{n \to \infty} \int_{\Omega} f_n \, d\xi.$$

Because of the discussion above we say a function is $\mathcal{F}$-integrable if it is the uniform limit of some $\mathcal{F}$-simple functions. The integral of an $\mathcal{F}$-integrable function f on a set B in $\mathcal{F}$ is defined by $\int_B f d\xi = \int_{\Omega} f 1_B d\xi$. It can be easily shown that any bounded $\mathcal{F}$-measurable function f on Ω is the uniform limit of a sequence of $\mathcal{F}$-simple functions and hence $\mathcal{F}$-integrable with respect to any bounded vector measure ξ.

The next proposition lists basic properties of the vector integral we just defined.

Proposition 5.7 *If the functions f and g are both $\mathcal{F}$-integrable, c is any scalar and B, C are any two disjoint members of $\mathcal{F}$ then*

(a) $$\int_B (cf+g)d\xi = c\int_B f\,d\xi + \int_B g\,d\xi;$$

(b) $$\int_{B\cup C} f\,d\xi = \int_B f\,d\xi + \int_C f\,d\xi;$$

(c) $$\left\| \int_B f(\lambda)d\xi \right\| \le \|f\|_\infty \|\xi\|(B).$$

Proof. For (a), pick two sequences $\{f_n\}$ and $\{g_n\}$ of simple functions that converge uniformly to f and g, respectively. Then it is easy to see that the sequence $(cf_n + g_n)$ converges to $cf+g$ uniformly and hence $\lim_n \int_B (cf_n + g_n)d\xi = \int_B (cf+g)d\xi$. Now because of the linearity of the integral on simple functions,

$$\int_B (cf_n + g_n)d\xi = c\int_B f_n\,d\xi + \int_B g_n\,d\xi.$$

Taking the limit as n goes to infinity on both sides, we get

$$\lim_n \int_B (cf_n+g_n)d\xi = \lim_n \left(c\int_B f_n d\xi + \int_B g_n d\xi\right) = c\lim_n \int_B f_n d\xi + \lim_n \int_B g_n d\xi$$

which clearly shows (a). Proof of parts (b) and (c) are similar and left to the reader. ∎

We will close this section with the following result.

Proposition 5.8 *Let ξ be a bounded vector measure on Borel subsets of $[0,2\pi)$ with values in some Hilbert space $\mathcal{H}$. If the scalar valued function F defined by $F([a,b),[c,d)) = \langle \xi([a,b)),\ \xi([c,d))\rangle$ can be extended to a measure on the Borel subsets of $[0,2\pi)^2$, then for any two bounded measurable functions $f,g : [0,2\pi) \to \mathbb{C}$ we have*

$$\left\langle \int_0^{2\pi} f\,d\xi,\ \int_0^{2\pi} g\,d\xi \right\rangle = \int_0^{2\pi}\int_0^{2\pi} fg\,dF.$$

Proof. If functions f and g are simple, then we can represent them as $f = \sum_{i=1}^m \alpha_i 1_{E_i}$ and $g = \sum_{i=1}^n \beta_i 1_{F_i}$ and write

$$\begin{aligned}
\left\langle \int_0^{2\pi} f\,d\xi,\ \int_0^{2\pi} g\,d\xi \right\rangle &= \left\langle \sum_{i=1}^m \alpha_i \xi E_i,\ \sum_{i=1}^n \beta_i \xi F_i \right\rangle \\
&= \sum_{i=1}^m \sum_{j=1}^n \alpha_i \beta_j \langle \xi E_i,\ \xi F_j\rangle \\
&= \sum_{i=1}^m \sum_{j=1}^n \alpha_i \beta_j\, F(E_i, F_j) = \int_0^{2\pi}\int_0^{2\pi} fg\,dF.
\end{aligned}$$

Now suppose f and g are any two bounded measurable functions. Then there are sequences f_n and g_n of simple functions that converge uniformly to f and g , respectively. According to the simple function case above we have

$$\left\langle \int_0^{2\pi} f_n\, d\xi, \int_0^{2\pi} g_n\, d\xi \right\rangle = \int_0^{2\pi}\int_0^{2\pi} f_n g_n\, dF, \qquad \text{for any } n \in \mathbb{N}.$$

Since the sequence $f_n g_n$ converges uniformly to fg if we take the limit of both sides of this equation, by virtue of the continuity of inner products as in Proposition 3.2, we arrive at (5.8). ∎

5.2 HARMONIZABLE SEQUENCES

Here we introduce harmonizable sequences and present their basic properties. The study of harmonizable processes was started by Loéve in 1948 [138] and subsequently by Cramér [36]. The weaker type of harmonizability seems to have originated with Rozanov in 1959 [200], but its characterization in terms of stationary dilations was completed in 1978 by Miamee and Salehi [155]. Before we define these two types of harmonizability we recall that a sequence $\{X_t\}$ of random variables on some probability space $(\Omega, \mathcal{F}, P)$ is said to be of second order if every element X_t is of finite variance, that is, $X_t \in L^2(\Omega, \mathcal{F}, P)$ for all $t \in \mathbb{Z}$. However, for the presentation here one can think of $\{X_t\}$ as being a sequence in an arbitrary Hilbert space $\mathcal{H}$.

Definition 5.4 *A random sequence X_t with autocovariance function $R(s,t)$ is called strongly harmonizable (harmonizable, in short) if there exists a scalar valued measure F of bounded variation on Borel subsets of $[0, 2\pi)^2$ such that*

$$R(s,t) = \int_0^{2\pi}\int_0^{2\pi} e^{i(s\lambda - t\theta)} F(d\lambda, d\theta) \tag{5.4}$$

and weakly harmonizable if there exists a bounded vector measure ξ on Borel subsets of $[0, 2\pi)$ such that

$$X_t = \int_0^{2\pi} e^{it\lambda} \xi(d\lambda). \tag{5.5}$$

F (respectively, ξ) is called the spectral (respectively, random) measure of X_t.

,

Lemma 5.1 *Every weakly harmonizable process X_t is bounded.*

Proof. Let X_t be a weakly harmonizable sequence so that (5.5) holds for some bounded random measure ξ. By virtue of properties of integrals with respect to such random measures given in Section 5.1, we get

$$\|X_t\| = \left\| \int_0^{2\pi} e^{it\lambda}\xi(d\lambda) \right\| \leq \|\xi\|([0, 2\pi)), \quad \text{for every integer } t. \qquad \blacksquare$$

Definition 5.5 *A sequence X_t in a Hilbert space $\mathcal{H}$ is said to have a stationary dilation if there exists a Hilbert space $\mathcal{K}$ containing $\mathcal{H}$ and a stationary sequence Z_t in $\mathcal{K}$ such that*

$$X_t = PZ_t; \; t \in \mathbb{Z},$$

where P is the orthogonal projection from $\mathcal{K}$ onto $\mathcal{H}$. If this is the case, the sequence Z_t is called a stationary dilation for X_t.

Addressing the interesting question of which sequences have stationary dilations, Abreu [1] proved the following result.

Proposition 5.9 *Every harmonizable process has a stationary dilation.*

Proof. Suppose $\{X_t\} \subset \mathcal{H}$ be a harmonizable process whose spectral measure is F. That is suppose

$$R(s,t) = \langle X_s, X_t \rangle = \int_0^{2\pi} \int_0^{2\pi} e^{i(s\lambda - t\theta)} F(d\lambda, d\theta), \quad s, t \in \mathbb{Z}.$$

The total variation $|F|$ of F is a finite symmetric Borel measure on $[0, 2\pi)^2$. Let μ be the finite positive measure defined on Borel subsets $\mathcal{B}[0, 2\pi)$ of $[0, 2\pi)$ by

$$\mu(B) = |F|(B \times [0, 2\pi)) = \frac{1}{2}\Big[|F|(B \times [0, 2\pi)) + |F|([0, 2\pi) \times B)\Big]$$

for any $B \in \mathcal{B}[0, 2\pi)$. Then for every continuous complex valued function f on $[0, 2\pi)$ we have

$$\int_0^{2\pi} f d\mu = \frac{1}{2}\int_0^{2\pi}\int_0^{2\pi} (f \times 1)\, d|F| + \frac{1}{2}\int_0^{2\pi}\int_0^{2\pi} (1 \times f)\, d|F| \tag{5.6}$$

and

$$0 \leq \int_0^{2\pi}\int_0^{2\pi} (f \times \overline{f})\, dF \leq \int_0^{2\pi}\int_0^{2\pi} |(f \times \overline{f})|\, d|F|.$$

But for each $\lambda, \theta \in [0, 2\pi)$ we have

$$|(f \times \overline{f})(\lambda, \theta)| = |f(\lambda)f(\theta)| \leq \frac{1}{2}[|f(\lambda)|^2 + |f(\theta)|^2].$$

Therefore

$$\begin{aligned}\int_0^{2\pi}\int_0^{2\pi}(f \times \overline{f})\,dF &\leq \int_0^{2\pi}\int_0^{2\pi}|f \times \overline{f}|\,d|F| \\ &\leq \frac{1}{2}\int_0^{2\pi}\int_0^{2\pi}(|f|^2 \times 1)\,d|F| + \frac{1}{2}\int_0^{2\pi}\int_0^{2\pi}(1 \times |f|^2)\,d|F|,\end{aligned}$$

which, by virtue of (5.6), yields

$$\int_0^{2\pi}\int_0^{2\pi}(f \times \overline{f})\,dF \leq \int_0^{2\pi}|f|^2 d\mu. \tag{5.7}$$

Now let $\widetilde{\mu}$ be the measure on the square $[0, 2\pi)^2$ which is concentrated on its diagonal and $\widetilde{\mu}(B, B) = \mu(B)$, for every $B \in \mathcal{B}[0, 2\pi)$, and define $\rho = \widetilde{\mu} - F$. Using (5.7) we get

$$\int_0^{2\pi}\int_0^{2\pi}(f \times \overline{f})\,d\rho \geq 0, \text{ for any continuous function } f \text{ on } [0, 2\pi).$$

Thus ρ is a nonnegative measure and hence it is the spectral measure of some harmonizable sequence Y_t. Recalling $\mathcal{H}_X(\infty) = \overline{\text{sp}}\{X_t : t \in \mathbb{Z}\}$ and $\mathcal{H}_Y(\infty) = \overline{\text{sp}}\{Y_t : t \in \mathbb{Z}\}$, we define $\mathcal{K}$ to be their direct sum $\mathcal{H}_X(\infty) \bigoplus \mathcal{H}_Y(\infty)$ and let

$$Z_t = X_t + Y_t, \quad \text{for each t } \in \mathbb{Z}.$$

Then

$$R_Z(s, t) = \langle Z_s, Z_t\rangle = \langle X_s + Y_s, X_t + Y_t\rangle = \langle X_s, X_t\rangle + \langle Y_s, Y_t\rangle.$$

Hence $\widetilde{\mu} = F + \rho$ is the covariance measure of Z_t and this implies Z_t is stationary. Finally, if $P : \mathcal{K} \to \mathcal{H}_X(\infty)$ denotes the orthogonal projection from $\mathcal{K}$ onto $\mathcal{H}_X(\infty)$, it is clear from the construction of Z_t that

$$X_t = PZ_t, \text{ for each } t \in \mathbb{Z}. \quad \blacksquare$$

The following proposition shows why harmonizable processes are sometimes called *strongly harmonizable.*

Proposition 5.10 *Every harmonizable sequence X_t is weakly harmonizable.*

Proof. Let $\{X_t\} \subset \mathcal{H}$ be a harmonizable sequence. By Proposition 5.9 this sequence has a stationary dilation $\{Z_t\}$ in some Hilbert space $\mathcal{K}$ containing $\mathcal{H}$: that is,

$$X_t = PZ_t, \quad t \in Z,$$

with P being the orthogonal projection $P : \mathcal{K} \to \mathcal{H}$. Let ζ be the orthogonally scattered random measure of the stationary sequence Z_t. Then integration theory with respect to vector measures, as presented in Section 5.1, and continuity of the projection P allow us to write

$$X_t = PZ_t = P\left(\int_0^{2\pi} e^{it\lambda} d\zeta\right) = \int_0^{2\pi} e^{it\lambda} dP\zeta = \int_0^{2\pi} e^{it\lambda} d\xi,$$

where the random measure ξ is defined by $\xi(B) = P\zeta(B)$, for each Borel subset B of $[0, 2\pi)$. It remains to show that ξ is of bounded semivariation. To see this, note that for any disjoint collection $\{B_i : 1 \le i \le n\}$ of Borel subsets of $[0, 2\pi)$ and any collection $\{\alpha_i : 1 \le i \le n\}$ of scalars with $|\alpha_i| \le 1$, we have

$$\begin{aligned}
\left\|\sum_{i=1}^n \alpha_i \xi(B_i)\right\|^2 &= \left\|\sum_{i=1}^n \alpha_i P\zeta(B_i)\right\|^2 = \left\|P\Big(\sum_{i=1}^n \alpha_i \zeta(B_i)\Big)\right\|^2 \\
&\le \left\|\sum_{i=1}^n \alpha_i \zeta(B_i)\right\|^2 = \left\langle \sum_{i=1}^n \alpha_i \zeta(B_i), \sum_{i=1}^n \alpha_i \zeta(B_i)\right\rangle \\
&= \sum_{i,j=1}^n \alpha_i \alpha_j \langle \zeta(B_i), \zeta(B_j)\rangle \\
&= \sum_{i=1}^n |\alpha_i|^2 \parallel \zeta(B_i) \parallel^2 \le \parallel \zeta([0, 2\pi)) \parallel^2 .
\end{aligned}$$

∎

Abreu [1] observed that the class of harmonizable processes fails to include all projections of stationary sequences. Seeking a characterization for sequences possessing a stationary dilation, Rozanov [200] showed that projection of any stationary process is weakly harmonizable. The converse of Rozanov's result was given by Miamee and Salehi [155]. These comments establish the following characterization.

Proposition 5.11 *A sequence X_t has stationary dilation if and only if it is weakly harmonizable.*

Niemi in a series of papers [168, 170–172] studied various related questions and proved the following general result.

Theorem 5.1 *If $\mathcal{H}$ is a Hilbert space, then for any bounded vector measure $\xi : \mathcal{H} \to \mathcal{F}$ there exists (1) a Hilbert space $\mathcal{K}$ containing $\mathcal{H}$ and (2), an orthogonally scattered vector measure $\zeta : \mathcal{F} \to \mathcal{K}$, such that*

$$\xi(E) = P\zeta(E), \text{ for any } E \in \mathcal{F}.$$

Here $P : \mathcal{K} \to \mathcal{H}$ stands for the orthogonal projection.

For more information on the subject of dilation the reader may want to check Rosenberg [196], Makagon and Salehi [144], Miamee [154,160], Chattergi [32], Mlak [164], Masani [153], and the references therein.

We close this section with the following results concerning boundedness of harmonizable sequences.

Corollary 5.11.1 *Every harmonizable process is bounded.*

5.3 LIMIT OF ERGODIC AVERAGE

Here we extend Proposition 4.2 to express the limit of

$$S_N(\lambda) = \frac{1}{2N+1} \sum_{t=-N}^{N} X_t e^{-i\lambda t} \tag{5.8}$$

in spectral terms.

Proposition 5.12 *If X_t is a harmonizable sequence having the random spectral measure ξ and corresponding spectral (correlation) measure F, then*

$$\lim_{N\to\infty} S_N(\lambda) = \xi(\{\lambda\}) \tag{5.9}$$

in the mean-square sense.

Proof. Using (1.14), we obtain

$$\begin{aligned} S_N(\lambda) &= \frac{1}{2N+1} \sum_{t=-N}^{N} \int_0^{2\pi} e^{i(\gamma-\lambda)t} \xi(d\gamma) \\ &= \int_0^{2\pi} d_N(\gamma - \lambda)\xi(d\gamma), \end{aligned}$$

where the Dirichlet kernel $d_N(x)$ is bounded and continuous and converges to $1_{\{0\}}(x)$. The convergence $\| S_N(\lambda) - \xi(\{\lambda\}) \|^2 \to 0$ of follows from

$$\langle S_N(\lambda), \xi(\{\lambda\})\rangle = \int_0^{2\pi} d_N(\gamma - \lambda)F(d\gamma, \lambda) \longrightarrow F(\{\lambda\}, \{\lambda\})$$

because $F(\cdot, \{\lambda\})$ is a (finite) measure in the first coordinate for each fixed λ, and from the fact that

$$\begin{aligned} \| S_N(\lambda) \|^2 &= \frac{1}{(2N+1)^2} \sum_{s=-N}^{N} \sum_{t=-N}^{N} R(s,t) e^{-i\lambda(s-t)} \\ &= \int_0^{2\pi} \int_0^{2\pi} d_N(\gamma_1 - \lambda) d_N(\gamma_2 - \lambda) F(d\gamma_1, d\gamma_2) \end{aligned}$$

converges to

$$\int_0^{2\pi} \int_0^{2\pi} 1_{\{\lambda\}}(\gamma_1) 1_{\{\lambda\}}(\gamma_2) F(d\gamma_1, d\gamma_2) = F(\{\lambda\}, \{\lambda\}). \quad \blacksquare$$

5.4 LINEAR TIME INVARIANT FILTERS

The process Y_t is said to be obtained from the process X_t by application of the *linear filter* $W = \{w_k(t) : t, k \in \mathbb{Z}\}$ if

$$Y_t = \sum_{k=-\infty}^{\infty} w_k(t) X_k, \quad \text{for } t \in \mathbb{Z}.$$

The coefficients $w_k(t)$ are called the *weights* of the filter. A filter $W = \{w_k(t) : t, k \in \mathbb{Z}\}$ is called a *linear time invariant (LTI) filter* if $w_k(t)$ depends only on $t - k$, that is, if $w_k(t) = w_{t-k}$. In this case we can write

$$Y_t = \sum_{k=-\infty}^{\infty} w_{t-k} X_k = \sum_{k=-\infty}^{\infty} w_k X_{t-k}. \tag{5.10}$$

The time invariant linear filter $W = \{w_k\}$ is said to be *causal* if

$$w_k = 0 \text{ for } k < 0,$$

which according to (5.10) means that we can express Y_t in the form

$$Y_t = \sum_{k=0}^{\infty} w_k X_{t-k}.$$

Examples .

1. The filter defined by $Y_t = aX_{-t}$, $t \in \mathbb{Z}$ is linear but not time invariant since the weights are $w_k(t) = a\delta_{t+k}$, which do not depend only on $t - k$.

2. The filter defined by $Y_t = a_{-1}X_{t+1} + a_0X_t + a_1X_{t-1}, t \in \mathbb{Z}$ is linear and time invariant but not causal unless a_{-1} happens to be zero.

3. The infinite moving average $Y_t = \sum_{j=0}^{\infty} \alpha_j \xi_{t-j}$ (where ξ_t is usually taken to be orthonormal) is an instance of a causal linear time invariant filter.

Definition 5.6 *A linear filter $W = \{w_k\}$ with absolutely summable weights*

$$\sum_{j=-\infty}^{\infty} |w_j| < \infty$$

is called stable.

Lemma 5.2 *If $W = \{w_k\}$ is a stable filter then the series $\sum_{k=-\infty}^{\infty} w_k e^{ik\lambda}$ converges point wise to some function $W(\lambda)$. This function $W(\lambda)$ is called its transfer function.*

Proof. Clearly for every λ the series $\sum_{k=-\infty}^{\infty} w_k e^{ik\lambda}$ is absolutely convergent, and hence convergent. ∎

Since a harmonizable sequence X_t is defined as an integral of functions $e^{i\lambda t}$ with respect to some random measures $\xi(d\lambda)$, namely,

$$X_t = \int_0^{2\pi} e^{it\lambda} \xi(d\lambda),$$

the familiar frequency domain interpretation of the effect of time invariant linear filtering for stationary sequences is retained.

Proposition 5.13 *Suppose X_t is a harmonizable sequence with random measure ξ and suppose $W = \{w_j\}$ is a stable time invariant linear filter; then for every t, the sum in*

$$Y_t = \sum_{j=-\infty}^{\infty} w_j X_{t-j} \tag{5.11}$$

converges in norm and thus defines the filtered sequence Y_t. Denoting $W(\lambda)$ as the transfer function (see Lemma 5.2) of the filter, the sequence Y_t is harmonizable with spectral representation

$$Y_t = \int_0^{2\pi} e^{it\lambda} W(\lambda) d\xi(\lambda) \tag{5.12}$$

and spectral measure defined for all $B \in \mathcal{B}([0, 2\pi)^2)$ by

$$F_Y(B) = \int_B W(\lambda)\overline{W(\theta)} F_X(d\lambda, d\theta). \tag{5.13}$$

Proof. By Lemma 5.1, $\{X_t\}$ is bounded. Therefore $M_X = \sup_t \|X_t\| < \infty$. Now if $0 < m < n$, then because of

$$\begin{aligned}\Big\|\sum_{j=m}^{n} w_j X_{t-j}\Big\|^2 &= \sum_{j=m}^{n}\sum_{k=m}^{n} w_j \overline{w_k}\langle X_{t-j}, X_{t-k}\rangle \\ &= \Big|\sum_{j=m}^{n}\sum_{k=m}^{n} w_j \overline{w_k}\langle X_{t-j}, X_{t-k}\rangle\Big| \\ &\le \sum_{j=m}^{n}\sum_{k=m}^{n} |w_j|\,|\overline{w_k}|\,|\langle X_{t-j}, X_{t-k}\rangle| \le M_X^2\Big[\sum_{j=m}^{n}|w_j|\Big]^2,\end{aligned}$$

the norm $\|\sum_{j=m}^{n} w_j X_{t-j}\|$ goes to zero as $m, n \to \infty$. The same argument applies to the case $-n < -m < 0$ and so by the Cauchy criterion the series $\sum_{j=-\infty}^{\infty} w_j X_{t-j}$ converges in norm to some random variable, which we call Y_t: that is,

$$\lim_{n\to\infty}\sum_{j=-n}^{n} w_j X_{t-j} = Y_t,\ \ t \in \mathbb{Z}. \tag{5.14}$$

To prove (5.12) and (5.13), we note that by Lemma 5.2

$$\lim_{n\to\infty}\sum_{k=-n}^{n} w_k e^{ik\lambda} = W(\lambda), \quad \text{for every } \lambda \in [0, 2\pi).$$

So applying the Dominated Convergence Theorem IV.10.10 in [53] with $f_n(\lambda) = \sum_{k=-n}^{n} w_k e^{ik\lambda}$, $f(\lambda) = W(\lambda)$, and $g(\lambda)$ being the constant function $g(\lambda) = \sum_{k=-\infty}^{\infty} |w_k|$, we get

$$\lim_{n\to\infty}\int_0^{2\pi}\sum_{-n}^{n} w_k e^{ik\lambda} d\xi(\lambda) = \int_0^{2\pi} W(\lambda)\xi(d\lambda). \tag{5.15}$$

Now letting n to go infinity on both sides of

$$\sum_{k=-n}^{n} w_k X_{t-k} = \int_0^{2\pi}\sum_{k=-n}^{n} w_k e^{ik\lambda}\xi(d\lambda)$$

and using (5.14) and (5.15), we get (5.12), as claimed. To prove the last assertion, according to (5.12) and (5.8) we can write

$$\begin{aligned} R_Y(s,t) = \langle Y_s, Y_t\rangle &= \Big\langle \int_0^{2\pi} e^{is\lambda} w(\lambda)\xi(d\lambda), \int_0^{2\pi} e^{it\theta} W(\theta)\xi(d\theta)\Big\rangle \\ &= \int_0^{2\pi}\int_0^{2\pi} e^{is\lambda - it\theta} W(\lambda)\overline{W(\theta)} F_X(d\lambda, d\theta),\end{aligned}$$

which, in conjunction with the uniqueness of Fourier inversion, completes the proof. ∎

PROBLEMS AND SUPPLEMENTS

5.1 Let $F(d\lambda)$ be a measure on $[0, 2\pi)$ and let $\tilde{F}(\lambda)$ be its associated distribution function defined on $[0, 2\pi)$ by

$$\tilde{F}(\lambda) = \begin{cases} 0, & \text{if } \lambda = 0 \\ F([0, \lambda)), & \text{if } 0 < \lambda < 2\pi; \end{cases}$$

show that for any continuous function $f(\lambda)$ on $[0, 2\pi)$,

$$\int_0^{2\pi} f(\lambda) F(d\lambda) = \int_0^{2\pi} f(\lambda) d\tilde{F}(\lambda).$$

5.2 Determine the transfer function of the time invariant linear filter with coefficients $w_0 = 1, w_1 = -2\alpha, w_2 = 1$, and $w_j = 0$ for $j \neq 0, 1, 2$. What value of α should be used for the filter to suppress sinusoidal oscillations of period 6?

5.3 Consider the process

$$X_t = A \cos\left(\frac{\pi t}{3}\right) + B \sin\left(\frac{\pi t}{3}\right) + Z_t + 2.5 Z_{t-1},$$

where A and B are uncorrelated mean zero random variables that are also uncorrelated with the white noise Z_t. Find

(a) the covariance function and the spectral distribution function of X_t;

(b) the covariance function and the spectral distribution function of the filtered process, $Y_t = X_t - 2X_{t-1} + X_{t-2}$.

5.4 Let $Z_\lambda, 0 \leq \lambda < 2\pi$ be an orthogonally scattered process with distribution function F_λ and suppose that $\psi \in L^2(F)$.

(a) Show that $W_F = \int_0^F \psi(\lambda) dZ_\lambda,\ F \in [0, 2\pi)$ is also orthogonal scattered and has spectral distribution function

$$G_F = \int_0^F |\psi(\lambda)|^2 dF_\lambda.$$

(b) Show that if $g \in L^2(G)$ then $g\psi \in L^2(F)$ and

$$\int_0^{2\pi} g(\lambda) dW_\lambda = \int_0^{2\pi} g(\lambda) \psi(\lambda) dZ_\lambda.$$

(c) Show that if $|\psi| > 0$, a.e., then

$$Z_F = \int_0^F \frac{1}{\psi(\lambda)} dW_\lambda.$$

5.5 Prove parts (b) and (c) of Proposition 5.7.

CHAPTER 6

FOURIER THEORY OF THE COVARIANCE

As in the stationary case it is possible and somewhat natural to develop a Fourier theory for the covariance without considering the representation of the process itself. In the stationary case, the Fourier theory for the covariance was complete as soon as it was understood that every stationary covariance was nonnegative definite and that the Herglotz theorem (Bochner's theorem for continuous parameter) implied that every stationary covariance was the Fourier transform of a positive measure. Not long afterward, Kolmogorov and Cramér developed the spectral theory for stationary processes. Kolmogorov's result used the spectral theory for unitary operators whereas Cramér developed, in a more direct fashion, the isomorphism that makes the process at time t correspond to a stochastic integral of the exponential function $e^{i\lambda t}$ with respect to a process of orthogonal increments (in other words, an orthogonally scattered vector measure).

In the case of PC processes the representation theory for the covariance and for the process has developed more or less simultaneously in the sense that Gladyshev gave results on process representation in his first (1961) paper

Periodically Correlated Random Sequences:Spectral Theory and Practice. By H.L. Hurd and A.G. Miamee

[77] on this topic. But here we will follow the route of the stationary case and develop the Fourier theory for the covariance and the representation for PC sequences that is a consequence of harmonizability, also due to Gladyshev. The clarification brought by unitary operators, treated in Chapter 7, will help in understanding Gladyshev's and other representations of PC sequences.

6.1 FOURIER SERIES REPRESENTATION OF THE COVARIANCE

Following the notation of Gladyshev [77], since $R(t+\tau, t)$ is periodic in the variable t with period T, it can be represented as a discrete Fourier series

$$R(t+\tau, t) = \sum_{k=0}^{T-1} B_k(\tau) e^{i2\pi kt/T}, \tag{6.1}$$

where

$$B_k(\tau) = \frac{1}{T} \sum_{t=0}^{T-1} e^{-i2\pi kt/T} R(t+\tau, t). \tag{6.2}$$

where, as shown in the following lemma, the representation (6.1) is a pointwise equality. In continuous time or if $R(t+\tau, t)$ is almost periodic in t, a Fourier series representation still exists but the sense of (6.1) may not be pointwise equality.

Lemma 6.1 (Discrete Fourier Series) *Let V be a vector space over $\mathbb{C}$. The map $D : V^n \mapsto V^n$ defined by*

$$Dv|_k = \widetilde{v}_k = \frac{1}{n} \sum_{j=0}^{n-1} v_j e^{i2\pi jk/n}, \quad k = 0, 1, ..., n-1 \tag{6.3}$$

is invertible (a bijection) and D^{-1} is given by

$$D^{-1}\widetilde{v}|_j = \sum_{k=0}^{n-1} \widetilde{v}_k e^{-i2\pi jk/n}, \quad j = 0, 1, ..., n-1. \tag{6.4}$$

Proof. Given $v \in V^n$, we shall show D is invertible by showing that (6.4) explicitly gives the inverse, that is, $D^{-1}Dv = v$. Let j' be any integer in $\{0, 1, \ldots, T-1\}$; then

$$\begin{aligned} D^{-1}Dv|_{j'} &= \frac{1}{n} \sum_{k=0}^{n-1} \sum_{j=0}^{n-1} v_j e^{(i2\pi jk - i2\pi j'k)/n} \\ &= \frac{1}{n} \sum_{j=0}^{n-1} v_j \sum_{k=0}^{n-1} e^{i2\pi(j-j')k)/n} \\ &= v_{j'}, \end{aligned} \tag{6.5}$$

where it is not difficult to show that

$$\sum_{k=0}^{n-1} e^{i2\pi(j-j')k/n} = \begin{cases} 0 & \text{if} \quad j \neq j', \\ n & \text{if} \quad j = j' \end{cases}. \quad \blacksquare$$

Thus the function $R(t+\tau,t)$ is completely represented by the collection of coefficient functions $\{B_k(\tau) : k = 0, 1, \ldots T-1\}$, and the sequence X_t is stationary if and only if $R(t+\tau,t) \equiv B_0(\tau)$ for all t.

We now turn our attention to properties of the coefficient functions $\{B_k(\tau) : k = 0, 1, \ldots T-1\}$.

First, the connection between nonnegative definite functions and second-order sequences also holds in the nonstationary case.

Proposition 6.1 *A necessary and sufficient condition that a complex sequence $\{R(s,t) : s,t \in \mathbb{Z}\}$ be the covariance of a random sequence is that it is nonnegative definite (we will henceforth say NND), meaning that for every n, every collection of complex constants $\{\alpha_1, \alpha_2, ..., \alpha_n\}$, and collection of times $\{t_1, t_2, ..., t_n\}$,*

$$\sum_{p=1}^{n}\sum_{q=1}^{n} \alpha_p \overline{\alpha_q} R(t_p, t_q) \geq 0. \tag{6.6}$$

The necessity is very simple and is omitted. Several proofs exist for the sufficiency; see, for example, Lemma 5.1 of Rozanov [201].

For the stationary case the covariance is a function of the difference of the arguments, (see Chapter 4) so

$$R(\tau) = \int_0^{2\pi} e^{i\lambda\tau} F(d\lambda) \tag{6.7}$$

where F is a nonnegative measure on the Borel subsets of $[0, 2\pi)$. Now we show that $B_0(\tau)$ is always the covariance of some stationary sequence; that is, it is always NND.

Proposition 6.2 *If X_t is a PC sequence, then*

$$B_0(\tau) = \frac{1}{T}\sum_{t=0}^{T-1} R(t+\tau, t) \tag{6.8}$$

is nonnegative definite.

Proof. To see that $B_0(\tau)$ is NND, take any n and any collection of complex numbers $\{\alpha_1, \alpha_2, \ldots, \alpha_n\}$ and integers $\{t_1, t_2, \ldots, t_n\}$ and consider

$$\sum_{p=1}^{n}\sum_{q=1}^{n} \alpha_p \overline{\alpha_q} B_0(t_p - t_q) = \sum_{p=1}^{n}\sum_{q=1}^{n} \alpha_p \overline{\alpha_q} \frac{1}{T}\sum_{t=0}^{T-1} R(t + t_p - t_q, t). \tag{6.9}$$

But for each fixed t_p, t_q, the transformation $u = t - t_q$ converts the last sum into

$$\sum_{u=-t_q}^{-t_q+T-1} R(u+t_p, u+t_q) = \sum_{u=0}^{T-1} R(u+t_p, u+t_q),$$

where the last equality occurs because we are summing a periodic sequence over exactly one period and so it does not matter where in the period the sum begins. Thus (6.9) becomes

$$\begin{aligned}\sum_{p=1}^{n}\sum_{q=1}^{n} \alpha_p\overline{\alpha_q} B_0(t_p - t_q) &= \sum_{p=1}^{n}\sum_{q=1}^{n} \alpha_p\overline{\alpha_q}\frac{1}{T}\sum_{u=0}^{T-1} R(u+t_p, u+t_q) \\ &= \frac{1}{T}\sum_{u=0}^{T-1}\sum_{p=1}^{n}\sum_{q=1}^{n} \alpha_p\overline{\alpha_q} R(u+t_p, u+t_q) \geq 0\end{aligned}$$

because

$$\sum_{p=1}^{n}\sum_{q=1}^{n} \alpha_p\overline{\alpha_q} R(u+t_p, u+t_q) \geq 0$$

follows from the application of (6.6) to the times $\{u+t_j\}_{j=1}^{n}$. ∎

An immediate application of Herglotz's theorem (Theorem 4.1) to the coefficient $B_0(\tau)$ yields

$$B_0(\tau) = \int_0^{2\pi} e^{i\lambda\tau} F_0(d\lambda) \tag{6.10}$$

for some nonnegative measure F_0. But what about the other coefficients $B_k(\tau)$? Are they Fourier transforms too? The answer is yes, but not necessarily with respect to nonnegative measures. We answer the question using the following characterization of sequences that are Fourier transforms of complex measures μ.

Proposition 6.3 *A bounded complex sequence $\{a_n : n \in \mathbb{Z}\}$ has representation*

$$a_n = \int_0^{2\pi} e^{i\lambda n} \mu(d\lambda) \tag{6.11}$$

for μ a complex measure on the Borel subsets of $[0, 2\pi)$ if and only if it satisfies the following boundedness condition: there exists a positive number M such that

$$\left|\sum_{j=1}^{m} \alpha_j a_j\right| \leq M \sup_{\lambda\in[0,2\pi)} \left|\sum_{j=1}^{m} \alpha_j e^{i\lambda j}\right| \tag{6.12}$$

for any positive integer m and any complex sequence $\{\alpha_j\}_{j=1}^{m}$.

Proof. If a_n is given by (6.11), then the boundedness condition holds with $M = \int_0^{2\pi} |d\mu|$. For the converse, suppose we are given a sequence $\{a_n\}$ that satisfies the boundedness. We will show that this condition expresses the boundedness of a linear functional on a dense subspace of $C[0, 2\pi)$, and hence extension to all of $C[0, 2\pi)$ gives the desired result. Let $\mathcal{M}$ be the space of finite linear combination of functions from $E = \{e^{i\lambda n}, n \in \mathbb{Z}\}$; that is, $\mathcal{M}$ is the set of trigonometric polynomials $t(\lambda) = \sum_{j=1}^{m} \alpha_j e^{i\lambda j}$. Clearly $\mathcal{M} \subset C[0, 2\pi)$. A linear functional may be defined on $\mathcal{M}$ by setting, for every $t \in \mathcal{M}$,

$$L(t) = L\left(\sum_{j=1}^{m} \alpha_j e^{i\lambda j}\right) = \sum_{j=1}^{n} \alpha_j a_j, \tag{6.13}$$

where $t(\lambda) = \sum_{j=1}^{m} \alpha_j e^{i\lambda j}$. We immediately see that the boundedness condition (6.12) expresses the boundedness of L on $\mathcal{M}$. But since $\mathcal{M}$ is dense in $C[0, 2\pi)$, a result from elementary Fourier series or from the Stone–Weierstrauss theorem, then L may be extended to $C[0, 2\pi)$ without increasing the norm. Hence from the Riesz representation theorem (see Rudin [202, Chapter 6]) for bounded linear functionals on $C[0, 2\pi)$ there exists a measure μ so that for $f \in C[0, 2\pi)$, the functional L is given by

$$L(f) = \int_0^{2\pi} f(\lambda)\mu(d\lambda) \tag{6.14}$$

and $\int_0^{2\pi} |d\mu| \leq M$. Observe that L evaluated on the elements of E produces

$$a_n = L(e^{i\lambda n}) = \int_0^{2\pi} e^{i\lambda n} \mu(d\lambda)$$

as required. ∎

For bounded linear functionals on $C[0, 1]$ the Riesz representation leads to Stieltjes integrals with respect to functions of bounded variation (see Riesz and Nagy [192, page 106, section 50]).

The characterization of Fourier transforms given in Proposition 6.3 is sometimes called the problem of trigonometric moments (see Riesz and Nagy [192, page 116]). Bochner [20] extended the characterization to bounded functions $f : \mathbb{R} \to \mathbb{C}$ and Eberlein to LCA groups, an account of which can be found in the book by Rudin [203].

Now we apply the preceding to the $B_k(\tau)$.

Proposition 6.4 *If X_t is PC-T, then for each $k = 0, 1, \ldots, T-1$ there exists a measure F_k on Borel subsets of $[0, 2\pi)$ such that*

$$\begin{aligned} B_k(\tau) &= \int_0^{2\pi} e^{i\lambda\tau} F_k(d\lambda), \quad \tau \in \mathbb{Z}; \\ \int_0^{2\pi} |F_k|(d\lambda) &\leq B_0(0) = \int_0^{2\pi} F_0(d\lambda). \end{aligned} \tag{6.15}$$

Proof. For general k we shall apply Proposition 6.3, as we have already remarked that Proposition 6.2 and Herglotz's theorem together imply (6.15) for $k = 0$. Hence it suffices to show that there exists a positive constant M for which

$$\left| \sum_{j=1}^{m} \alpha_j B_k(\tau_j) \right| \leq M \sup_{\lambda \in [0, 2\pi)} \left| \sum_{j=1}^{m} \alpha_j e^{i\lambda\tau_j} \right|$$

for any m, complex sequence $\{\alpha_j : 1 \leq j \leq m\}$, and integer sequence $\{\tau_j : 1 \leq j \leq m\}$.

Substituting (6.2) in the left-hand side of the previous expression leads to

$$\begin{aligned} S &= \left| \sum_{j=1}^{m} \alpha_j B_k(\tau_j) \right| = \left| \sum_{j=1}^{m} \alpha_j \frac{1}{T} \sum_{t=0}^{T-1} R(t+\tau_j, t) e^{-i2\pi kt/T} \right| \\ &= \frac{1}{T} \left| \sum_{t=0}^{T-1} \sum_{j=1}^{m} \alpha_j R(t+\tau_j, t) e^{-i2\pi kt/T} \right| \\ &= \frac{1}{T} \left| \sum_{t=0}^{T-1} \sum_{j=1}^{m} \alpha_j E\{X_{t+\tau_j} \overline{X_t}\} e^{-i2\pi kt/T} \right| \\ &= \frac{1}{T} \left| \sum_{t=0}^{T-1} E\left\{ \left[\sum_{j=1}^{m} \alpha_j X_{t+\tau_j} \right] \overline{X_t} \right\} e^{-i2\pi kt/T} \right| \\ &\leq \frac{1}{T} \sum_{t=0}^{T-1} \left[E \left| \sum_{j=1}^{m} \alpha_j X_{t+\tau_j} \right|^2 \right]^{1/2} \left[E|X_t|^2 \right]^{1/2} \end{aligned}$$

where the preceding inequality follows from the Cauchy–Schwarz inequality applied to second order random variables. Application of the Cauchy–Schwarz inequality to finite sequences yields

$$S^2 \leq \frac{1}{T} \sum_{t=0}^{T-1} E \left| \sum_{j=1}^{m} \alpha_j X_{t+\tau_j} \right|^2 \frac{1}{T} \sum_{t=0}^{T-1} E\, |X_t|^2 . \tag{6.16}$$

But

$$
\begin{aligned}
\sum_{t=0}^{T-1} E\Big|\sum_{j=1}^{m} \alpha_j X(t+\tau_j)\Big|^2 &= \sum_{t=0}^{T-1}\sum_{k=1}^{m}\sum_{j=1}^{m} \alpha_k \overline{\alpha}_j R(t+\tau_k, t+\tau_j) \\
&= T\sum_{k=1}^{m}\sum_{j=1}^{m} \alpha_k \overline{\alpha}_j B_0(\tau_k - \tau_j) \\
&= T\int_0^{2\pi} \Big|\sum_{j=1}^{m} \alpha_j e^{i\lambda\tau_j}\Big|^2 F_0(d\lambda) \qquad (6.17)
\end{aligned}
$$

and

$$
\sum_{t=0}^{T-1} E\,|X_t|^2 = \sum_{t=0}^{T-1} R(t,t) = TB_0(0) = T\int_0^{2\pi} F_0(d\lambda).
$$

Using these facts in (6.16) leads finally to

$$
S^2 \le \int_0^{2\pi} \Big|\sum_{j=1}^{m} \alpha_j e^{i\lambda\tau_j}\Big|^2 F_0(d\lambda) \int_0^{2\pi} F_0(d\lambda)
$$

and so

$$
S \le \sup_{\lambda\in[0,2\pi)} \Big|\sum_{j=1}^{m} \alpha_j e^{i\lambda\tau_j}\Big| \int_0^{2\pi} F_0(d\lambda), \qquad (6.18)
$$

so that $\int_0^{2\pi} F_0(d\lambda)$ serves for the constant M, and this same constant serves for all k. ∎

Note that the inequality (6.15) says only that $|F_k|([0,2\pi)) \le F_0([0,2\pi))$, the total mass of $|F_k|$ is dominated by that of F_0, not absolute continuity ($|F_k| \ll F_0$), which requires $|F_k|(A) \le F_0(A)$ for all Borel sets A. It will be shown later using the Cauchy–Schwarz inequality that there is a sense in which F_k is absolutely continuous with respect to F_0.

To see that the measures F_k need not be nonnegative, consider any properly PC sequence X_t for which $R(t,t)$ is constant with respect to t. For then there must exist some τ and $k > 0$ for which $B_k(\tau) \ne 0$, meaning $|F_k([0,2\pi))| \ne 0$. But $R(t,t)$ constant means that $F_k([0,2\pi)) = 0$ for all $k > 0$ because

$$
B_k(0) = \int_0^{2\pi} F_k(d\lambda). \qquad (6.19)
$$

Examples of such processes were given in Sections 2.1.4, 2.2.6, and 2.2.7.

The reader may wish to show that Proposition 6.4 is not necessarily true if the underlying covariance $R(s,t)$ is replaced with an arbitrary $f(s,t) =$

$f(s+T,t+T)$. The following result, due to Gladyshev [77], is a statement in the spirit of Khintchine's theorem that is related to this problem.

Proposition 6.5 (Gladyshev) *Suppose the sequence $\{B_k(\tau), \tau \in \mathbb{Z}\}_{k=0}^{T-1}$ of coefficient functions is determined by (6.2) from some $R(s,t)$ that satisfies $R(s,t) = R(s+T, t+T)$. Then $R(s,t)$ is NND if and only if for any n, any set of complex numbers $\{\alpha_1, \alpha_2, \ldots, \alpha_n\}$, any set of times $\{t_1, t_2, \ldots, t_n\}$, and any set of indices $\{k_1, k_2, \ldots, k_n \in [0, 1, ..., T-1]\}$, one has*

$$\sum_{p=1}^{n}\sum_{q=1}^{n} \alpha_p \bar{\alpha}_q \beta^{k_p k_q}(t_p - t_q) \geq 0, \tag{6.20}$$

where $\beta^{jk}(\tau) = B_{k-j}(\tau)e^{i2\pi j\tau/T}$.

Proof. Suppose R is NND so that (6.6) holds; then evaluation of (6.20) gives

$$\begin{aligned} S &= \sum_{p=1}^{n}\sum_{q=1}^{n} \alpha_p \bar{\alpha}_q \beta^{k_p k_q}(t_p - t_q) \\ &= \sum_{p=1}^{n}\sum_{q=1}^{n} \alpha_p \bar{\alpha}_q B_{k_q - k_p}(t_p - t_q) e^{i2\pi k_p (t_p - t_q)/T} \\ &= \sum_{p=1}^{n}\sum_{q=1}^{n} \alpha_p \bar{\alpha}_q e^{i2\pi k_p (t_p - t_q)/T} \frac{1}{T}\sum_{t=0}^{T-1} R(t + t_p - t_q, t) e^{-i2\pi t (k_q - k_p)/T}. \end{aligned} \tag{6.21}$$

In the last sum of the preceding line, setting $t' = t - t_q$ yields

$$\sum_{t'=-t_q}^{-t_q+T-1} R(t' + t_p, t' + t_q) e^{-i2\pi (t' + t_q)(k_q - k_p)/T} =$$

$$\sum_{t'=0}^{T-1} R(t' + t_p, t' + t_q) e^{-i2\pi (t' + t_q)(k_q - k_p)/T}$$

because for any p, q the summand is periodic in t' with period T and the sum is over one full period. Hence the sum can be started anywhere and the same value will be obtained. So finally we obtain

$$S = \frac{1}{T}\sum_{t=0}^{T-1}\left[\sum_{p=1}^{n}\sum_{q=1}^{n} R(t + t_q, t + t_p)\alpha_p e^{i2\pi k_p (t + t_p)/T} \overline{\alpha}_q e^{-i2\pi k_q (t + t_q)/T}\right] \tag{6.22}$$

and the quantity in brackets is nonnegative for each $t = 0, 1, \ldots, T-1$.

To show the converse, we are given (6.20) and wish to show that (6.6) holds for arbitrary n, and any collection of complex numbers $\{\alpha_1, \alpha_2, \ldots, \alpha_n\}$, and any collection of integers $\{t_1, t_2, \ldots, t_n\}$. Using (6.1) we first obtain

$$\begin{aligned} S &= \sum_{p=1}^{n}\sum_{q=1}^{n} \alpha_p \bar{\alpha}_q R(t_p, t_q) \\ &= \sum_{p=1}^{n}\sum_{q=1}^{n} \alpha_p \bar{\alpha}_q \sum_{k=0}^{T-1} B_k(t_p - t_q) e^{i2\pi k t_q/T}. \end{aligned}$$

By observing the way $B_k(\tau)$ is formed it is clear that for k outside the base set $\{0, 1, ..., T-1\}$ the relation $B_k(\tau) = B_{k+T}(\tau)$ holds for every k and τ. For any arbitrary periodic sequence $h_k = h_{k+T}$ the truth of

$$\sum_{k=0}^{T-1} h_k = \frac{1}{T}\sum_{k=0}^{T-1}\sum_{k'=0}^{T-1} h_{k-k'} \tag{6.23}$$

applied to $B_k(t_p - t_q)$ yields

$$\begin{aligned} S &= \frac{1}{T}\sum_{p=1}^{n}\sum_{q=1}^{n} \alpha_p \bar{\alpha}_q \sum_{k=0}^{T-1}\sum_{k'=0}^{T-1} B_{k'-k}(t_p - t_q) e^{i2\pi t_q (k'-k)/T} \\ &= \frac{1}{T}\sum_{p=1}^{n}\sum_{q=1}^{n} \alpha_p \bar{\alpha}_q \sum_{k=0}^{T-1}\sum_{k'=0}^{T-1} \beta^{kk'}(t_p - t_q) e^{i2\pi (k' t_q - k t_p)/T}. \end{aligned} \tag{6.24}$$

We now claim that the preceding quadruple sum can be put into the form

$$\frac{1}{T}\sum_{p'=1}^{nT}\sum_{q'=1}^{nT} \xi_{p'} \bar{\xi}_{q'} \beta^{k_{p'} k_{q'}}(t_{p'} - t_{q'}), \tag{6.25}$$

which can be recognized as the form of (6.20) and the result $S \geq 0$ follows. To do this we first note that every $p' \in \{1, 2, ..., nT\}$ can be written as $p' = kn + p$ for $k \in \{0, 1, ..., T-1\}$ and $p \in \{1, 2, ..., n\}$. The identifications

$$\begin{aligned} \xi_{p'} &= \alpha_p e^{-i2\pi k t_p}, \\ k_{p'} &= k, \\ t_{p'} &= t_p \end{aligned}$$

transform the last sum in (6.24) into (6.25) and the claim is proved. ∎

Note that condition (6.20) means that $\{\beta^{jk}(\tau)\}$ for $j, k \in \{0, 1, \ldots T-1\}$ is a cross-covariance matrix of a T-variate stationary sequence. Gladyshev [77] used this to argue that the $B_k(\tau)$ are Fourier transforms. We have chosen our particular path because it introduces us to the Riesz theorem about linear functionals, which leads us to similar results both in the continuous time and almost periodic cases [102, 106].

6.2 HARMONIZABILITY OF PC SEQUENCES

Gladyshev [77] also showed that all PC sequences are (strongly) harmonizable, which, in view of Definition 5.4, means that

$$R(s,t) = \int_0^{2\pi} \int_0^{2\pi} e^{i(s\lambda_1 - t\lambda_2)} F(d\lambda_1, d\lambda_2)$$

where F is a finite measure on $\mathcal{B}([0,2\pi)^2)$. The proof of this relation for covariances of PC sequences seems almost obvious, since using (6.15) in (6.1) produces

$$R(s,t) = \sum_{k=0}^{T-1} \int_0^{2\pi} e^{i\lambda(s-t)+i2\pi kt/T} F_k(d\lambda), \tag{6.26}$$

a representation of $R(s,t)$ by a finite sum of one dimensional Fourier transforms. We only need to see how to express (6.26) in the form (5.4). We get a clue about making this transformation from the following result, in which we *assume* $R(s,t)$ is the autocovariance function of a harmonizable sequence (see Definition 5.4).

Proposition 6.6 *If autocovariance $R(s,t)$ is a Fourier transform of a measure of bounded variation F, then $R(s,t) = R(s+T, t+T)$ if and only if the support of F is contained in the union $S_T = \cup_{k=-T+1}^{T-1} S_k$ of $2T-1$ diagonal lines*

$$S_k = \{(\lambda_1, \lambda_2) \in [0,2\pi) \times [0,2\pi) : \lambda_2 = \lambda_1 - 2\pi k/T\}. \tag{6.27}$$

Proof. If $\text{supp}(F) \subset S_T$, then (5.4) becomes

$$R(s,t) = \int\int_{S_T} e^{is\lambda_1 - it\lambda_2} F(d\lambda_1, d\lambda_2) = R(s+T, t+T) \tag{6.28}$$

because $e^{i(s+T)\lambda_1 - i(t+T)\lambda_2} = e^{is\lambda_1 - it\lambda_2}$, whenever $(\lambda_1, \lambda_2) \in S_T$.

Conversely if $R(s,t) = R(s+T, t+T)$, then for every natural number N,

$$\begin{aligned} R(s,t) &= \frac{1}{(2N+1)} \sum_{k=-N}^{N} R(s+kT, t+kT) \\ &= \int_0^{2\pi} \int_0^{2\pi} e^{i(s\lambda_1 - t\lambda_2)} D_N[T(\lambda_1 - \lambda_2)] F(d\lambda_1, d\lambda_2), \end{aligned} \tag{6.29}$$

where

$$D_N(\sigma) = \frac{\sin[(N+\frac{1}{2})\sigma]}{2(N+\frac{1}{2})\sin(\sigma/2)} \tag{6.30}$$

is a sequence of bounded continuous functions with $D_N(j2\pi) = 1, j \in \mathbb{Z}$ and converging to 0 for $\sigma \neq 2\pi\mathbb{Z}$. Hence the Lebesgue convergence theorem implies (6.28) holds for every s, t and the invertibility of Fourier transforms (or the uniqueness of the measure) implies that the support of F must be contained in S_T. ∎

Now we have a target for the rearrangement of the finite collection of measures on lines given by (6.26). The proof of the following proposition is just a demonstration of the rearrangement. It gives us the correspondence between the complex measures F_k and the measure F, which we now know to be supported (concentrated) on a finite number of lines.

Proposition 6.7 *The covariance of every PC sequence is a Fourier transform of a measure with bounded total variation. In other words, every PC sequence is harmonizable.*

Proof. We only need to show that (6.26) can be rearranged to yield (6.28). First, (6.26) can be rewritten as

$$R(s,t) = \int_0^{2\pi}\int_0^{2\pi} e^{ixt+iy(s-t)}\rho(dx,dy), \tag{6.31}$$

where the support of the measure $\rho(\cdot)$, as shown in Figure 6.1 for $T=5$, is the set $\{(x,y) : 0 \le y < 2\pi,\ x = 2\pi k/T, 0 \le k \le T-1\}$ of T vertical lines of length 2π and spacing $2\pi/T$. The total variation of ρ is given by

$$|\rho| = \sum_{k=0}^{T-1}\int_0^{2\pi} |F_k(d\lambda)| \le T\int_0^{2\pi} F_0(d\lambda) < \infty.$$

The transformation $\lambda_1 = y$ and $\lambda_2 = y - x$ in (6.31) produces

$$R(s,t) = \int_{\lambda_1=0}^{2\pi}\int_{\lambda_2=\lambda_1}^{\lambda_1+2\pi} e^{i\lambda_1 s - i\lambda_2 t}\rho'(d\lambda_1, d\lambda_2). \tag{6.32}$$

The support of ρ' is a family of T diagonal lines given by $\lambda_2 = \lambda_1 - 2\pi k/T$ for $\lambda_1 \in [0, 2\pi)$ as illustrated in Figure 6.2. This follows from the fact that the support of ρ is T vertical lines with spacing $2\pi/T$. For example, the line segment $\{(x,y) : x = 0, 0 \le y < 2\pi\}$ is transformed to $\{(\lambda_1, \lambda_2) : \lambda_2 = \lambda_1, 0 \le \lambda_1 < 2\pi\}$.

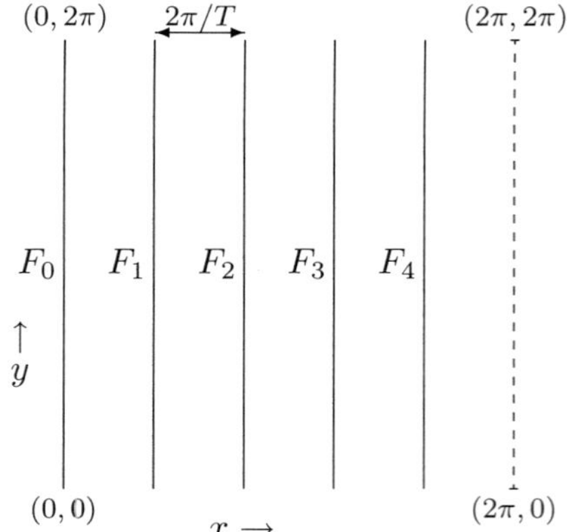

Figure 6.1 Support of ρ in $[0, 2\pi) \times [0, 2\pi)$ for a PC-T sequence with $T = 5$.

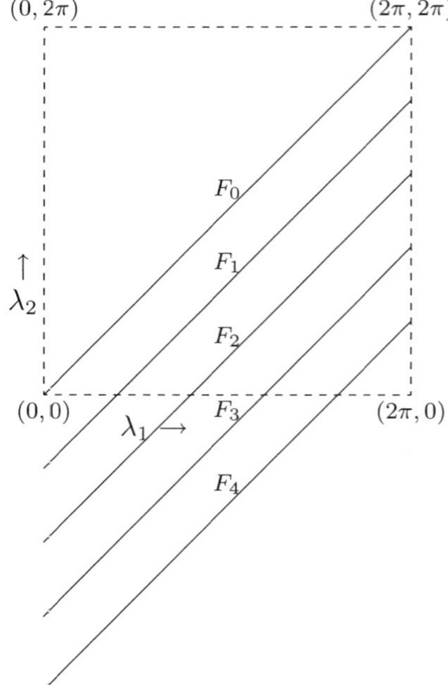

Figure 6.2 Support of ρ' under the transformation $\lambda_1 = y$ and $\lambda_2 = y - x$.

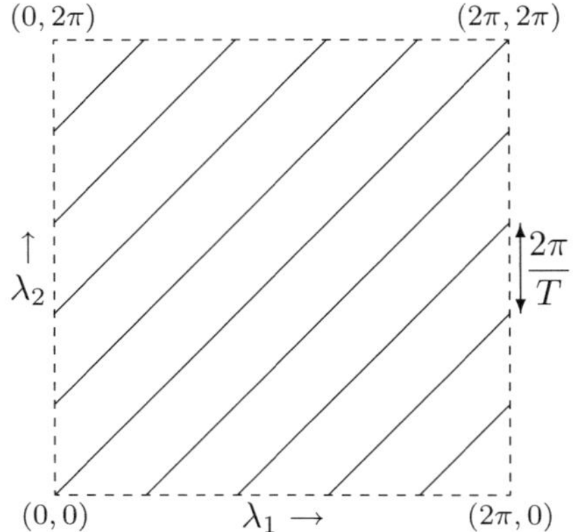

Figure 6.3 Support of $F = \rho'$: λ_2 is taken (mod 2π).

The final step to put (6.31) into the desired form (6.28) is to replace λ_2 in (6.31) with $\lambda_2(\text{mod } 2\pi)$, noting that the exponentials are invariant under these shifts (see Figure 6.3). The transformation $\lambda_2(\text{mod } 2\pi)$ translates the support of ρ', where $\lambda_2 < 0$, upward by 2π to fill in the square $[0, 2\pi) \times [0, 2\pi)$. So we finally see that the measure F is merely a rearrangement of ρ having the same (finite) variation as ρ. ∎

It follows from the sequence of transformations that

$$B_k(\tau) = \int \int_{S_k} e^{i\lambda_1 \tau} F(d\lambda_1, d\lambda_2) = \int_0^{2\pi} e^{i\lambda\tau} F_k(d\lambda) \tag{6.33}$$

and so from the uniqueness of Fourier transforms we conclude that we can identify the measure F_k with the restriction of F to the set S_k. An application of the inversion formula (see Loève, [139, page 199] or Brockwell and Davis, [28, Theorem 4.9.1])

$$F_k([a, b)) = \lim_{N\to\infty} \frac{1}{2\pi} \sum_{\tau=-N}^{N} \frac{e^{-i\tau a} - e^{-i\tau b}}{i\tau} B_k(\tau)$$

to (6.33) makes precise the relation

$$F_k(\lambda_0 + d\lambda) = F\left(\lambda_0 + d\lambda, \lambda_0 - \frac{2\pi k}{T} + d\lambda\right). \tag{6.34}$$

The set S_1 for $T = 5$ is illustrated in Figure 6.4.

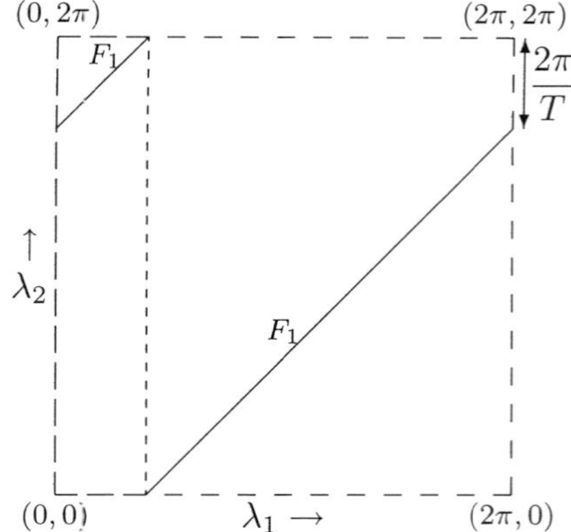

Figure 6.4 Support set S_1 for $T = 5$.

Harmonizability of X_t. Since PC sequences are (strongly) harmonizable and thus, by Proposition 5.10, also weakly harmonizable, we have

$$X_t = \int_0^{2\pi} e^{i\lambda t}\, \xi(d\lambda), \tag{6.35}$$

where the covariance in the increments of ξ were described by Proposition 6.6. This same representation will be proved again in Chapter 7 via the explicit construction of the random measure ξ.

Now we show that the representation (6.35) for a PC-T sequence X_t leads to another connection to multivariate stationary sequences, but in the frequency domain.

Proposition 6.8 (Gladyshev) *A second order sequence X_t is PC-T if and only if there exists a T-variate stationary sequence $\{Z_t^k\}_{k=0}^{T-1}, t \in \mathbb{Z}$ such that*

$$X_t = \sum_{k=0}^{T-1} Z_t^k e^{i2\pi kt/T}, \quad \textit{for every } t \in \mathbb{Z}. \tag{6.36}$$

Proof. If X_t has the representation (6.36), then

$$\langle X_s, X_t\rangle = \sum_{k=0}^{T-1}\sum_{j=0}^{T-1} \langle Z_s^k, Z_t^j\rangle e^{i2\pi(ks-jt)/T} = \langle X_{s+T}, X_{t+T}\rangle \tag{6.37}$$

because for every k, k' the covariances $\langle Z_s^k, Z_t^{k'}\rangle$ depend only on $s - t$.

Conversely, if X_t is PC with period T, then from (6.35),

$$\begin{aligned} X_t &= \sum_{k=0}^{T-1} \int_{2k\pi/T}^{2(k+1)\pi/T} e^{i\lambda t}\xi(d\lambda) \\ &= \sum_{k=0}^{T-1} \int_0^{2\pi/T} e^{i(\gamma+2\pi k/T)t}\xi(d\gamma + 2\pi k/T) \\ &= \sum_{k=0}^{T-1} e^{i2\pi kt/T} \int_0^{2\pi/T} e^{i\gamma t}\xi^k(d\gamma), \end{aligned} \tag{6.38}$$

where $\xi^k(\cdot)$ is defined on the Borel sets of $[0, 2\pi/T)$ by

$$\xi^k(A) = \xi(A + \frac{2\pi k}{T}). \tag{6.39}$$

The identification

$$Z_t^k = \int_0^{2\pi/T} e^{i\gamma t}\xi^k(d\gamma) \tag{6.40}$$

gives the result since the random measures ξ^k are orthogonally scattered. That is, if A and B are subsets of $[0, 2\pi/T)$, then

$$\mathcal{F}^{pq}(A \cap B) = E\{\xi^p(A)\overline{\xi^q(B)}\} = 0$$

whenever $A \cap B = \emptyset$ because the translated sets

$$A + \frac{2\pi p}{T} \text{ and } B + \frac{2\pi q}{T}$$

also do not intersect, and this is true for every $p, q = 0, 1, \ldots, T-1$. Figure 6.5 may help in the perception of this fact. ∎

The following proposition, also due to Gladyshev [77], relates the matrix valued cross spectral measures (distribution) $\mathbf{F}$ for the multivariate stationary sequence $\{[\mathbf{X}_n]^p = X_n^p = X_{nT+p} : 0 \le p \le T-1\}$ with $\mathcal{F}$ of the multivariate stationary sequence $\{Z_t^k : 0 \le k \le T-1\}$. Here we note that $\mathbf{F}$ and $\mathcal{F}$ are both $T \times T$, but $\mathbf{F}$ is defined on the Borel sets of $[0, 2\pi)$, whereas $\mathcal{F}$ is defined on the Borel sets of $[0, 2\pi/T)$.

Proposition 6.9 *If X_t is PC-T, then*

$$\mathbf{F}(d\lambda) = T\mathbf{V}(\lambda)\mathcal{F}(d\lambda/T)\mathbf{V}^{-1}(\lambda), \tag{6.41}$$

where $\mathbf{V}(\lambda)$ is a unitary matrix (a map $\mathbb{C}^T \to \mathbb{C}^T$) whose (p,k)th element is given by

$$v^{pk}(\lambda) = \frac{1}{\sqrt{T}}e^{i2\pi pk/T + i\lambda p/T}, \tag{6.42}$$

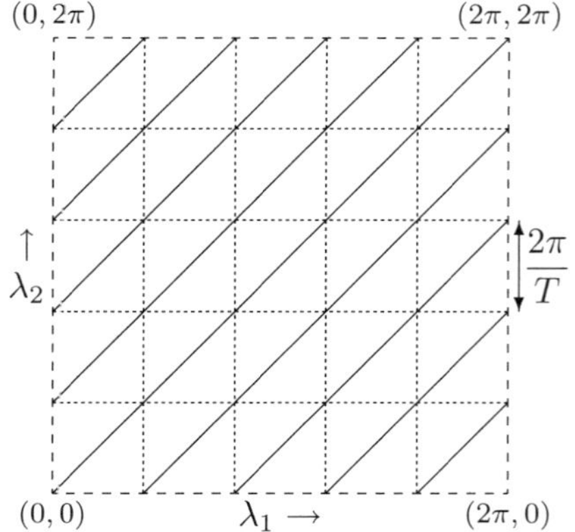

Figure 6.5 Partition of $[0, 2\pi) \times [0, 2\pi)$ into subsquares of side $2\pi/T$. The cross spectral measure $\mathcal{F}^{pq}(\cdot)$ is obtained from the diagonal of subsquare pq by $\mathcal{F}^{pq}(A) = F(A + 2\pi p/T, A + 2\pi q/T)$.

where p, k *are both in the set* $\{0, 1, \ldots, T-1\}$.

Proof. For the T-variate stationary sequence $X_n^p = X_{p+nT}, n \in \mathbb{Z},\ p = 0, 1, ..., T-1$, we can write

$$\langle X_s^p, X_t^{p'} \rangle = \int_0^{2\pi} e^{i(s-t)\lambda} F^{pp'}(d\lambda). \tag{6.43}$$

On the other hand, using representation (6.36) of Proposition 6.8 we can write

$$\begin{aligned}\langle X_s^p, X_t^{p'} \rangle &= \sum_k \sum_{k'} e^{i2\pi pk/T} e^{-i2\pi p'k'/T} \int_0^{2\pi/T} e^{i\gamma(p+sT-p'-tT)} \mathcal{F}^{kk'}(d\gamma) \\ &= \sum_k \sum_{k'} e^{i2\pi pk/T} e^{-i2\pi p'k'/T} \int_0^{2\pi} e^{i\lambda(p-p')/T + i\lambda(s-t)} \mathcal{F}^{kk'}\left(\frac{d\lambda}{T}\right),\end{aligned}$$

where we changed variables by $\lambda = T\gamma$. Comparing the last display with expression (6.43) and applying the uniqueness of Fourier transforms together yield

$$\begin{aligned}\mathbf{F}^{pp'}(A) &= \int_A \mathbf{F}^{pp'}(d\lambda) \\ &= T \int_A \sum_k \sum_{k'} v^{pk}(\lambda) \overline{v}^{p'k'}(\lambda) \mathcal{F}^{kk'}\left(\frac{d\lambda}{T}\right).\end{aligned} \tag{6.44}$$

Since the functions $v^{pk}(\lambda)$ are all continuous functions, setting $A = [a, b)$ with $b - a$ sufficiently small, we can write

$$\mathbf{F}^{pp'}([a,b)) = \sum_k \sum_{k'} v^{pk}(\lambda^{(kk')})\overline{v}^{p'k'}(\lambda^{(kk')})\mathcal{F}^{kk'}([a,b)/T),$$

where $a \leq \lambda^{(kk')} < b$. We abbreviate this by

$$\mathbf{F}^{pp'}(d\lambda) = T\sum_k \sum_{k'} v^{pk}(\lambda)\overline{v}^{p'k'}(\lambda)\mathcal{F}^{kk'}\left(\frac{d\lambda}{T}\right), \tag{6.45}$$

which by taking all the p, p' gives (6.41). ∎

The invertibility and continuity of $\mathbf{V}(\lambda)$ for $\lambda \in [0, 2\pi)$ means (6.41) can also be expressed as

$$\mathcal{F}(d\lambda) = \frac{1}{T}\mathbf{V}^{-1}(T\lambda)\mathbf{F}(Td\lambda)\mathbf{V}(T\lambda). \tag{6.46}$$

Corollary 6.9.1 *The complex matrix measures* $\mathbf{F}$ *and* $\mathcal{F}$ *are mutually absolutely continuous in the sense that for every Borel set* A, $|\mathbf{F}^{pp'}|(A) = 0,\ p, p' = 0, 1, \ldots, T-1 \iff |\mathcal{F}^{kk'}|(A/T) = 0,\ k, k' = 0, 1, \ldots, T-1.$

Proof. If $|\mathcal{F}^{kk'}|(A/T) = 0,\ k, k' = 0, 1, \ldots, T-1$, then

$$\begin{aligned} |\mathbf{F}^{pp'}|(A) &\leq T\int_A \sum_k \sum_{k'} \max_\lambda |v^{pk}(\lambda)\overline{v}^{p'k'}(\lambda)||\mathcal{F}^{kk'}|\left(\frac{d\lambda}{T}\right) \\ &\leq T\sum_k \sum_{k'} \int_A |\mathcal{F}^{kk'}|\left(\frac{d\lambda}{T}\right) = T\sum_k \sum_{k'} |\mathcal{F}^{kk'}(A/T)| \qquad (6.47) \\ &= 0 \qquad (6.48) \end{aligned}$$

for $p, p' = 0, 1, \ldots, T-1$. The reverse implication follows from (6.46). ∎

We will say that the complex matrix measure $\mathbf{F}$ is absolutely continuous with respect to Lebesgue measure μ and write $|\mathbf{F}| \ll \mu$ if $|\mathbf{F}|^{pp'} \ll \mu$ for $p, p' = 0, 1, \ldots, T-1$. Similarly, $|\mathcal{F}| \ll \mu$ will mean $|\mathcal{F}|^{kk'}(A/T) \ll \mu$ for $k, k' = 0, 1, \ldots, T-1$. We have immediately from Corollary 6.9.1 that $|\mathbf{F}| \ll \mu$ if and only if $|\mathcal{F}| \ll \mu$. Thus if either of these hold we have

$$\frac{d\mathbf{F}^{pp'}}{d\mu} = T\sum_k \sum_{k'} v^{pk}(\lambda)\overline{v}^{p'k'}(\lambda)\frac{d\mathcal{F}^{kk'}(\cdot/T)}{d\mu} \tag{6.49}$$

or, more compactly, setting $f_{\mathbf{X}}(\lambda) = d\mathbf{F}/d\mu$ and $f_Z(\lambda/T) = d\mathcal{F}(\cdot/T)/d\mu$, we have

$$f_{\mathbf{X}}(\lambda) = TV(\lambda) f_{\mathbf{Z}}\Big(\frac{\lambda}{T}\Big) V^{-1}(\lambda) \tag{6.50}$$

and the inverse whose form is expressed by (6.46).

See the supplements at the end of this chapter for some discussion of Proposition 6.9 in terms of the spectral processes of $Y(n)$ and $Z_p(t)$.

6.3 SOME PROPERTIES OF $B_k(\tau)$, F_k, AND F

From (6.2) it is clear that $B_k(\tau)$ is periodic in the index k, that is, $B_k(\tau) = B_{k+T}(\tau)$. In addition,

$$\begin{aligned} B_k(-\tau) &= \frac{1}{T}\sum_{t=0}^{T-1} R(t-\tau, t) e^{-i2\pi kt/T} \\ &= \frac{1}{T}\sum_{t=0}^{T-1} \overline{R(t, t-\tau)} e^{-i2\pi kt/T} \\ &= \frac{1}{T}\sum_{s=-\tau}^{-\tau+T-1} \overline{R(s+\tau, s)} e^{-i2\pi k(s+\tau)/T} \\ &= \overline{B_{-k}(\tau)} e^{-i2\pi k\tau/T} = \overline{B_{T-k}(\tau)} e^{-i2\pi k\tau/T}, \end{aligned} \tag{6.51}$$

so there is a symmetry about the index $(T-1)/2$; in other words, (6.51) shows that the entire collection of coefficient functions $\{B_k(\tau) : k = 0, 1, \dots, T-1\}$ is determined by the collection $\{B_k(\tau) : k = 0, 1, \dots, [(T-1)/2]\}$. This also means that the spectral measures $\{F_k : k = 0, 1, \dots, T-1\}$ are determined by the collection $\{F_k : k = 0, 1, \dots, [(T-1)/2]\}$.

This conclusion may also be drawn from the Hermitian property of F (see Chapter 5)

$$F([a,b),[c,d)) = E\{\xi([a,b))\overline{\xi([c,d))}\} = \overline{F([c,d),[a,b))]} \tag{6.52}$$

for any rectangle $[a,b) \times [c,d)$ in $[0,2\pi) \times [0,2\pi)$. In the case of PC sequences, for which the support of F is contained in $2T-1$ diagonal lines, we conclude that the measure on the kth line below the main diagonal is determined by the conjugate of the kth line above the main diagonal. And since each spectral measure F_k may be identified with two pieces appropriately spliced from F, the kth line below the main diagonal and the $(T-k)$th line above the main diagonal, the conjugate symmetry of F implies that the $\{F_k : k = 0, \dots, T-1\}$ are determined entirely by F restricted to $\lambda_2 \le \lambda_1$ or from the preceding remark regarding splicing, by the collection $\{F_k : k = 0, 1, \dots, [(T-1)/2]\}$.

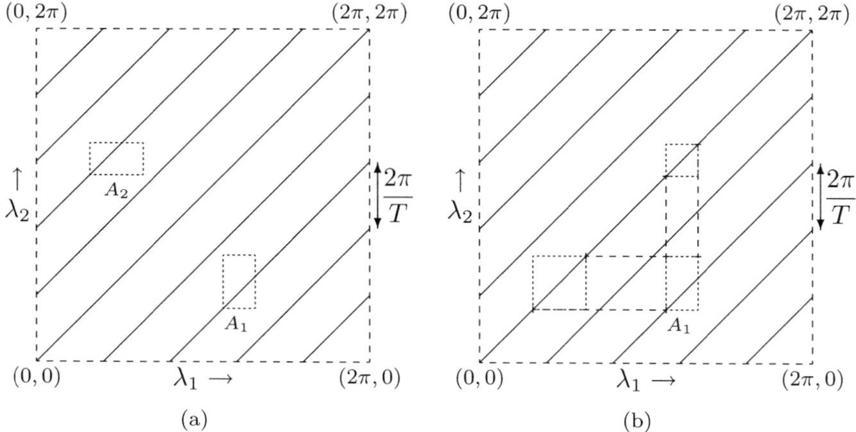

Figure 6.6 Geometric relationship between F and F_k for a PC sequence. $I_1 = [a, b), I_2 = [c, d)$. (a) Hermitian symmetry of F implies $F(A_2) = \overline{F(A_1)}$ for $A_1 = I_1 \times I_2$ and $A_2 = I_2 \times I_1$. (b) Cauchy–Schwarz inequality implies $|F([a, b) \times [c, d))|^2 \leq F([a, b)^2)F([c, d)^2)$.

The main diagonal measure F_0 has a domination property arising from the Cauchy–Schwarz inequality applied to ξ,

$$|E\{\xi([a, b))\overline{\xi([c, d))}\}|^2 \leq E\{|\xi([a, b))|^2\}E\{|\xi([c, d))|^2\},$$

which can also be written

$$|F([a, b) \times [c, d))|^2 \leq F([a, b) \times F([a, b))F([c, d) \times [c, d)) \tag{6.53}$$

as illustrated in Figure 6.6(b). Taking $[c, d) = [a - 2k\pi/T, b - 2k\pi/T)$ produces

$$\left|F_k\Big([a, b)\Big)\right| \leq F_0\Big([a, b)\Big)F_0\Big([a - \frac{2k\pi}{T}, b - \frac{2k\pi}{T})\Big). \tag{6.54}$$

This immediately implies that if F_0 is absolutely continuous with respect to Lebesgue measure, then F_k is also, and its density satisfies

$$|f_k(\lambda)|^2 \leq f_0(\lambda)f_0(\lambda - 2k\pi/T). \tag{6.55}$$

But also if F has a point mass at (λ_1, λ_2) then the diagonal must have point masses at (λ_1, λ_1) and (λ_2, λ_2).

For additional conclusions that may be drawn when X_t is a *real* PC sequence, see the supplements to this chapter.

6.4 COVARIANCE AND SPECTRA FOR SPECIFIC CASES

The notation and definitions just completed are applied here to some specific examples.

6.4.1 PC White Noise

Periodically correlated white noise might also be called an uncorrelated PC sequence, thus providing an example of a situation where the terminology *cyclostationary* works better.

Definition 6.1 *If ϵ_t is a normalized white noise (see Definition 4.6) and $\sigma(t) = \sigma(t+T)$ is a nonnegative periodic sequence, then*

$$X_t = \sigma(t)\epsilon_t \tag{6.56}$$

is called PC white noise.

It is easy to compute that

$$R(s,t) = \sigma^2(t)\delta_{s-t}, \tag{6.57}$$

which leads to

$$\begin{aligned} B_k(\tau) &= \frac{1}{T}\sum_{t=0}^{T-1} R(t+\tau,t)e^{-i2\pi kt/T} \\ &= \delta_\tau \frac{1}{T}\sum_{t=0}^{T-1} \sigma^2(t)e^{-i2\pi kt/T} \\ &= \delta_\tau B_k(0) = \frac{B_k(0)}{2\pi}\int_0^{2\pi} e^{i\lambda\tau}\,d\lambda. \end{aligned} \tag{6.58}$$

So the spectral densities for PC white noise are simply

$$f_k(\lambda) = \frac{B_k(0)}{2\pi} \quad \text{for } 0 \le \lambda < 2\pi.$$

Note if the variance is constant, $\sigma^2(t) \equiv \sigma^2$, then

$$B_k(\tau) = \begin{cases} \sigma^2 & k = 0 \text{ and } \tau = 0 \\ 0 & \text{otherwise} \end{cases},$$

showing that the nonstationary part (corresponding to $k \neq 0$) of F is null.

Taking the other extreme, suppose

$$\sigma^2(t) = \begin{cases} \sigma^2 & \text{if} \quad t = 0 \\ 0 & \text{if} \quad t = 1, 2, \ldots, T-1 \end{cases}; \tag{6.59}$$

then we find $B_k(\tau) = \sigma^2\delta_\tau/T$ and hence the spectral densities are all the same:

$$f_k(\lambda) = \frac{\sigma^2}{2\pi T} \tag{6.60}$$

for all $k = 0, 1, \ldots, T-1$.

In view of 6.60, we can see that each element of the matrix spectral distribution function $\mathcal{F}^{pp'}$ is comprised only of the absolutely continuous part, which is

$$f_Z^{pp'}(\lambda) = \frac{\sigma^2}{2\pi T},$$

and so

$$f_Z(\lambda) = \frac{\sigma^2}{2\pi T}\begin{bmatrix} 1 & 1 & \ldots & 1 \\ 1 & 1 & \ldots & 1 \\ \vdots & \vdots & \vdots & \vdots \\ 1 & 1 & \ldots & 1 \end{bmatrix}, \quad 0 \le \lambda < 2\pi/T. \tag{6.61}$$

Clearly $f_Z(\lambda)$ is of rank 1 and since $\mathbf{V}(\lambda)$ is unitary for every λ, then from (6.50), $f_{\mathbf{X}}(\lambda)$ is of rank 1 for every λ. Anticipating the application of multivariate (innovation) rank (see Section 4.4.3) to PC sequences (see Chapter 8), this particular case (6.59) is called *rank-1 PC-T white noise.*

Note that the PC-T white noise is actually a special case of $X_t = f_t Y_t$ for f_t periodic and Y_t stationary and where we identify $f_t^2 = \sigma^2(t)$. By convention $\sigma(t)$ is the positive square root of $\sigma^2(t)$.

6.4.2 Products of Scalar Periodic and Stationary Sequences

If $X_t = f_t Y_t$ for f_t periodic and Y_t stationary, then from (2.7) the simple result $R_X(s,t) = f_s\overline{f_t} \cdot R_Y(s-t)$ leads to

$$\begin{aligned} B_k(\tau) &= \frac{1}{T}\sum_{t=0}^{T-1} R_X(t+\tau, t)e^{-i2\pi kt/T} \\ &= R_Y(\tau)\frac{1}{T}\sum_{t=0}^{T-1} f_{t+\tau}\overline{f_t}e^{-i2\pi kt/T} \\ &= R_Y(\tau)\sum_{p=0}^{T-1} F_p\overline{F_{p-k}}e^{i2\pi k\tau/T}, \end{aligned} \tag{6.62}$$

where F_p is the Fourier coefficient

$$F_p = \frac{1}{T}\sum_{t=0}^{T-1} f_t e^{-i2\pi tp/T}$$

with

$$f_t = \sum_{p=0}^{T-1} F_p e^{i2\pi tp/T}.$$

If the sequence Y_t has an absolutely continuous spectrum

$$R_Y(\tau) = \int_0^{2\pi} e^{i\lambda\tau} f_Y(\lambda) d\lambda,$$

then from (6.62) and using standard Fourier relationships we obtain

$$f_k(\lambda) = \sum_{p=0}^{T-1} f_Y(\lambda - 2\pi p/T) F_p \overline{F_{p-k}}, \tag{6.63}$$

where in (6.63) we must use $F_p = F_{p+T}$ and the argument of f_Y must be taken (mod 2π).

6.5 ASYMPTOTIC STATIONARITY

Definition 6.2 *For a second order sequence X_t, we define*

$$R_{\text{avg}}(\tau) = \lim_{N\to\infty} \frac{1}{2N+1} \sum_{t=-N}^{N} E\{X_{t+t_0+\tau}\overline{X_{t+t_0}}\} \tag{6.64}$$

whenever the limit exists. If $R_{\text{avg}}(\tau)$ is defined for all $\tau \in \mathbb{Z}$, then we say the sequence X_t is asymptotically stationary.

It is not hard to see that the limit is independent of t_0 at any τ for which the limit exists and if the limit exists for all τ, then the function $R_{\text{avg}}(\tau)$ must be NND and hence a covariance.

Proposition 6.10 *Every PC sequence is asymptotically stationary and*

$$R_{\text{avg}}(\tau) = B_0(\tau). \tag{6.65}$$

Proof. The proof follows from the fact that for any positive integer k,

$$s_k = \frac{1}{kT} \sum_{t=-kT}^{kT-1} E\{X_{t+\tau}\overline{X_t}\} = B_0(\tau)$$

and so for any N, choosing $k^* = [(2N+1)/T]$ produces

$$\epsilon_N = \left| \frac{1}{2N+1} \sum_{t=-N}^{N} E\{X_{t+\tau}\overline{X_t}\} - s_{k^*} \right| = O(1/N). \quad \blacksquare$$

Proposition 6.11 *Every harmonizable sequence is asymptotically stationary and*

$$R_{\text{avg}}(\tau) = \int\int_{\Delta} e^{i\lambda_1 \tau} F(d\lambda_1, d\lambda_2). \tag{6.66}$$

Proof. The partial sum

$$\frac{1}{2N+1} \sum_{t=-N}^{N} E\{X_{t+\tau}\overline{X_t}\} = \int_0^{2\pi} \int_0^{2\pi} D_N(\lambda_1 - \lambda_2) e^{i\lambda_1 \tau} F(d\lambda_1, d\lambda_2), \tag{6.67}$$

where $D_N(\cdot)$, given by (6.30), converges as $N \to \infty$ to the right-hand side of (6.66) from the same argument used in Proposition 6.6. ∎

Proposition 6.12 *If F is the spectral measure of a harmonizable sequence and Δ is the diagonal of $[0, 2\pi) \times [0, 2\pi)$, then the diagonal measure defined by $F_\Delta(A) = F(\Delta \cap A \times A)$ satisfies $F_\Delta(A) \geq 0$ for any measurable subset A of $[0, 2\pi)$.*

Proof. For any n the diagonal Δ can be covered by n disjoint squares $\delta_{jn} \times \delta_{jn}$ of side $2\pi/n$,

$$\Delta \subset \Delta_n = \bigcup \delta_{jn} \times \delta_{jn},$$

in a way that $\Delta_n \downarrow \Delta$. Since $F(\delta_{jn} \times \delta_{jn} \cap A \times A) = E\{|\xi(\delta_{jn} \cap A)|^2\} \geq 0$ and $\Delta_n \cap A \times A \downarrow \Delta \cap A \times A$, then by the continuity of measure $F(\Delta_n \cap A \times A) \to F(\Delta \cap A \times A)$, and thus $F(\Delta \cap A \times A) \geq 0$. ∎

In particular, if X_t is stationary it is clearly asymptotically stationary and $R_{\text{avg}}(\tau)$ is the same as the covariance function of X_t. And we also recover the facts that were proved independently in Proposition 6.10: PC-T sequences are asymptotically stationary with $R_{\text{avg}}(\tau) = B_0(\tau)$ and F_Δ is F_0. For some additional facts, including some early references, on asymptotic stationarity, see [38, 42, 121, 126, 178].

6.6 LEBESGUE DECOMPOSITION OF F

Since F is a measure on the Borel sets of $[0, 2\pi)^2$, from its Lebesgue decomposition we can write

$$F = F^{ac} + F^s, \tag{6.68}$$

where F^{ac} is absolutely continuous with respect to two-dimensional Lebesgue measure μ^2 and F^s is singular with respect to μ^2. The first observation is that for a stationary or PC-T sequence, F is singular with respect to μ^2 because $\mu^2(S_T) = 0$.

Since for any harmonizable sequence it is always true that $F^{ac}(\Delta) = 0$ for the diagonal Δ, we conclude that $F(\Delta) > 0$ if and only if $F^s(\Delta) > 0$.

To get a sense of the meaning of the absolutely continuous part, we will say that a second order sequence X_t is *transient* whenever the time averaged variance is zero, or more explicitly whenever

$$R_{\text{avg}}(0) = \lim_{N\to\infty} \frac{1}{2N+1} \sum_{t=-N}^{N} E\{|X_t|^2\} = 0. \tag{6.69}$$

Proposition 6.13 *If X_t is harmonizable with spectral measure F, then the following are equivalent:*

(a) *X_t is transient;*

(b) *$F(\Delta) = 0$;*

(c) *$F = F^{ac}$.*

Proof. The result follows from (6.66) and the remarks above. ∎

See the supplements for more on this issue.

6.7 THE SPECTRUM OF m_t

An important source of singular discrete components in the spectrum of a PC sequence is the mean function $m_t = E\{X_t\}$, where $|m_t| \le E\{|X_t|\} < \infty$ because $X_t \in L^2(\Omega, \mathcal{F}, P)$. If a PC-T sequence X_t has a nonzero mean then we can write $X_t = X_t' + m_t$, where $m_t = m_{t+T}$ and X_t' is PC-T but with zero mean. Thus

$$\begin{aligned} R_X(s,t) &= E\{[X_s' + m_s]\overline{[X_t' + m_t]}\} &(6.70)\\ &= R_{X'}(s,t) + m_s\overline{m_t} \\ &= \int_0^{2\pi}\int_0^{2\pi} e^{i(s\lambda_1 - t\lambda_2)} F_{X'}(d\lambda_1, d\lambda_2) &(6.71)\\ &\quad + \sum_{j=0}^{T-1}\sum_{k=0}^{T-1} \widetilde{m}_j \overline{\widetilde{m}_k} e^{i2\pi(js-kt)/T} \end{aligned}$$

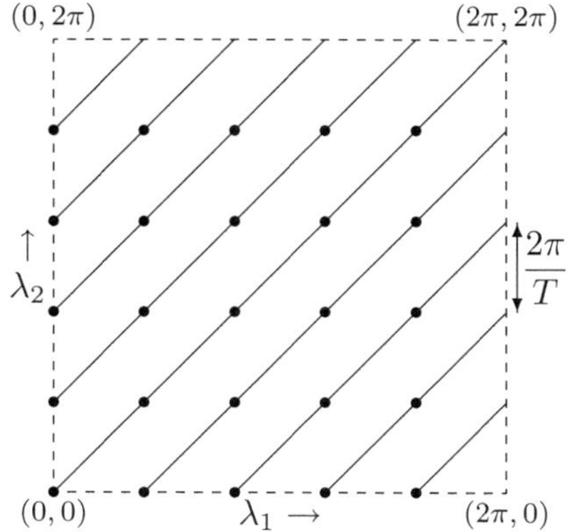

Figure 6.7 Possible locations of spectral atoms of F produced by the periodic mean m_t for $T = 5$.

and the periodicity $m_t = m_{t+T}$ permits the discrete Fourier series representation

$$m_t = \sum_{k=0}^{T-1} \widetilde{m}_k e^{i2\pi kt/T} \tag{6.72}$$

with scalar coefficients

$$\widetilde{m}_k = \frac{1}{T} \sum_{t=0}^{T-1} m_t e^{-i2\pi kt/T}. \tag{6.73}$$

It follows easily that the spectral measure associated with $R_X(s,t)$ can be expressed as

$$F_X = F_{X'} + \sum_{j=0}^{T-1} \sum_{k=0}^{T-1} \widetilde{m}_j \overline{\widetilde{m}_k} \delta(\lambda_1 - j2\pi/T, \lambda_2 - k2\pi/T), \tag{6.74}$$

where $F_{X'}$ is the spectral measure associated with the covariance, and $\delta(a,b) = 1$ if $a = 0$ and $b = 0$ and otherwise $\delta(a,b) = 0$.

So the point masses produced by m_t are at the points $(2\pi j/T, 2\pi k/T)$ with mass $\widetilde{m}_j \overline{\widetilde{m}_k}$. The possible location of the point masses attributed to the periodic mean for a PC-5 sequence are illustrated in Figure 6.7.

Now suppose X_t is PC-T but its periodic mean is null and we are studying the spectral measure for the *covariance* of X_t. Can there still remain point masses in the spectrum of X_t, that is, in F_X? The answer is yes. To give a simple example, suppose

$$X_t = Ae^{i\lambda_a t} + Be^{i\lambda_b t}, \tag{6.75}$$

where A and B are zero mean second order random variables. This sequence has $m_t = E\{X_t\} = 0$ and covariance

$$\begin{aligned} R_X(s,t) &= E\{|A|^2\}e^{i\lambda_a(s-t)} + E\{A\overline{B}\}e^{i\lambda_a s - i\lambda_b t} \\ &+ E\{B\overline{A}\}e^{i\lambda_b s - i\lambda_a t} + E\{|B|^2\}e^{i\lambda_b(s-t)}. \end{aligned}$$

If $E\{A\overline{B}\} = 0$, then $R_X(s,t)$ depends only on $s-t$ so X_t is stationary and hence PC-T for every integer $T \geq 1$. More to the point, if $E\{A\overline{B}\} \neq 0$, the sequence is still PC-T (i.e., $R_X(s+T, t+T) = R_X(s,t)$) provided $\lambda_a - \lambda_b = 2j\pi/T$ for some integer j. Extending this, a PC-T sequence can have a countable number of harmonic terms in its singular discrete part

$$X_t^{sd} = \sum_{j=-\infty}^{\infty} A_j e^{i\lambda_j t} \tag{6.76}$$

provided

$$S_{sd} = \{(\lambda_j, \lambda_k) : E\{A_j\overline{A_k}\} \neq 0\} \subset S_T \tag{6.77}$$

and

$$\sum_{(j,k):(\lambda_j,\lambda_k)\in S_{sd}} |E\{A_j\overline{A_k}\}| < \infty.$$

That is, the frequency pairs corresponding to nonzero correlations must all lie in the set S_T. The condition of Hermitian symmetry is seen in the simple statement

$$\overline{E\{A_j\overline{A_k}\}} = E\{A_k\overline{A_j}\}$$

and that of the Schwarz inequality

$$|E\{A_j\overline{A_k}\}|^2 \leq E\{|A_j|^2\}E\{|A_k|^2\}.$$

In order for X_t^{sd} to be stationary it is necessary and sufficient for $E\{A_j\overline{A_k}\} = 0$ for $j \neq k$.

6.8 EFFECTS OF COMMON OPERATIONS ON PC SEQUENCES

This section contains a treatment of the most common operations that are often used to precondition time series or that serve as components of more complicated algorithms.

6.8.1 Linear Time Invariant Filtering

Given a random sequence X_t and a nonrandom sequence $w_n, n \in \mathbb{Z}$ called *filter coefficients*, a new (filtered) sequence is formed by

$$Y_t = \sum_{n\in\mathbb{Z}} w_n X_{t-n} \tag{6.78}$$

provided the sum converges in mean-square sense. If $w_n = 0$ for $n < 0$ the filter is called *causal*: Y_t depends only on the input X_s for $s \le t$. If $\sum_{n\in\mathbb{Z}} |w_n| < \infty$, the filter is called *stable*. Since $R_X(t,t)$ is bounded for PC sequences, the finiteness of

$$\sum_{n\in\mathbb{Z}}\sum_{m\in\mathbb{Z}} |w_n||w_m||R_X(t-n,t-m)|$$
$$\le \sum_{n\in\mathbb{Z}}\sum_{m\in\mathbb{Z}} |w_n||w_m| \max_{t=0,1,\ldots,T-1} |R_X(t,t)| < \infty$$

along with the Cauchy criterion suffices for the existence of the sum Y_t and then

$$\begin{aligned} R_Y(s,t) &= \sum_{n\in\mathbb{Z}}\sum_{m\in\mathbb{Z}} w_n\overline{w_m}R_X(s-n,t-m) \\ &= \sum_{n\in\mathbb{Z}}\sum_{m\in\mathbb{Z}} w_n\overline{w_m}R_X(s+T-n,t+T-m) = R_Y(s+T,t+T), \end{aligned}$$

showing that the PC property is preserved by stable linear filtering. Causality is not required for the existence of the sum (6.78) defining Y_t, but in practical situations filters are typically causal.

The harmonizability of PC sequences permits a useful and helpful interpretation of the effects of linear time invariant filtering of PC sequences.[1] In Chapter 5 it was shown that the resulting sequence Y_t (6.78) from the filtering of a harmonizable sequence X_t by a stable linear filter can also be expressed as

$$Y_t = \int_0^{2\pi} e^{i\lambda t}\, W(\lambda)\xi(d\lambda), \tag{6.79}$$

where

$$W(\lambda) = \sum_{n\in\mathbb{Z}} w_n e^{i\lambda n} \tag{6.80}$$

is the Fourier transform of the filter weights. The covariance of Y_t then becomes

$$R_Y(s,t) = \int_0^{2\pi}\int_0^{2\pi} e^{i\lambda_1 s - i\lambda_2 t} W(\lambda_1)\overline{W(\lambda_2)}F_X(d\lambda_1,d\lambda_2), \tag{6.81}$$

where F_X is the spectral measure for the sequence X_t. When X_t is PC-T so the support of F_X is contained in S_T, (6.81) describes how the filter response $W(\lambda)$ modifies F_X and hence the output covariance. For example, Figure 6.8(a) presents the frequency response $W(\lambda)$ for a lowpass filter having 8

[1]This interpretation may also be utilized in continuous time, where PC processes are not necessarily harmonizable.

real valued coefficients that were determined by a least-square algorithm[2] and for which the cutoff design frequency is $\lambda_c = 0.30\pi$. Figure 6.8(b) shows a

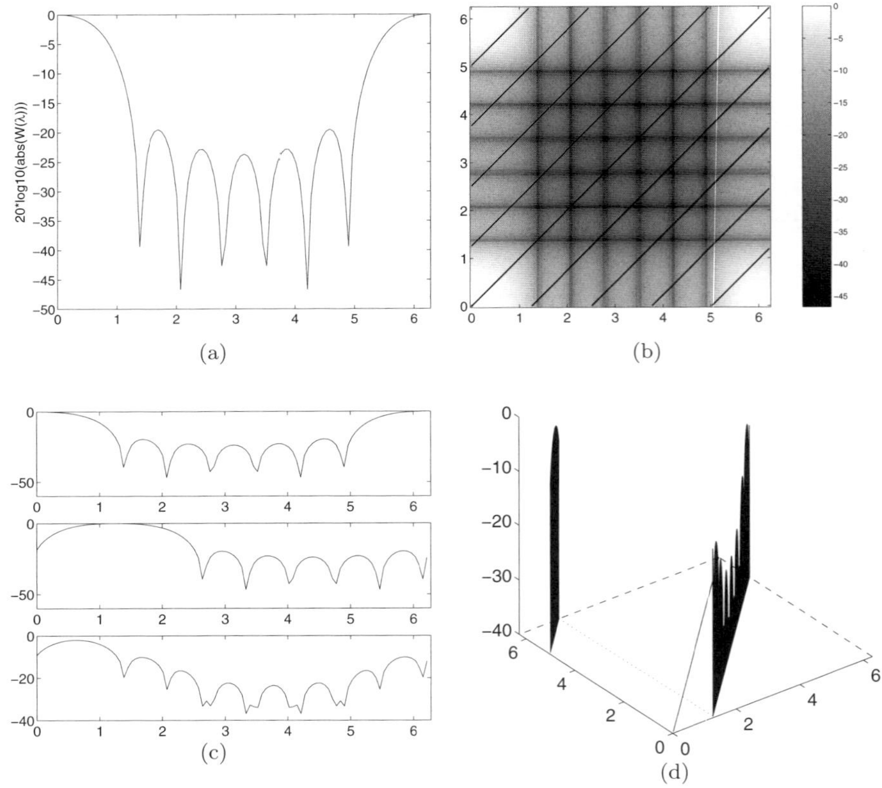

Figure 6.8 Effects of filtering a PC-5 sequence by an 8 coefficient lowpass filter with $\lambda_c = 0.30\pi$. (a) Frequency response $W(\lambda)$. (b) $10\log_{10}|W(\lambda_1)\overline{W(\lambda_2)}|$ image plot. (c) (Top) $20\log_{10}|W(\lambda)|$; (middle) $20\log_{10}|W(\lambda - 2\pi/T)|$; (bottom) $10\log_{10}|W(\lambda)\overline{W(\lambda - 2\pi/T)}|$. (d) $10\log_{10}|W(\lambda_1)\overline{W(\lambda_2)}|$ with $\lambda_2 = \lambda_1 - 2\pi/T$.

greyscale image of $10\log 10|W(\lambda_1)\overline{W(\lambda_2)}|$ with S_T for $T = 5$ overlaid in black. This image, whose diagonal is the response of Figure 6.8(a), shows that only the very low frequencies survive the filtering and the contributions of the off-diagonal measures, where the product $W(\lambda_1)\overline{W(\lambda_2)}$ is small, are strongly

[2]This particular set of coefficients was determined by the function `firls` from the MATLAB Signal Processing Toolbox.

suppressed. Real coefficients produce the symmetry $W(\lambda) = \overline{W(2\pi - \lambda)}$ and hence the large values of $W(\lambda_1)\overline{W(\lambda_2)}$ in the upper left and lower right corners. The plots of Figure 6.8(c) illustrate the values of $W(\lambda_1)\overline{W(\lambda_2)}$ along the $k = 1$ support line : $\lambda_2 = \lambda_1 - 2\pi/T$: the top trace is $20\log_{10}|W(\lambda)|$, the middle is $20\log_{10}|W(\lambda - 2\pi/T)|$, and the bottom is $10\log_{10}|W(\lambda)\overline{W(\lambda - 2\pi/T)}|$. Figure 6.8(d) shows the densities $f_k^Y(\lambda)$ produced by application of the same filter to PC-5 rank 1 white noise, for which all the densities are equal to the same constant.

Note that an ideal lowpass filter that passes only frequencies in $0 \leq \lambda < \lambda_c$ would completely suppress the PC structure provided $\lambda_c < 2\pi/T$ (in this example, $T = 5$). This result makes intuitive sense; if a PC sequence is smoothed by a filter having sufficiently long memory, then the nonstationary fluctuations will be smoothed out.

Before continuing with further illustrations we shall formalize the statement of effects of linear time invariant filtering on a PC sequence.

Proposition 6.14 *Suppose X_t is PC-T having spectral measures $F_k^X, k = 0, 1, \ldots T-1$, and $\{w_n, n \in \mathbb{Z}\}$ are the coefficients of a stable linear filter. Then the spectral measures of the filtered sequence Y_t are given by*

$$F_k^Y(A) = \int_A W(\lambda)\overline{W(\lambda - 2\pi k/T)}F_k^X(d\lambda), \tag{6.82}$$

where $W(\lambda)$ is given by (6.80).

Proof. Graphically, almost by inspection. But, using Proposition 5.13 and (6.33), we obtain

$$\begin{aligned} B_k^Y(\tau) &= \int_0^{2\pi} e^{i\lambda\tau}F_k^Y(d\lambda) \\ &= \int\int_{S_k} e^{i\lambda_1\tau}W(\lambda_1)\overline{W(\lambda_2)}F_X(d\lambda_1, d\lambda_2)) \\ &= \int_0^{2\pi} e^{i\lambda\tau}W(\lambda)\overline{W(\lambda - 2\pi k/T)}F_k^X(d\lambda) \end{aligned} \tag{6.83}$$

and then by recalling that $S_k = \{(\lambda_1, \lambda_2) : \lambda_2 = \lambda_1 - 2\pi k/T\}$ and always taking λ_1 and λ_2 modulo 2π. ∎

Note that the effect of LTI filtering on the main diagonal, which represents the "average" stationary spectrum, is exactly that of filtering a stationary sequence,

$$B_0^Y(\tau) = \int_0^{2\pi} e^{i\lambda\tau}|W(\lambda)|^2F_0^X(d\lambda).$$

In the case of LTI filtering (with real coefficients) as illustrated in Figure 6.9, the extent to which the off-diagonal (nonstationary) part of F_X is passed depends on where the passband intersects its reflection about $\lambda = \pi$. If the

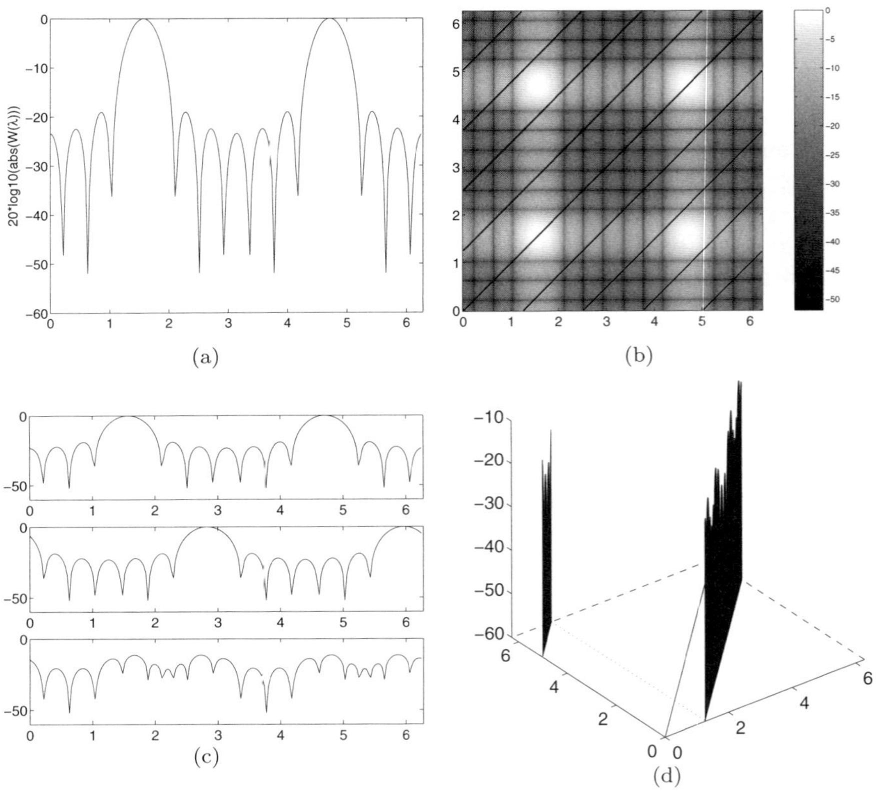

Figure 6.9 Effects of filtering a PC-5 sequence by a 12 coefficient bandpass filter with band edges $(\lambda_a, \lambda_b) = (0.4\pi, 0.6\pi)$ (a) Frequency response $W(\lambda)$. (b) $10\log_{10}|W(\lambda_1)\overline{W(\lambda_2)}|$ image plot. (c) (Top) $20\log_{10}|W(\lambda)|$; (middle) $20\log_{10}|W(\lambda - 2\pi/T)|$; (bottom) $10\log_{10}|W(\lambda)\overline{W(\lambda - 2\pi/T)}|$. (d) $10\log_{10}|W(\lambda_1)\overline{W(\lambda_2)}|$ with $\lambda_2 = \lambda_1 - 2\pi/T$.

area of high response covers a support line, then the nonstationary spectrum in this region will contribute to the output. It may be seen from Figure 6.9(b) that the area of high response is not directly over a support line. The bottom trace of Figure 6.9(c) shows that the response along the $k = 1$ support line is substantially smaller than the response on the diagonal.

In a subsequent section we will extend this analysis to compute the effect of periodically time varying (PTV) filters.

6.8.2 Differencing

In many applications (such as in economics or meteorology) the observed time series contains a trend term proportional to time t or a very low frequency fluctuation that appears as a trend in short series. In this case it is common practice to produce a new sequence

$$Y_t = X_t - X_{t-1} = (1 - B)X_t,$$

where B is the back shift operator defined by $X_{t-1} = BX_t$.

A nonrandom periodic component is sometimes called a *periodic trend* and in this case differencing with a one period lag

$$Y_t = X_t - X_{t-T} = (1 - B^T)X_t$$

will completely suppress (eliminate) from X_t any additive periodic function with period T. The T-point differences are also used to remove stochastic periodic terms produced by models having roots on the unit circle at $z_j = e^{i\lambda_j}$, where $\lambda_j = 2j\pi/T, j = 0, 1, \ldots, T-1$.

Since both of these operations are LTI filters, their effect on harmonizable sequences and, in particular, on PC sequences can be understood using (6.81) and Proposition 6.14.

In the case of first differences with lag 1, the filter coefficient sequence is $w_0 = 1$ and $w_1 = -1$ and the resulting frequency response is simply computed by (6.80) to be

$$W(\lambda) = 1 - e^{i\lambda}.$$

Figure 6.10 presents the resulting $20\log_{10}|W(\lambda)|$ relative to the maximum $(W(\pi) = 2)$ along with the grayscale image of $10\log_{10}|W(\lambda_1)\overline{W(\lambda_2)}|$, also relative to its maximum. The filter response at $\lambda = 0$ is null and exceeds 0.707 of the maximum in the region $\pi/2 \le \lambda < \pi$. However, for low frequencies, $0 < \lambda < \pi/2$, the filter has large suppressive effects. So the entire lower part of the spectrum of a harmonizable (including stationary) sequence will be seriously affected by first differencing. Hence, if one thinks that the low frequencies carry some important information, more thought should be given to the design of the detrending filter. The support set S_5 for a PC-5 sequence is overlaid in the usual manner to show how the filter response would affect the spectrum.

Figure 6.11 presents the corresponding displays for the first difference for lag 5. In this case $W(2j\pi/5) = 0$ for $j = 0, 1, \ldots, 5$ and so $W(\lambda_1)\overline{W(\lambda_2)} = 0$ at the points $(\lambda_1, \lambda_2) = (2j\pi/T, 2k\pi/T)$ for $j, k \in \{0, 1, \ldots, 5\}$ and where $T = 5$.

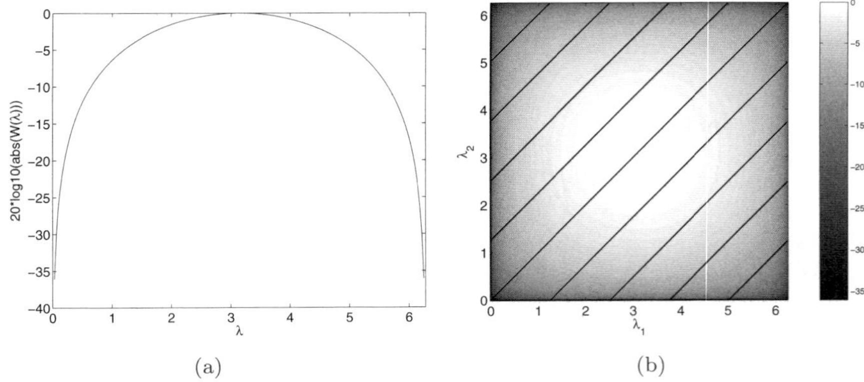

Figure 6.10 Frequency response for first difference with lag 1, $\phi(B) = 1-B$. (a) Frequency response $20\log_{10}|W(\lambda)|$. (b) Outer product frequency response $10\log_{10}|W(\lambda_1)\overline{W(\lambda_2)}|$. S_T for $T=5$ is overlaid in black.

So any discrete components at these points will be completely suppressed in the output, where recall that discrete components can be produced by the periodic mean or by random periodic components. Of course, any other discrete components in the original spectrum will be passed but with their amplitudes modified by the filter response. Figure 6.11(b) shows that on the central part of each small square $[2j\pi/T, 2(j+1)\pi/T) \times [2k\pi/T, 2(k+1)\pi/T)$ the measure on the diagonal lines does not experience much suppression, but near the points $(\lambda_1, \lambda_2) = (2j\pi/T, 2k\pi/T)$ the measure is substantially affected. As in the lag 1 case, this argues for methods that do not affect so much of the spectral covariance measure F_X in such a significant manner.

6.8.3 Random Shifts

We have already noted several ways in which periodically correlated processes have a very close connection to stationary processes. In this section we show the connection in yet another way: periodically nonstationary and periodically correlated sequences are essentially those sequences that can be made stationary by an independent random time shift.

The motivation for this problem comes from engineering problems in which periodic functions arise and one wishes to treat the periodic function as a stationary process. Typically this is done because treating the periodic function as a stationary process makes calculations concerning the spectrum easier, and in some cases the average spectrum conveys enough information to solve the engineering problem at hand. And one is tempted by the argument that

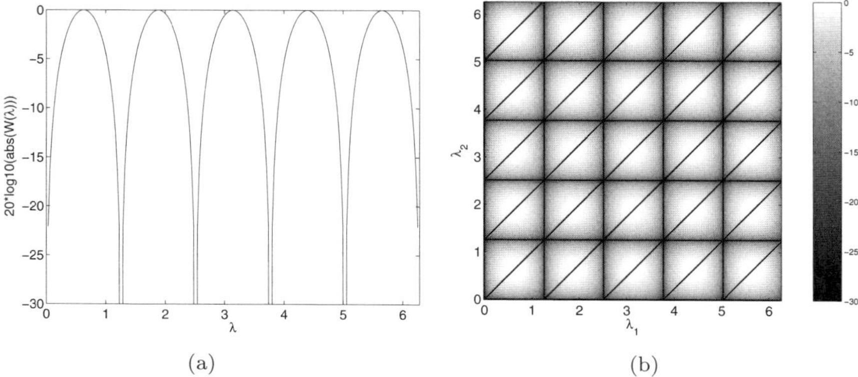

Figure 6.11 Frequency response for first difference with lag 5, $\phi(B) = 1-B^5$. (a) Frequency response $20\log_{10}|W(\lambda)|$. (b) Outer product frequency response $10\log_{10}|W(\lambda_1)\overline{W(\lambda_2)}|$. S_T for $T=5$ is overlaid in black.

the exact time origin is unknown anyway and so we may as well consider it as random and uniformly distributed over the period of the periodic function. Fredrick Beutler [24] studied this problem for nonrandom functions in some detail. He showed that if $f : \mathbb{R} \to \mathbb{R}, f(t) = f(t+T)$ is Borel measurable and Θ is a real random variable uniformly distributed on $\{0, 1, \ldots, T\}$, then $Y_t(\omega) = f(t + \Theta(\omega))$ is strictly stationary. He also showed that the uniform distribution for Θ is not necessarily the only distribution that can make $Y_t(\omega)$ stationary. The connection to PC processes (see Hurd [103]) and sequences is a little more complicated because the nonrandom function $f(\cdot)$ is replaced with random function X_t. See Gardner [65] for a treatment that also includes certain almost PC processes.

In discrete time, the shifted sequence can be explicitly written as

$$Y_t(\omega) = X_{t+\Theta(\omega)}(\omega) \tag{6.84}$$

so that in forming $Y_t(\omega)$ the randomness of the process $X_t(\omega)$ is mixed up by the random shift $\Theta(\omega)$. Given that both X and Θ are defined on the same probability space $(\Omega, \mathcal{F}, P)$, we come immediately to this question: How big does $\mathcal{F}$ need to be to ensure that $Y_t(\omega)$ is $\mathcal{F}$ measurable for each fixed t? If we denote $\mathcal{F}_X$ to be the sigma-field induced by X (the smallest sigma-field containing the ω-sets giving the finite dimensional distributions) and $\mathcal{F}_\Theta$ to be the sigma-field induced by Θ, then we will see below that it is enough for $\mathcal{F}$ to contain the join $\mathcal{F}_X \bigvee \mathcal{F}_\Theta$. Recall $\mathcal{F}_X \bigvee \mathcal{F}_\Theta$ is the smallest sigma-field containing sets of the form $A \cap B$ for $A \in \mathcal{F}_X$ and $B \in \mathcal{F}_\Theta$.

Although Θ could generally be taken to have range $\mathbb{Z}$, for our current problem it suffices to consider $\Theta : \Omega \mapsto \{0, 1, ..., T-1\}$. To see that $\mathcal{F}_X \bigvee \mathcal{F}_\Theta$ is indeed sufficient for our problem, let

$$S_j = \{\omega : \Theta(\omega) = j\} = \Theta^{-1}(j),\ j = 0, 1, ..., T-1 \tag{6.85}$$

so that

$$\Theta(\omega) = \sum_{j=0}^{T-1} 1_{S_j}(\omega) j, \tag{6.86}$$

and this leads at once to

$$\begin{aligned} Y_t(\omega) &= X_{t+j}(\omega),\ \omega \in S_j,\ j = 0, 1, \ldots, T-1 \\ &= \sum_{j=0}^{T-1} 1_{S_j}(\omega) X_{t+j}(\omega). \end{aligned} \tag{6.87}$$

So now it is clear that for each fixed t, $Y_t(\omega)$ is a sum of products of $\mathcal{F}_X$ measurable functions with $\mathcal{F}_\Theta$ measurable functions; that is, $Y_t(\omega)$ is $\mathcal{F}_X \bigvee \mathcal{F}_\Theta$ measurable.

Now if X_t is independent of Θ in the sense that $\mathcal{F}_X$ and $\mathcal{F}_\Theta$ are independent sigma-fields, then for any t and Borel set A,

$$\begin{aligned} Pr[Y_t \in A] &= \sum_{j=0}^{T-1} Pr[\Theta = j \cap X_{t+j} \in A] \\ &= \sum_{j=0}^{T-1} Pr[\Theta = j] Pr[X_{t+j} \in A] \\ &= \sum_{j=0}^{T-1} \mu_j Pr[X_{t+j} \in A], \end{aligned} \tag{6.88}$$

where

$$\mu_j = Pr[\Theta = j] = P(S_j).$$

This may immediately be extended to obtain for every n, every collection of times $t_1, t_2, ..., t_n$ and Borel sets $A_1, A_2, ..., A_n$

$$\begin{aligned} &Pr[Y_{t_1} \in A_1, Y_{t_2} \in A_2, ..., Y_{t_n} \in A_n] \\ &\quad = \sum_{j=0}^{T-1} \mu_j Pr[X_{t_1+j} \in A_1, X_{t_2+j} \in A_2, ..., X_{t_n+j} \in A_n]. \end{aligned} \tag{6.89}$$

The finite dimensional distributions of Y_t are just μ-weighted time averages of the finite dimensional distributions of X_t.

The same thing hold for moments. For example, if X_t is second order, then

$$\begin{aligned} E\{Y_{t+\tau}\overline{Y_t}\} &= \int_\Omega \sum_{j=0}^{T-1}\sum_{k=0}^{T-1} 1_{S_j}(\omega)1_{S_k}(\omega)X_{t+\tau+j}(\omega)\overline{X_{t+k}(\omega)}P(d\omega) \\ &= \sum_{j=0}^{T-1}\sum_{k=0}^{T-1} \int_\Omega 1_{S_j}(\omega)1_{S_k}(\omega)X_{t+\tau+j}(\omega)\overline{X_{t+k}(\omega)}P(d\omega) \\ &= \sum_{j=0}^{T-1} \mu_j E\{X_{t+\tau+j}\overline{X_{t+j}}\}, \end{aligned} \tag{6.90}$$

where the last equality results from $S_j \cap S_k = \emptyset$ for $j \neq k$ and the independence of X with Θ.

After the following simple lemma we will be prepared to state the main result.

Lemma 6.2 *A sequence $p : \mathbb{Z} \to \mathbb{C}$ is periodic with period T; that is, $p_t = p_{t+T}$ if and only if the sequence*

$$\widetilde{p}_t = \frac{1}{T}\sum_{j=0}^{T-1} p_{t+j} \tag{6.91}$$

is constant with respect to the variable t.

Proof. If $p_t = p_{t+T}$, then $\widetilde{p}_t$ will not depend on t because it is a uniform average over exactly one period; it does not matter where in the period one begins the sum. Conversely, if $\widetilde{p}_t$ does not depend on t, then

$$0 = \widetilde{p}_{t+1} - \widetilde{p}_t = p_{t+T} - p_t$$

and so p_t must be periodic with period T. ∎

Proposition 6.15 *If X_t is periodically nonstationary with period T and Θ is an integer valued random variable, uniformly distributed on $\{0, 1, ..., T-1\}$ and independent of X_t, then $Y_t = X_{t+\Theta}$ is strictly stationary. Conversely, if $Y_t = X_{t+\Theta}$ is strictly stationary for some Θ uniformly distributed on $\{0, 1, ..., T-1\}$ and independent of X_t, then X_t is periodically nonstationary with period T.*

Proof. Both statements are applications of Lemma 6.2 to (6.89). ∎

We state the following separately because of our focus on the second order case.

Proposition 6.16 *If X_t is PC with period T and Θ is an integer valued random variable, uniformly distributed on $\{0, 1, ..., T-1\}$ and independent of X_t, then $Y_{t+\Theta}$ is wide sense stationary and its covariance is $B_0(\tau)$. Conversely, if $Y_{t+\Theta}$ is wide sense stationary for some Θ uniformly distributed on $\{0, 1, ..., T-1\}$ and independent of X_t, then X_t is PC with period T.*

Proof. Both statements are applications of Lemma 6.2 to (6.90). In the first claim, (6.2) is used to see that the covariance of Y is $B_0(\tau)$ (which is NND by Proposition 6.2). ∎

When X_t is harmonizable the covariance $E\{Y_s\overline{Y_t}\}$ can be expressed nicely in terms of the spectral measure F_X.

Proposition 6.17 *If X_t is a harmonizable sequence and $Y_t = X_{t+\Theta}$, where $\Theta : \Omega \mapsto \{0, 1, \ldots, T-1\}$ is independent of X_t, then*

$$R_Y(s,t) = \int_0^{2\pi}\int_0^{2\pi} e^{i\lambda_1 s - i\lambda_2 t}\Phi_\Theta(\lambda_1 - \lambda_2)F_X(d\lambda_1, d\lambda_2) \tag{6.92}$$

where $\Phi_\Theta(u) = E\{e^{i\Theta u}\} = \sum_{j=0}^{T-1} e^{iju}\mu_j$ is the characteristic function of the random variable Θ.

Proof. Since $E\{X_{s+\Theta}\overline{X_{t+\Theta}} \mid \Theta = \Theta_0\} = R_X(s + \Theta_0, t + \Theta_0)$, then setting $\mu_j = Pr[\Theta = j]$,

$$\begin{aligned} R_Y(s,t) &= \sum_{j=0}^{T-1} \mu_j R_X(s+j, t+j) \qquad (6.93) \\ &= \sum_{j=0}^{T-1} \mu_j \int_0^{2\pi}\int_0^{2\pi} e^{i\lambda_1 s - i\lambda_2 t} e^{ij(\lambda_1 - \lambda_2)} F_X(d\lambda_1, d\lambda_2) \\ &= \int_0^{2\pi}\int_0^{2\pi} e^{i\lambda_1 s - i\lambda_2 t} \sum_{j=0}^{T-1} \mu_j e^{ij(\lambda_1 - \lambda_2)} F_X(d\lambda_1, d\lambda_2). \end{aligned}$$

The preceding line is easily identified as (6.92). ∎

In the special case when θ is uniformly distributed on $\{0, 1, \ldots, T-1\}$, we obtain

$$\Phi_\Theta(u) = e^{i(T-1)u/2}\frac{\sin(Tu/2)}{T\sin(u/2)}.$$

Note that $\Phi_\Theta(\lambda_1 - \lambda_2) = 0$ whenever $\lambda_2 = \lambda_1 + 2\pi k/T$ for any integer $k \neq 0$. Hence the main diagonal of S_T is preserved but all the off-diagonal components of S_T are *removed* by the random time shift. This gives another way to view the direct claim of Proposition 6.16 and shows that if the support of F_X is null on some of the off-diagonals S_k, then there are distributions other than uniform that will cause $X_{t+\Theta}$ to be stationary (see [24, 65, 103]).

6.8.4 Sampling

If X_t is a random sequence and we form $Y_t = X_{kt}$ for k a positive integer and $t \in \mathbb{Z}$, how are the covariance and spectral properties of Y_t related to those of the original process X_t? The question can be answered for arbitrary harmonizable sequences with random spectral measure $\xi(\cdot)$ and spectral covariance measure F. For then by (6.35),

$$Y_t = \int_0^{2\pi} e^{i\lambda kt}\, \xi(d\lambda), \tag{6.94}$$

which is an integral of exponential weights with respect to the random measure ξ. But $e^{i(\lambda+2\pi/k)kt} = e^{i\lambda kt}$ implies that the exponential weights are the same for all $t \in \mathbb{Z}$ for any frequencies that differ by an integer multiple of $2\pi/k$. Hence we can combine the parts of ξ that differ by multiples of $2\pi/k$, resulting in

$$Y_t = \int_0^{2\pi/k} e^{i\lambda kt} \sum_{j=0}^{k-1} \xi\Big(d\lambda + \frac{2j\pi}{k}\Big), \tag{6.95}$$

which, after the transformation $\gamma = k\lambda$, may finally be written as

$$Y_t = \int_0^{2\pi} e^{i\gamma t}\, \xi'(d\gamma), \tag{6.96}$$

where $\xi'(d\gamma) = \sum_{j=0}^{k-1} \xi(d\gamma/k + 2j\pi/k)$. The interpretation of (6.96) is that sampling by a factor of k does not preserve the original weighting of the random amplitudes by the exponentials, but causes the random amplitudes corresponding to frequencies that differ by an integer multiple of $2\pi/k$ to be weighted identically. This confounding of frequencies is usually called *aliasing*.

As for the covariance of Y_t, one can repeat the preceding steps or compute directly from (6.96) to reach

$$R_Y(s,t) = \int\int e^{i(s\lambda_1 - t\lambda_2)} F_Y(d\lambda_1, d\lambda_2), \tag{6.97}$$

where

$$F_Y(d\lambda_1, d\lambda_2) = \sum_{j=0}^{k-1} \sum_{j'=0}^{k-1} F_X(d\lambda_1/k + 2j\pi/k, d\lambda_2/k + 2j'\pi/k). \tag{6.98}$$

To illustrate using a sampling factor of two ($k = 2$), the spectrum in each subsquare in Figure 6.12(a) is shifted onto the principal square and then rescaled to fill the entire square $[0, 2\pi)^2$ in Figure 6.12(b). This illustrates also the idea that if the original sequence X_t had been lowpass filtered prior to sampling, then the spectrum in the "upper" quadrants would have been greatly suppressed so their effect on the sum (6.98) would have been small.

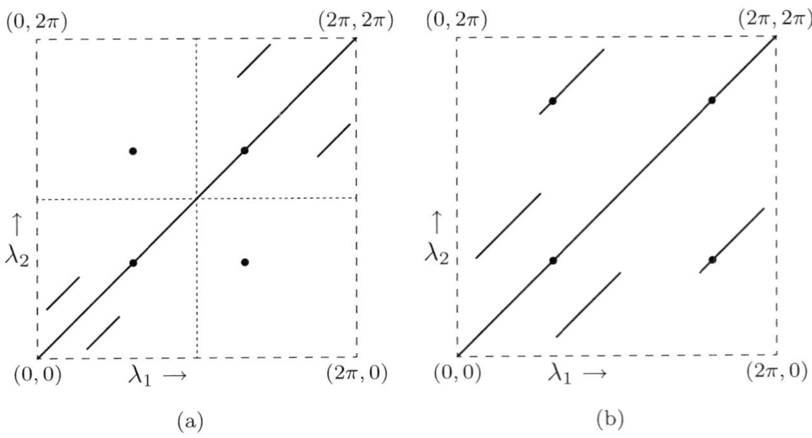

Figure 6.12 Aliasing effects due to sampling with $k = 2$. (a) Spectral support for PC-8 sequence. (b) Resulting F_Y produced by sampling with $k = 2$.

To determine whether $Y_t = X_{kt}$ is PC or stationary we examine $R_Y(s,t) = R_X(ks, kt)$ to find the smallest positive δ for which $R_Y(s+\delta, t+\delta) = R_Y(s,t)$ will be true (nontrivially) for all integers s, t. Since

$$R_Y(s + \delta, t + \delta) = R_X(ks + k\delta, kt + k\delta),$$

the previous equality will be satisfied whenever $k\delta = nT$, which has solutions δ, n since k and T are integers. The smallest δ giving a solution is paired with the least positive integer n that makes $\delta = nT/k$ a positive integer. If T and k are relatively prime, then $n = k$ and the period of the new sequence is again T. If T/k is an integer, then the period of the new sequence is $\delta = T/k$ (in the new index set). For example, if $T = 4$ and $k = 2$, then by choosing $n = 1$ we get Y_t to be PC with period $\delta = 2$. But if $T = 5$ and $k = 2$, then we need to take $n = 2$ in order to see that Y_t is again PC with period $\delta = 5$. Figure 6.13 illustrates the spectral effects of sampling a PC-4 sequence by a factor of $k = 2$ and $k = 4$. Figure 6.13(a) shows the support of the original F_X and the support of F_Y after the factor of 2 sampling is presented in Figure

6.13(b). An important case is when $k = T$, for then $\delta = 1$ so the sampled sequence is stationary. In this case all of the measures on the diagonals in the small squares of Figure 6.13(c) are added together to form the measure F_Y of Figure 6.13(d).

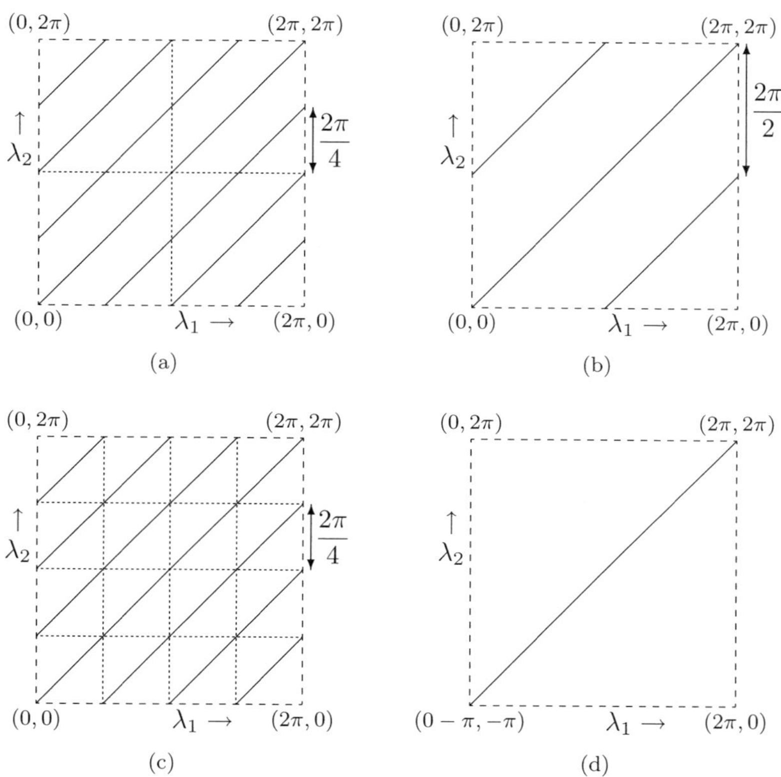

Figure 6.13 Result of sampling a PC-4 sequence by factors of 2 and 4. Small dashed lines define the subsets that are aliased. (a) Support set S_4 for a PC-4 sequence. (b) Support set for F_Y resulting from sampling by a factor of $k = 2$. (c) Support set S_4 for a PC-4 sequence. (d) Support set for F_Y resulting from sampling by a factor of $k = 4$.

An interesting application of filtering and sampling is the procedure called *aggregation*. Given a random sequence X_t that has a periodic structure with period T, it is sometimes of interest to inquire about the sequence obtained by summing (or perhaps averaging) the values of X_t over one period. For example, if t signified a monthly index and $T = 12$, then one might have interest in the yearly aggregate.

Definition 6.3 *The T-sample aggregate of a process X_t is*

$$Y_n = \sum_{p=0}^{T-1} X_{nT+p}. \tag{6.99}$$

The effects of aggregation may now be understood in terms of LTI filtering and sampling. Precisely, the sequence Y_n may be seen as the T-point sampling of the filtered sequence $Y_t = \sum_{n\geq 0} X_{t-n}w_n$, where $\{w_n\}$ has finite uniform weights, $w_n = 1$ for $n = 0, 1, \ldots, T-1$ and $w_n = 0$ for $n \geq T$. This yields a lowpass type of filter whose frequency response is illustrated in Figure 6.14 for $T = 12$. Note that the frequency response has zeros at $2\pi k/T$ so the periodic

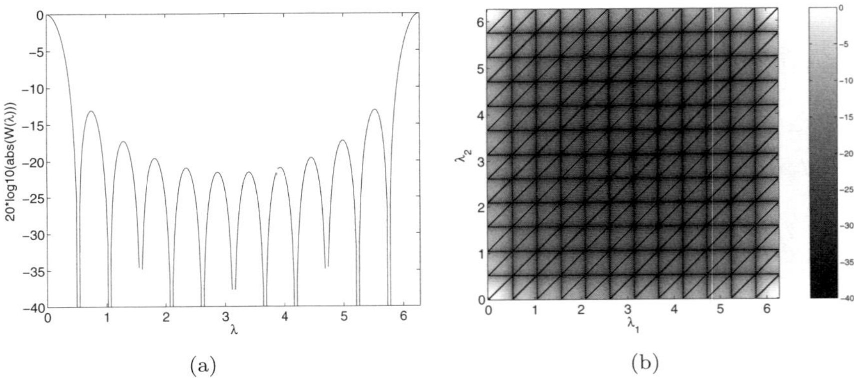

Figure 6.14 Frequency response for filter with uniform weights $w_n = 1$ for $n = 0, 1, \ldots, N-1$ for $N = 12$. The PC-12 support set is overlaid in black. (a) Frequency response $20\log_{10}|W(\lambda)|$. (b) Outer product frequency response $10\log_{10}|W(\lambda_1)\overline{W(\lambda_2)}|$.

components with period T are removed and near $\lambda = \pi$, the suppression of frequencies is 20 dB (an order of magnitude). The support lines for a PC-12 sequence are overlaid in black.

Figure 6.14 shows the spectral densities that result from filtering PC-12 white noise by the uniform weight filter with $N = 12$. Recall that if X_t is PC-12 then the filter output Y_t will generally be PC-12 but the 12-point sampling of Y_t will be stationary and its spectrum is the sum of all the diagonal measures of the $T \times T = 144$ subsquares, but scaled to fill $[0, 2\pi) \times [0, 2\pi)$. However, the original spectral covariance measure F_X will be greatly suppressed on all the squares except those corresponding to low frequencies $\lambda_1, \lambda_2 < 2\pi/12 \pmod{2\pi}$.

6.8.5 Bandshifting

Bandshifting refers to forming the product sequence

$$Y_t = X_t e^{i\lambda_s t}, \tag{6.100}$$

an operation we can easily understand in the context of harmonizable sequences. For then it follows from (6.35) that

$$Y_t = \int_0^{2\pi} e^{i(\lambda+\lambda_s t)} \xi(d\lambda) = \int_0^{2\pi} e^{i\gamma t} \xi'(d\gamma),$$

where the meaning of

$$\xi'(d\gamma) = \xi(d\gamma - \lambda_s)$$

should be clear (for $0 \le \gamma < 2\pi$ take $[\gamma - \lambda_s] \pmod{2\pi}$). So the random spectral measure ξ' for Y_t is that of X_t shifted upward by λ_s. This is often called *complex bandshifting* because of multiplication by the complex exponential rather than by sines or cosines. The covariance of Y_t is simply computed to be (even for nonharmonizable sequences)

$$R_Y(s,t) = R_X(s,t) e^{i\lambda_s(s-t)}, \tag{6.101}$$

from which it is easy to see that complex bandshifting a stationary or a PC-T sequence by any shift frequency λ_s leaves the respective result stationary or PC-T. This may also be understood from the spectral covariance measure. Indeed, for any harmonizable sequence

$$\begin{aligned} R_Y(s,t) &= \int\int e^{i(\lambda_1-\lambda_s)s-(\lambda_2-\lambda_s)t} F_X(d\lambda_1, d\lambda_2) \\ &= \int\int e^{i(\gamma_1 s - \gamma_2 t)} F_X(d\gamma_1 - \lambda_s, d\gamma_2 - \lambda_s), \end{aligned} \tag{6.102}$$

so

$$F_Y(d\gamma_1, d\gamma_2) = F_X(d\gamma_1 - \lambda_s, d\gamma_2 - \lambda_s) \tag{6.103}$$

and, as above, $d\gamma_1 - \lambda_s$ and $d\gamma_2 - \lambda_s$ are to be taken mod 2π. The effect of the bandshifting by frequency λ_s is to shift the measure F along the main diagonal upward by λ_s. Since frequency is taken mod 2π, the amplitude associated with a frequency $\gamma_0 > 2\pi$ will now appear at $\gamma_0 - 2\pi$, a phenomenon sometimes called *wrapping*. To illustrate, we consider the weighted measure $W(\lambda_1)\overline{W(\lambda_2)}F(d\lambda_1, d\lambda_2)$ produced by the lowpass LTI filter with cutoff frequency $\lambda_c = 0.15\pi$ whose response is illustrated in Figure 6.15(b). Figures 6.15(c) and 6.15(d) show the effect of bandshifting by $\lambda_s = 0.2\pi$ and 0.5π respectively. Note the "wrapping" effect.

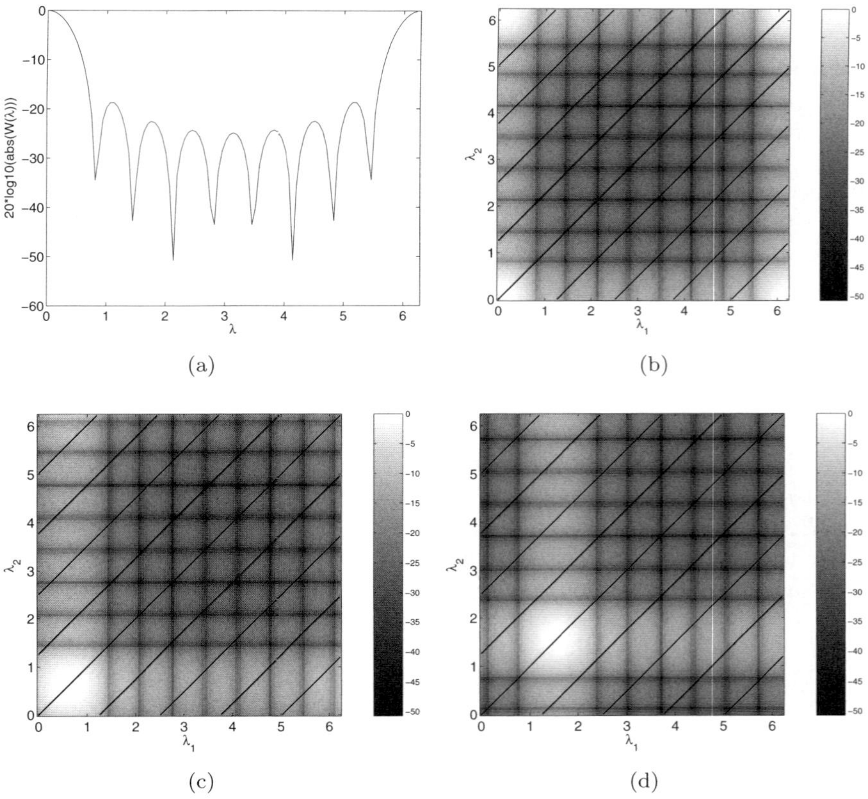

Figure 6.15 Effect of bandshifting and lowpass filtering. (a) Frequency response $W(\lambda)$ for a 8 coefficient lowpass filter with cutoff $\lambda_c = 0.15\pi$. (b) $10\log|W(\lambda_1)\overline{W(\lambda_2)}|$. (c) Effect of bandshifting by $\lambda_s = 0.2\pi$. (d) Effect of bandshifting by $\lambda_s = 0.5\pi$.

6.8.6 Periodically Time Varying (PTV) Filters

If in (6.78) the filter coefficients $w_n, n \in \mathbb{Z}$ are replaced by a collection $w_n(t), n \in \mathbb{Z}$ satisfying $w_n(t) = w_n(t+T)$ for every $n \in \mathbb{Z}$, the filtered sequence is formed by

$$Y_t = \sum_{n\in\mathbb{Z}} w_n(t) X_{t-n} \tag{6.104}$$

provided the sum converges in mean square sense. If $w_n(t) = 0$ for all t and $n < 0$ the filter is called *causal*: Y_t depends only on the input X_s for $s \le t$. If $\sum_{n\in\mathbb{Z}} |w_n(t)| < \infty$ for $t = 0, 1, \ldots, T-1$, the filter is called *stable*; in this case note $\sum_{n\in\mathbb{Z}} \max_{t=0,1,\ldots,T-1} |w_n(t)| < \infty$. If a PC-T sequence is filtered by a stable PTV filter, also with period T, then since $R_X(t,t)$ is bounded,

$$\sum_{n\in\mathbb{Z}}\sum_{n'\in\mathbb{Z}} |w_n(t)||w_{n'}(t)||R_X(t-n,t-n')| < \infty, \tag{6.105}$$

ensuring the existence of the sum Y_t. Furthermore,

$$\begin{aligned} R_Y(s,t) &= \sum_{n\in\mathbb{Z}}\sum_{n'\in\mathbb{Z}} w_n(s)\overline{w_{n'}(t)}R_X(s-n,t-n') \\ &= \sum_{n\in\mathbb{Z}}\sum_{n'\in\mathbb{Z}} w_n(s+T)\overline{w_{n'}(t+T)}R_X(s+T-n,t+T-n') \\ &= R_Y(s+T,t+T), \end{aligned}$$

showing that the PC-T property is preserved by stable PTV filtering of the same period T.

As for the case of LTI filtering, we can use the harmonizability of PC sequences to express and interpret the covariance Y_t. First we define for stable PTV filters,

$$W(t,\lambda) = \sum_{n=-\infty}^{\infty} e^{-i\lambda n} w_n(t), \tag{6.106}$$

so the sum in (6.104) becomes

$$\begin{aligned} Y_t &= \sum_{n\in\mathbb{Z}} w_n(t) \int_0^{2\pi} e^{i\lambda(t-n)}\xi(d\lambda) \\ &= \int_0^{2\pi} \sum_{n\in\mathbb{Z}} w_n(t)e^{-i\lambda n}e^{i\lambda t}\xi(d\lambda) \\ &= \int_0^{2\pi} W(t,\lambda)e^{i\lambda t}\xi(d\lambda). \end{aligned} \tag{6.107}$$

The interchange of integral and sum is justified essentially by (6.105) and Fubini's theorem.

Second, the covariance of Y_t then becomes

$$R_Y(s,t) = \int_0^{2\pi}\int_0^{2\pi} e^{i\lambda_1 s - i\lambda_2 t} W(s,\lambda_1)\overline{W(t,\lambda_2)}F_X(d\lambda_1,d\lambda_2), \tag{6.108}$$

where F_X is the spectral covariance measure for the sequence X_t.

A little more interpretation may be obtained by expressing $w_n(t)$ by its discrete Fourier series

$$w_n(t) = \sum_{k=0}^{T-1} e^{i2\pi tk/T} w_n^k \tag{6.109}$$

with

$$w_n^k = \frac{1}{T} \sum_{t=0}^{T-1} e^{i2\pi tk/T} w_n(t), \tag{6.110}$$

where it is clear that $\sum_n |w_n^k| < \infty$ for $k = 0, 1, \ldots, T-1$. Then the expression for Y_t becomes

$$\begin{aligned} Y_t &= \sum_{k=0}^{T-1} e^{i2\pi kt/T} \sum_{n \in \mathbb{Z}} w_n^k X_{t-n} \\ &= \sum_{k=0}^{T-1} e^{i2\pi kt/T} \int_0^{2\pi} e^{i\lambda t} W^k(\lambda) \xi(d\lambda), \end{aligned} \tag{6.111}$$

showing that the action of a PTV filter is represented by the filtering of X_t by T different LTI filters and then bandshifting the kth filter output by $2\pi k/T$ and summing the result. This finally gives the alternative expression

$$\begin{aligned} R_Y(s,t) &= \sum_{k=0}^{T-1} \sum_{l=0}^{T-1} \int_0^{2\pi} \int_0^{2\pi} e^{i(\lambda_1 + 2\pi k/T)s - i(\lambda_2 + 2\pi l/T)t} \\ & \quad W^k(\lambda_1) \overline{W^l(\lambda_2)} F_X(d\lambda_1, d\lambda_2). \end{aligned} \tag{6.112}$$

We note that PTV filtering of PC sequences can also be understood by blocking X_t into vectors of length T. This lifts the problem to one of filtering stationary T-variate sequences by a time invariant filter having matrix coefficients.

For further discussion of PTV filtering applied to PC sequences (and continuous time PC processes) see [59]. The design of PTV filters is addressed in [184], where related references are given.

PROBLEMS AND SUPPLEMENTS

6.1 If X_t is stationary, show that $X_t + X_{2t}$ is harmonizable and describe the most general form of support of its spectral measure F. What happens if X_t is PC-T?

6.2 Denote $\{\lambda_j : j = 1, 2, \ldots\}$ as the set of rational numbers in $[0, 2\pi)$. Let $\{A_j : j = 1, 2, \ldots\}$ be a sequence of orthogonal second order random variables with $\sum_j E\{|A_j|^2\} < \infty$. The sequence $X_t = \sum_j A_j e^{i\lambda_j t}$ is a stationary sequence whose discrete spectrum is dense in $[0, 2\pi)$. Use this idea to construct a PC-T sequence whose discrete spectrum is dense in S_T.

6.3 In the case that X_t is a *real* PC sequence, $R(s,t)$ is real and so

$$B_k(-\tau) = \overline{B_{-k}(\tau)} e^{-i2\pi k\tau/T} = \overline{B_{T-k}(\tau)} e^{-i2\pi k\tau/T} \tag{6.113}$$

becomes

$$B_k(-\tau) = B_{-k}(\tau)e^{-i2\pi k\tau/T} = B_{T-k}(\tau)e^{-i2\pi k\tau/T}.$$

Appealing directly to (6.2) we obtain further that

$$\text{Re } B_k(\tau) = \text{Re } B_{T-k}(\tau) \text{ and } \text{Im } B_k(\tau) = -\text{Im } B_{T-k}(\tau) \tag{6.114}$$

so

$$B_k(\tau) = \overline{B_{T-k}(\tau)}. \tag{6.115}$$

Expressions (6.51) and (6.115) together lead to

$$B_k(-\tau) = B_k(\tau)e^{-i2\pi k\tau/T}. \tag{6.116}$$

So if we define a modified coefficient function by

$$\widetilde{B}_k(\tau) = B_k(\tau)e^{-i\pi k\tau/T}, \tag{6.117}$$

it may be seen that $\widetilde{B}_k(\tau)$ is even, $\widetilde{B}_k(-\tau) = \widetilde{B}_k(\tau)$. An application of the inversion formula (see Loève [139, page 199], Brockwell and Davis [28, page 151]),

$$\widetilde{F}_k([a,b)) = \lim_{N\to\infty} \frac{1}{2\pi} \sum_{\tau=-N}^{N} \frac{e^{-i\tau a} - e^{-i\tau b}}{i\tau} \widetilde{B}_k(\tau),$$

leads to a condition of symmetry about the point $\lambda = \pi$ expressed by

$$\widetilde{F}_k([a,b)) = \widetilde{F}_k([2\pi - b, 2\pi - a)) \tag{6.118}$$

for any interval $[a,b) \subset [0,2\pi)$. From (6.116) it also follows that $\widetilde{F}_k([a,b)) = F_k([a+\pi/T, b+\pi/T))$, where the meaning of the latter is obtained by extending F_k periodically with period 2π over the entire real line. For $k = 0$ this symmetry exists even for the complex case because $B_0(\tau) = \overline{B_0(\tau)}$ and F_0 is real. However, for $k \neq 0$, we evidently need X_t to be real in order to conclude that $\widetilde{B}_k(\tau)$ is even. We also note that (6.118) leads to a condition of symmetry about the point $\lambda = 0$ expressed by

$$\widetilde{F}_k([a,b)) = \widetilde{F}_k([-b,-a)).$$

This may also be seen as the consequence of using a different exponential sequence in (6.117), specifically

$$\widetilde{\widetilde{B}}_k(\tau) = B_k(\tau)e^{i\pi\tau - i\pi k\tau/T}$$

to obtain an even $\widetilde{B}_k(\tau)$.

It is also of interest to observe the meaning of these symmetries in terms of the measure F. Beginning with the measure ρ from which F was constructed, we note that the symmetries about $y = \pi$ and $y = 0$ in the measures F_k become symmetries about the lines $\lambda_2 = 2\pi - \lambda_1$ and $\lambda_2 = -\lambda_1$ in Figure 6.2. These lines become the antidiagonal of the square $[0, 2\pi) \times [0, 2\pi)$ under the transformation that takes μ_ξ into F. Thus when X_t is real, the measure F is symmetric about the antidiagonal $\lambda_2 = 2\pi - \lambda_1$. But as we have already noted, there is also the symmetry about the main diagonal given by $F([a, b), [c, d)) = \overline{F}([c, d), [a, b))$. Thus F is evidently determined completely on the square $[0, 2\pi) \times [0, 2\pi)$ by its values on any one of the four triangles formed by the diagonal, the antidiagonal, and the boundaries of the square.

6.4 Show that if the limit $R_{\text{avg}}(\tau) = \lim_{N\to\infty} \frac{1}{2N+1} \sum_{t=-N}^{N} E\{X_{t+\tau}\overline{X_t}\}$ exists for every $\tau \in \mathbb{Z}$, then $R_{\text{avg}}(\tau)$ is NND.

6.5 Compute the densities $f_k(\lambda)$ for $X_t = f_t Y_t$ with Y_t stationary white noise and with $f_t = 1 + m\cos(2\pi t/T)$, where m is a real parameter (usually $m \leq 1$).

6.6 To get a version of Proposition 6.9 in terms of the spectral processes of $Y(n)$ and $Z_p(t)$, first re-express (6.38) as

$$\begin{aligned} Y_p(n) = X_t &= \sum_{k=0}^{T-1} \int_{k2\pi/T}^{(k+1)2\pi/T} e^{i\lambda t} \xi(d\lambda) \\ &= \sum_{k=0}^{T-1} \int_0^{2\pi/T} e^{i(\gamma+2\pi/T)t} \xi(d\gamma + 2\pi k/T) \\ &= \sum_{k=0}^{T-1} e^{ik2\pi t/T} \int_0^{2\pi/T} e^{i\gamma t} \xi(d\gamma + 2\pi k/T) \\ &= \int_0^{2\pi} e^{i\gamma t} \eta_p(d\gamma), \end{aligned}$$

so considering the random value $\eta_p(\{0\})$ gives

$$\eta_p(\{0\}) = \sum_{k=0}^{T-1} e^{ik2\pi p/T} \xi(\{2\pi k/T\}).$$

Under conditions for WLLN, $\eta_p(\{0\}) = m(p)$, the mean at time p and $\xi(\{2\pi k/T\}) = m_k$, the kth Fourier coefficient in the Fourier series of $m(t)$.

6.7 Verify that the matrix $\mathbf{V}(\lambda)$ (6.42) is unitary for all $0 \leq \lambda < 2\pi$.

6.8 If $F(\Delta) > 0$, we will say that the sequence is *not transient* or *continuing.* If $F(\Delta) > 0$ and $F_c = 0$, we will say the sequence is *purely not transient* or *purely continuing (persistent).* Stationary sequences, PC-T sequences, and some others are purely continuing. They have no inherent transient part. Of course, the sum sequence

$$X_t = Y_t + Z_t \tag{6.119}$$

for Y_t PC-T and Z_t transient, with X_t and Y_t orthogonal sequences, is neither transient nor purely continuing, it is just continuing.

The measures F_k can be further decomposed,

$$F_k = F_k^{ac} + F_k^s, \tag{6.120}$$

where $F_k^{ac} \ll \mu$ and $F_k^s \perp \mu$ and this leads to the further decomposition

$$F_k = F_k^{sd} + F_k^{sc} + F_k^{ac}, \tag{6.121}$$

where F_k^{sd} is the *singular discrete* part consisting of point masses, F_k^{sc} is the *singular continuous* part consisting of F_k^s less the point masses, and F_k^{ac} is absolutely continuous with respect to Lebesgue measure.

The component F_k^{ac} can be expressed in terms of a density

$$F_k^{ac}(A) = \int_A f_k(\lambda)d\lambda, \tag{6.122}$$

which we will call the *spectral density* of the PC sequence. If F_k^s is null, then

$$B_k(\tau) = \int_0^{2\pi} e^{i\lambda\tau} f_k(\lambda)d\lambda.$$

It is possible to further develop the decomposition of the singular part of F,

$$F^s = F^{sc} + F^{sd1} + F^{sd2}$$

where F^{sd2} is concentrated on a countable set of $[0, 2\pi)^2$, F^{sd1} is concentrated on a countable set of lines in $[0, 2\pi)^2$, and F^{sc} is the remainder. Note the remainder is singular continuous with respect to the specific other singular parts F^{sd1} and F^{sd2}. Since $|F_k|$ is a finite measure, F_k^{sd} can contain no more than a countable number of point masses and hence F^{sd} can contain no more than a countable set of point masses. Of course, this could have been concluded directly from the singular discrete part of F^s.

CHAPTER 7

REPRESENTATIONS OF PC SEQUENCES

As we mentioned previously, PC sequences arise from the mixing of stationarity and periodicity. This is very clearly seen, we believe, through the unitary operator of a PC sequence and the representations that follow. At the most basic level, the connection of the PC structure to a group of shift operators was first mentioned by Ogura [174] as a motivation for the development of spectral representations (that are much like those of Gladyshev). The explicit use of the unitary operator was introduced in [111] for continuous time PC processes; its role in almost PC processes was explored in [112]. In this chapter we show the existence of this operator and develop the representations that follow from it. We then use the spectral theory to prove Gladyshev's representation theorem (Proposition 6.8) but in a broader context than the original proof. The spectral theorem also permits the explicit construction of the random spectral measure ξ appearing in the harmonizable representation. We will also see that, not surprisingly, the role of the unitary operator for PC sequences is a straightforward extension of its role in the stationary case.

Periodically Correlated Random Sequences:Spectral Theory and Practice. By H.L. Hurd and A.G. Miamee

7.1 THE UNITARY OPERATOR OF A PC SEQUENCE

Reviewing from Chapter 3, a unitary operator on a Hilbert space $\mathcal{H}$ is a linear operator U from $\mathcal{H}$ onto $\mathcal{H}$ for which $\langle Ux, Uy\rangle = \langle x, y\rangle$ for every $x, y \in \mathcal{H}$; that is, unitary operators are linear and preserve inner products. Every unitary operator can be written as an integral with respect to a spectral measure (see Theorem 3.6)

$$U = \int_0^{2\pi} e^{i\lambda} E(d\lambda). \tag{7.1}$$

Proposition 7.1 *A second order stochastic sequence X_t is PC-T if and only if there exists a unitary operator U on its time domain $\mathcal{H}_X$ (see Definition 4.2) such that*

$$X_{t+T} = UX_t, \text{ for every } t \in \mathbb{Z}. \tag{7.2}$$

Proof. In view of the discussion in Chapter 4, we assume $1 \in \mathcal{H}_X$. If there exists a unitary U for which (7.2) is true, then the mean and correlation functions of X_t satisfy $m(t) = m(t+T)$ and $R(s,t) = R(s+T, t+T)$ for every $s, t \in \mathbb{Z}$.

Conversely, if X_t is PC with period T, then for $\mathbf{z} = \sum_{j=1}^n a_j X_{t_j}$ in $\mathcal{L}_X = \text{sp}\{X_t, t \in \mathbb{Z}\}$, define the operator U by $U\mathbf{z} = \sum_{j=1}^n a_j X_{t_j+T}$. It is not difficult to show (see Problem 7.5 at the end of the chapter) that this extension of U to $\mathcal{L}_X$ is linear, well defined, and preserves inner product. It is also easy to see that U,as a map from $\mathcal{L}_X$ to $\mathcal{L}_X$, is onto. It then follows from the continuity of U that it extends to an operator on the closure $\mathcal{H}_X = \overline{\mathcal{L}_X}$. One can finally verify that this extension is unitary. ∎

So for PC-T sequences, shifts of length T are unitary. If X_t is a PC sequence and U is a shift operator for $T = 1$, then X_t is stationary.

Recall from Proposition 4.8 that multivariate stationary sequences have the property that there is a single unitary operator $U_{\mathbf{X}}$ that acts as the shift operator for each component of $\mathbf{X}_n$. It should be clear that if $\mathbf{X}_n$ is the T-variate stationary sequence arising from T-blocks of a PC-T sequence, then $U = U_{\mathbf{X}}$.

The existence of U leads to another characterization of PC sequences that becomes very useful for obtaining representations of PC processes.

Proposition 7.2 *A second order sequence X_t is PC-T if and only if there exists a unitary operator V and a periodic sequence P_t taking values in $\mathcal{H}_X$ for which*

$$X_t = V^t P_t \tag{7.3}$$

for every $t \in \mathbb{Z}$.

Proof. If X_t is given by (7.3) then

$$\begin{aligned}\langle X_s, X_t\rangle &= \langle V^s P_s, V^t P_t\rangle \\ &= \langle V^T U V^s P_{s+T}, V^T V^t P_{t+T}\rangle \\ &= \langle X_{s+T}, X_{t+T}\rangle, \end{aligned} \tag{7.4}$$

so X_t is PC.

Conversely, if X_t is PC-T, the spectral representation of U given by (7.1) provides a means to construct a unitary operator by

$$V = \int_0^{2\pi} \exp(i\lambda/T)E(d\lambda), \tag{7.5}$$

where $V^T = U$. In other words, V is a Tth root of U. If we take $P_t = V^{-t}X_t$ we can easily verify that (7.3) holds and P_t is periodic because

$$\begin{aligned}\|P_{t+T} - P_t\| &= \|V^{-t-T}[X_{t+T}] - V^{-t}[X_t]\| \\ &= \|V^{-T}[X_{t+T}] - X_t\| \\ &= 0,\end{aligned}$$

by virtue of (7.2). ∎

The expression (7.3) is a clear contrast to the stationary case, in which $P_t \equiv X_0$, a fixed random variable. We also note that the characterization is not unique as $\widetilde{V} = V\exp(i2\pi/T)$ and $\widetilde{P}(t) = P(t)\exp(-i2\pi t/T)$ yield another unitary operator and periodic function that also solve (7.3). The spectral theorem shows that any Borel measurable function $f_U(\lambda)$ with $|f_V(\lambda)| = 1$ and $f_U^T(\lambda) = \exp(i\lambda)$ will yield an operator that is not necessarily unitary, but with $V^T = U$. The operator V defined by (7.5) is, however, unitary and we shall call it the *principal* Tth root of U.

7.2 REPRESENTATIONS BASED ON THE UNITARY OPERATOR

Representations for X_t follow naturally from (7.3) by simultaneously employing various representations for V and P. We begin by proving again the representation of Gladyshev (Proposition 6.8).

7.2.1 Gladyshev Representation

Proposition 7.3 (Gladyshev) *A second order sequence X_t is PC-T if and only if there exists a T-variate stationary sequence $\mathbf{Z}_t = [Z_t^j]$ such that*

$$X_t = \sum_{j=0}^{T-1} Z_t^j e^{i2\pi jt/T}. \tag{7.6}$$

Proof. If X_t has the stated representation, it is clearly PC with period T. Conversely, if X_t is PC with period T, then (7.6) follows from Proposition 7.2 and the representation of the periodic function P_t by its (finite) Fourier series

$$P_t = \sum_{j=0}^{T-1} \widetilde{P}_j e^{i2\pi jt/T}, \tag{7.7}$$

where $\widetilde{P}_k \in \mathcal{H}_X$, and then making the identification

$$Z_t^j = V^t \widetilde{P}_j. \tag{7.8}$$

The Z_t^j are jointly stationary because they are orbits of different starting vectors $\widetilde{P}_j$ under the same unitary operator V. ∎

As noted in Chapter 6, Gladyshev actually stated a result in which the jointly stationary sequences Z_t^j are "band limited" to the interval $[0, 2\pi/T)$. This corresponds to taking V to be the *principal* root of U as we have done in (7.5). Then the change of variable $\gamma = \lambda/T$ in (7.5) produces

$$V = \int_0^{2\pi/T} e^{i\gamma} dE(T\gamma). \tag{7.9}$$

Since V is now back in the canonical integral form (see [2]), it is clear that the spectral support of V is contained in the interval $[0, 2\pi/T)$ and so

$$Z_t^j = \int_0^{2\pi/T} e^{i\gamma t} dE(T\gamma)[\widetilde{P}_j].$$

The representation (7.6) with V^t acting on the terms $\widetilde{P}_j e^{i2\pi jt/T}$ provides additional meaning to the term *spectral redundancy* that is used in the description of cyclostationary processes (see Gardner [70]). It shows that the correlations or redundancy that appears in the spectrum of X_t is caused by the action of the one unitary operator on possibly T different vectors.

Proposition 1.1 made the simple observation that every PC-T sequence X_t may be viewed as a T-variate stationary sequence $\mathbf{X}_n = [X_n^j]$, where $X_n^j = X_{j+nT}, n \in \mathbb{Z}$ for $j = 0, 1, ..., T-1$. Proposition 7.3, also by Gladyshev, has just shown the important fact that every PC-T sequence may be represented, via (7.6) by a T-variate stationary sequence $\mathbf{Z}_t$ whose components are of bandwidth $2\pi/T$.

The next result is essentially the same as Proposition 7.3 but relies on a different proof and gives a minimal representation for X_t.

7.2.2 Another Representation of Gladyshev Type

Proposition 7.4 *A second order sequence X_t is PC-T if and only if there exist a q-variate stationary sequence $\mathbf{Z}_t = [Z_t^j]$ and q scalar periodic sequences $\{f_t^j = f_{t+T}^j, j = 1, 2, ..., q\}$ with $q \leq T$, such that*

$$X_t = \sum_{j=1}^{q} Z_t^j f_t^j. \tag{7.10}$$

Proof. If X_t has the stated representation, it is clearly PC with period T. Conversely, if X_t is PC with period T, then we first note that $\mathcal{L}_P = \text{sp}\{P_t : t = 0, 1, ..., T-1\}$ has dimension $q = \dim \mathcal{L}^P \leq T$. Let $\{\xi_j : j = 1, 2, \ldots, q\}$ be an orthonormal basis for $\mathcal{L}^P$, and so

$$P_t = \sum_{j=1}^{q} \langle P_t, \xi_j \rangle \xi_j. \tag{7.11}$$

Now the claim (7.10) follows from Proposition 7.2 and then making the identifications

$$Z_t^j = V^t \xi_j \tag{7.12}$$

and

$$f_t^j = \langle P_t, \xi_k \rangle = \langle P_{t+T}, \xi_k \rangle = f_{t+T}^j. \quad \blacksquare \tag{7.13}$$

Finally, we turn to spectral integral representations for PC sequences. The first is a direct application of the spectral representation (7.1) for unitary operators and the representation (7.11) for P_t.

7.2.3 Time-Dependent Spectral Representation

Proposition 7.5 *A second order sequence X_t is PC-T if and only if there exists a time dependent random spectral measure $\xi(\cdot, t) = \xi(\cdot, t+T)$ on the Borel subsets of $[0, 2\pi)$ that is orthogonally scattered (in the sense that $\langle \xi(A, s), \xi(B, t) \rangle = 0$ for every $s, t \in \mathbb{Z}$ whenever $A \cap B = \emptyset$) and such that*

$$X_t = \int_0^{2\pi} e^{i\lambda t} \xi(d\lambda, t), \quad \textit{for all } t \in \mathbb{Z}. \tag{7.14}$$

Remark. This is in contrast to the harmonizable representation in which the measure is independent of t but is not orthogonally scattered.

Proof. If X_t has the form (7.14) with $\xi(\cdot, t) = \xi(\cdot, t+T)$ and $\langle \xi(A, s), \xi(B, t) \rangle =$

0 whenever $A \cap B = \emptyset$, then since $\langle \xi(A,s), \xi(B,t) \rangle = \langle \xi(A \cap B, s), \xi(A \cap B, t) \rangle$, we set

$$\begin{aligned} F_{st}(A \cap B) &= \langle \xi(A,s), \xi(B,t) \rangle \\ &= \langle \xi(A, s+T), \xi(B, t+T) \rangle \\ &= F_{s+T,t+T}(A \cap B) \end{aligned} \tag{7.15}$$

for all $s, t \in \mathbb{Z}$ and Borel sets A, B of $[0, 2\pi)$. For fixed s, t,

$$\begin{aligned} |F_{st}(A)| &= |\langle \xi(A,s), \xi(A,t) \rangle| \\ &\leq \| \xi(A,s) \| \| \xi(A,t) \| \\ &= \sqrt{F_{s,s}(A) F_{t,t}(A)}. \end{aligned}$$

Furthermore, $F_{t,t}(\cdot) \geq 0$ and $F_{st}(\cdot)$ inherits countable additivity from $\xi(\cdot, t)$. Thus $F_{st}(\cdot)$ is a (finite) complex measure and

$$\begin{aligned} \langle X_s, X_t \rangle &= \left\langle \int_0^{2\pi} e^{i\lambda_1 s} \xi(d\lambda_1, s), \int_0^{2\pi} e^{i\lambda_2 t} \xi(d\lambda_2, t) \right\rangle \\ &= \int_0^{2\pi} \int_0^{2\pi} e^{i(\lambda_1 s - \lambda_2 t)} \langle \xi(d\lambda_1, s), \xi(d\lambda_2, t) \rangle \\ &= \int_0^{2\pi} e^{i\lambda(s-t)} F_{st}(d\lambda) \\ &= \int_0^{2\pi} e^{i\lambda(s+T-t-T)} F_{s+T,t+T}(d\lambda) = \langle X_{s+T}, X_{t+T} \rangle. \end{aligned} \tag{7.16}$$

Note that a representation (7.16) for the correlation of a PC-T sequence is obtained as a by-product.

Conversely, if X_t is PC-T, we use the spectral representation (7.5) for the unitary operator V^t to obtain

$$X_t = V^t[P_t] = \int_0^{2\pi} e^{i\lambda t/T} E(d\lambda)[P(t)] = \int_0^{2\pi} e^{i\lambda t/T} \xi(d\lambda, t), \tag{7.17}$$

where the *time dependent spectral measure* $\xi(\cdot, t)$ is defined through the application of the spectral measure appearing in (7.1) to the vector P_t. To be precise, for any Borel A, set

$$\xi(A, t) = E(A)[P_t],$$

from which it follows directly that $\xi(A, t) = \xi(A, t+T)$. It is also easy to see that $\xi(\cdot, t)$ and $\xi(\cdot, s)$ are mutually orthogonally scattered in the sense

$$\langle \xi(A, t), \xi(B, s) \rangle = 0,$$

whenever $A \cap B = \emptyset$. ∎

Oscillatory PC Sequences. A random sequence X_t is called *oscillatory* (see Priestley [185]) if there is an orthogonally scattered random spectral measure $\xi(\cdot)$ on the Borel sets of $[0, 2\pi)$ and a function $a(t, \lambda)$ for which

$$X_t = \int_0^{2\pi} e^{i\lambda t} a(t,\lambda)\xi(d\lambda), \tag{7.18}$$

where $a(t, \cdot) \in L^2(F)$ for all $t \in \mathbb{Z}$ and F is the spectral measure associated with $\xi(\cdot)$. Oscillatory processes are generally nonstationary but clearly include the stationary processes by taking $a(t, \lambda) \equiv 1$. It follows that the correlation of an oscillatory sequence has the representation

$$R(s,t) = \int_0^{2\pi} e^{i\lambda(s-t)} a(s,\lambda)\overline{a(t,\lambda)}F(d\lambda). \tag{7.19}$$

If $a(t, \lambda) = a(t + T, \lambda)$ for all $t \in \mathbb{Z}$ and $\lambda \in [0, 2\pi)$, then it is clear that

$$R(s+T, t+T) = \int_0^{2\pi} e^{i\lambda(s+T-t-T)} a(s+T,\lambda)\overline{a(t+T,\lambda)}F(d\lambda) = R(s,t).$$

So X_t is PC-T. In this case the spectral measure $F_{st}(\cdot)$ defined in (7.15) is given here by

$$F_{st}(A) = \int_A a(s,\lambda)\overline{a(t,\lambda)}F(d\lambda).$$

The characterization of oscillatory PC sequences is an open problem.

7.2.4 Harmonizability Again

Although Proposition 6.7 shows (in the same manner as Gladyshev) that all PC sequences are harmonizable, we will now make the argument again using the unitary operator. We will do this by explicitly giving the random measure ξ and then arguing that its spectral correlation measure F is of bounded variation.

Proposition 7.6 *Every PC-T sequence X_t is strongly harmonizable.*

Proof. In order to express ξ explicitly, we use the spectral representations (7.5) and (7.8) in relation (7.6) to obtain

$$\begin{aligned} X_t &= \sum_{p=0}^{T-1} \int_0^{2\pi/T} e^{it\lambda} E(Td\lambda)[\widetilde{P}_p] \\ &= \sum_{p=0}^{T-1} \int_{2p\pi/T}^{2(p+1)\pi/T} e^{it\eta} E(Td\eta - 2p\pi)[\widetilde{P}_p]. \end{aligned} \tag{7.20}$$

Now for any interval $A \subset [0, 2\pi)$, if we define

$$\xi(A) = \sum_{p=0}^{T-1} E(TA - 2p\pi)[\widetilde{P}_p], \tag{7.21}$$

then (7.20) can be read as (1.14). Here it is important to remember that the projection valued measure $E(\cdot)$ is defined on the Borel sets of $[0, 2\pi)$.

The measure ξ may be viewed in the following manner. First note we can write

$$A = \bigcup_{p=0}^{T-1} A \cap \Delta_p, \quad \text{where } \Delta_p = [2p\pi/T, 2(p+1)\pi/T). \tag{7.22}$$

Then $E(TA - 2p\pi)[\widetilde{P}_p]$ is the action of $E(T(A \cap \Delta_p))$ on vector $\widetilde{P}_p$, where $T(A \cap \Delta_p)$ is taken modulo 2π. Thus the correlation (or redundancy [70]) in the spectral measure arises from the repeated use of the same projection measure on different vectors $\widetilde{P}_p$.

The spectral (correlation) measure is determined by first defining it on intervals A and B in $[0, 2\pi)$ by

$$F(A, B) = \langle \xi(A), \xi(B) \rangle = \sum_{p=0}^{T-1} \sum_{p'=0}^{T-1} F^{pp'}(A, B), \tag{7.23}$$

where

$$F^{pp'}(A, B) = \langle E(TA - 2\pi p)[\widetilde{P}_p], E(TB - 2\pi p')[\widetilde{P}_{p'}] \rangle. \tag{7.24}$$

The countable additivity and boundedness is inherited from the projection valued measure E; the measure defined on intervals may then be extended in the usual way to the Borel sets of $[0, 2\pi) \times [0, 2\pi)$. But (7.24) for A, B intervals gives $F^{pp'}(A, B) = 0$ unless

$$(TA - 2p\pi) \cap (TB - 2p'\pi) \neq \emptyset. \tag{7.25}$$

This leads to the conclusion that the support set of F must be contained in $S_T = \{(\lambda_1, \lambda_2) : \lambda_2 = \lambda_1 - 2\pi k/T, k \in [-(T-1), T-1]\}$ intersected with the square $[0, 2\pi) \times [0, 2\pi)$.

The finiteness of the spectral correlation measure can be established from the finiteness of each of the measures $F^{pp'}(A, B)$. Indeed, from the Schwarz inequality

$$|F^{pp'}(A, B)|^2 \leq F^{pp}(A, A)\, F^{p'p'}(B, B) \tag{7.26}$$

and from (7.23) it can be seen that $F^{pp}(A, A)$ is a nonnegative spectral measure defined on the Borel sets of the rectangles $\Delta^p \times \Delta^p$ defined in (7.22) with $F^{pp}(\Delta^p, \Delta^p) \leq \|\widetilde{P}^p\|_{L^2}^2$, and similarly for $F^{p'p'}(B, B)$. Hence each $F^{pp'}(A, B)$ has finite variation and since $F(A, B)$ is just a finite sum of these parts, it too has finite variation. ∎

7.2.5 Representation Based on Principal Components

We now show that a representation of the type given in Proposition 7.4 may be obtained from the principal component analysis of the covariance $R(s,t)$. This idea appeared in a report by K. L. Jordan [123] and was given a thorough treatment for continuous time by Gardner and Franks [64]. More recently, some finer points, also for continuous time, have been given by Hurd and Kallianpur [111] and Makagon [145]. For continuous time, the representations are based essentially on Mercer's representation of NND compact operators on $L^2[0,1]$ (see [192]); this leads to the well known Karhunen–Loève expansion. It has appeared again recently in the context of EOFs (*empirical orthogonal functions*) for cyclostationary processes (see papers by Kim et al. [130–132]).

If $R(s,t)$ is the correlation of a PC-T sequence, the restriction of $R(s,t)$ to the principal square

$$\mathbf{R}_0 = [R(s,t) : (s,t) \in \{0,1,\ldots,T-1\}^2] \tag{7.27}$$

is a Hermitian NND matrix (which we assume to be of full rank) and so there exists a matrix Φ whose jth column is $\{\phi_j(t) : t = 0,1,\ldots,T-1\}$ and for which

$$\mathbf{R}_0 = \Phi\boldsymbol{\Sigma}\Phi^*, \tag{7.28}$$

where

$$\boldsymbol{\Sigma} = \begin{bmatrix} \sigma_0^2 & 0 & \ldots & 0 \\ 0 & \sigma_1^2 & \ldots & 0 \\ \vdots & \vdots & \vdots & \vdots \\ 0 & 0 & \ldots & \sigma_{T-1}^2 \end{bmatrix}. \tag{7.29}$$

That is, the columns of Φ are right eigenvectors of $\mathbf{R}_0$ with eigenvalues σ_j^2, so that

$$\sigma_j^2\phi_j = \mathbf{R}_0\phi_j,\ j = 0,1,\ldots,T-1,$$

or more explicitly,

$$\sigma_j^2\phi_j(s) = \sum_{t=0}^{T-1}[\mathbf{R}_0]_{s,t}\phi_j(t),\ s,j = 0,1,...,T-1. \tag{7.30}$$

Furthemore, the eigenvectors $\{\phi_j : j = 0,1,...,T-1\}$ are orthonormal

$$\sum_{t=0}^{T-1}\phi_j(t)\overline{\phi_k(t)} = \delta_{j-k}, \tag{7.31}$$

which may be written

$$\Phi^*\Phi = I.$$

These facts mean that $R(s,t)$ may be represented in the principal square by

$$R(s,t) = \sum_{j=0}^{T-1} \sigma_j^2 \phi_j(s)\overline{\phi_j(t)}. \tag{7.32}$$

Using the block notation (see (1.10)) the T-block $\mathbf{X}_0 = [X_0, X_1, \ldots, X_{T-1}]$ (treated as a column vector) is transformed by the linear transformation

$$\eta_0 = \Phi^* \mathbf{X}_0 \tag{7.33}$$

into a vector η_0 of uncorrelated random variables,

$$E\{\eta_0\eta_0^*\} = E\{\Phi^*\mathbf{X}_0(\Phi^*\mathbf{X}_0)^*\} = \Phi^*\mathbf{R}_0\Phi = \mathbf{\Sigma}.$$

We shall denote η_0 as the *principal component* vector of block number 0. Multiplying both sides of (7.33) by Φ and using $\Phi^*\Phi = \Phi\Phi^* = I$ yields

$$\mathbf{X}_0 = \Phi\eta_0$$

or in detail

$$X_t = [\mathbf{X}_0]_t = \sum_{j=0}^{T-1} \phi_j(t)[\eta_0]_j, \quad \text{for } t = 0, 1, \ldots, T-1. \tag{7.34}$$

The PC-T property $R(s,t) = R(s+T, t+T)$ implies that the very same Φ will diagonalize

$$\mathbf{R}_n = [R(s,t) : (s,t) \in \{nT, nT+1, \ldots, nT+T-1\}^2] \tag{7.35}$$

for every $n \in \mathbb{Z}$. That is,

$$\mathbf{R}_n = \Phi\Sigma\Phi^*, \tag{7.36}$$

or in other words, the eigenvectors $\phi_j(t)$ solve

$$\sigma_j^2\phi_j(s) = \sum_{t=0}^{T-1} R_n(s,t)\phi_j(t) \tag{7.37}$$

for $s, j = 0, 1, ..., T-1$ and hence

$$R(s,t) = \sum_{j=0}^{T-1} \sigma_j^2 \phi_j(s-nT)\overline{\phi_j(t-nT)}, \tag{7.38}$$

for $(s,t) \in \{nT, nT+1, \ldots, (n+1)T-1)\} \times \{nT, nT+1, \ldots, (n+1)T-1)\}$ (i.e., in the nth square) and for every $n \in \mathbb{Z}$.

From these facts we conclude that for each *fixed* n, the random vector of principal components

$$\eta_n = \Phi^* \mathbf{X}_n \tag{7.39}$$

given in detail by

$$[\eta_n]_j = \sum_{t=0}^{T-1} X_{t+nT} \overline{\phi_j(t)} \tag{7.40}$$

are orthogonal

$$\begin{aligned} \langle [\eta_n]_j, [\eta_n]_k \rangle &= \sum_{s=0}^{T-1} \sum_{t=0}^{T-1} \langle X_{s+nT} X_{t+nT} \rangle \overline{\phi_j(s+NT)} \phi_k(t+nT) \\ &= \sum_{s=0}^{T-1} \overline{\phi_j(s+nT)} \sigma_k^2 \phi_k(s+nT) \\ &= \sigma_j^2 \delta_{j-k} \end{aligned} \tag{7.41}$$

and Var $([\eta_n]_j) = \sigma_j^2$.

Furthermore, the principal component sequences $[\eta_n]_j$ (indexed on n) are jointly stationary because they are all propagated by the same unitary operator U,

$$\begin{aligned} [\eta_{n+1}]_j &= \sum_{t=0}^{T-1} X_{t+(n+1)T} \overline{\phi_j(t)} \\ &= \sum_{t=0}^{T-1} U X_{t+nT} \overline{\phi_j(t)} \\ &= U[\eta_n]_j. \end{aligned} \tag{7.42}$$

Stationarity of the vector sequence η_t could also have been concluded by writing out the expression for $\langle [\eta_n]_j, [\eta_m]_k \rangle$ in terms of X_t and noting that the PC-T property (1.7) implies that the inner product depends only on $n - m$. And even though the principal component vectors η_n have orthogonal components for fixed n it is not necessarily true that $\langle [\eta_n]_j, [\eta_m]_k \rangle = 0$ for $m \neq n$.

Finally, using $\Phi^*\Phi = \Phi\Phi^* = I$ yields as before

$$\mathbf{X}_n = \Phi \eta_n$$

or in detail (using $t = k + nT$ with $0 \le k < T-1$)

$$X_t = [\mathbf{X}_n]_k = \sum_{j=0}^{T-1} \phi_j(k) [\eta_n]_j \tag{7.43}$$

for $t = nT, nT+1, \ldots, nT+T-1$. This representation has wide use and interpretation in communications engineering [64, 73], where it is called the *translation series representation (TSR)*. By extending $\phi(t)$ periodically, $\phi_j(t) = \phi_j(t+T)$, then (7.43) is sometimes written

$$X_t = \sum_{j=0}^{T-1} \phi_j(t)[\eta_n]_j, \quad n = \lfloor t/T \rfloor,$$

making it clear that the "pulses" $\phi_j(t)$ are the same in each period with the randomness (the message) entering via the amplitude vector η_n.

Here we ask how the preceding remarks change when $\mathbf{R}_0$ is rank deficient, let us say rank $\mathbf{R}_0 = r < T$. Then $T-r$ eigenvalues $\sigma_j^2 = \text{Var}\left([\eta_n]_j\right)$ in (7.29) will be null. In this case Φ and Σ in representations (7.28) and (7.36) can be taken to be of dimension $T \times r$ and $r \times r$. Alternatively, the dimensions can remain unchanged because whatever vectors ϕ_j are used in the columns corresponding to $\sigma_j = 0$ will be ignored.

As for uniqueness, all vectors corresponding to distinct eigenvalues will be unique (recall we always assume $\sum_{t=0}^{T-1} |\phi_j(t)|^2 = 1$). But there will be ambiguity whenever there are eigenvalues σ_j^2 with multiplicity greater than one. Let us say that $\sigma_{j_1}^2 = \sigma_{j_2}^2 = \cdots = \sigma_{j_m}^2 = \sigma^2$. Then any orthonormal set of vectors that span the eigenspace $\mathcal{L}^{\sigma^2} = \{\mathbf{X} \in \mathbb{C}^T : \mathbf{R}_0\mathbf{X} = \sigma^2\mathbf{X}\}$ will serve as normalized eigenvectors.

See the supplements for discussion of the case when X_t is stationary.

7.3 MEAN ERGODIC THEOREM

Generally, ergodic theorems address convergence of the sum

$$S_N = \frac{1}{2N+1} \sum_{t=-N}^{N} X_t, \tag{7.44}$$

where X_t is the orbit of a dynamical system or, in our case, a random sequence. When X_t is stationary we *expect* from

$$E\{S_N\} = \frac{1}{2N+1} \sum_{t=-N}^{N} E\{X_t\} = \mu$$

that $S_N \to \mu$, and when this is true, the sequence is called *ergodic*. The mean ergodic theorem addresses the mean square convergence of S_N to μ and pointwise ergodic theorems address the convergence for each $\omega \in \Omega$. A more complete treatment of ergodicity for *cyclostationary* sequences, including both pointwise and mean-square theory, is given by Boyles and Gardner [23] who also give a good set of references to related work. Also see Honda [98]. Here we will give only a few results for PC sequences that are connected to the mean ergodic theorem. We will use the spectral theory for harmonizable sequences from Section 5.3 for which we can express the limit of a more general form of (7.44), namely,

$$S_N(\lambda) = \frac{1}{2N+1} \sum_{t=-N}^{N} X_t e^{-i\lambda t}, \tag{7.45}$$

which was also given in (5.8). Note that if X_t is PC-T then $F(\{(\lambda, \lambda\})) = F_0(\{\lambda\})$, the F_0 measure of the point λ where F_0 is defined in (6.26). Similarly, the limit of $S_N(\lambda)$ is $\xi(\{\lambda\})$, the random measure of the point λ. See Loève [139, page 474] for similar results when $\xi(\lambda)$ is interpreted as the random spectral process and $F(\lambda_1, \lambda_2)$ as the two dimensional spectral distribution function function.

For a PC sequence with possibly nonzero mean m_t, we defined (in Chapter 6) $\widetilde{m}_k$ as the kth Fourier coefficient of m_t,

$$\widetilde{m}_k = \frac{1}{T} \sum_{t=0}^{T-1} m_t e^{-i2\pi kt/T}. \tag{7.46}$$

Now we can observe that the quantity

$$E\{S_N(\lambda)\} = \frac{1}{2N+1} \sum_{t=-N}^{N} E\{X_t\} e^{-i\lambda t}$$

converges to $\widetilde{m}_k$ for $\lambda = 2\pi k/T, k = 0, 1, \dots, T-1$, and otherwise converges to 0. So it is natural to say that a PC-T sequence is mean ergodic whenever $S_N(2\pi k/T) \to \widetilde{m}_k$, in mean-square sense for all $k = 0, 1, \dots, T-1$.

Proposition 7.7 *A PC-T sequence is mean ergodic if and only if*

$$F_0(\{2\pi k/T\}) = F(\{2\pi k/T\}, \{2\pi k/T\}) = |\widetilde{m}_k|^2, \tag{7.47}$$

for all $k = 0, 1, \dots, T-1$.

Proof. Let $m_t = m_{t+T}$ be the periodic mean of the PC-T sequence X_t and write $X_t = X_t' + m_t$, where X_t' has zero mean. Denoting $S_N'(\lambda)$ as (5.8) evaluated for X_t', we see that

$$S_N(2\pi k/T) = S_N'(2\pi k/T) + \frac{1}{2N+1} \sum_{t=-N}^{N} m_t e^{-i2\pi kt/T}$$

converges to $\xi(\{2\pi k/T\}) = \xi'(\{2\pi k/T\}) + \widetilde{m}_k$. Continuing to follow the stationary case, since

$$F_0(\{2\pi k/T\}) = E\{|\xi(\{2\pi k/T\})|^2\} = E\{|\xi'(\{2\pi k/T\})|^2\} + |m_k|^2,$$

we obtain a mean ergodic result $S_N(2\pi k/T) \to \widetilde{m}_k$ for every k, in mean-square sense, if and only if

$$F_0(\{2\pi k/T\}) - |\widetilde{m}_k|^2 = E\{|\xi'(\{2\pi k/T\})|^2\} = 0$$

for every k. ∎

Thus we get mean ergodicity if and only if the diagonal measure $F_0(\{2\pi k/T\})$ is exactly large enough to account for $\widetilde{m}_k$, meaning that there is no random component contributing positive variance to $F_0(\{2\pi k/T\})$. This would be expected from the multivariate stationary case, but we make it precise in the next paragraph.

The mean function m_t can be viewed as the mean of the tth component of the blocked vector sequence $\mathbf{X}_n$ defined in (1.10). That is, $\mathbf{m} = E\{\mathbf{X}_n\} = (m_0, m_1, \ldots, m_{T-1})'$. Thus we can examine the weak law for $\mathbf{X}_n$ and relate it to the preceding paragraphs. Denoting $\mathbf{F}$ as the matrix spectral measure of $\mathbf{X}_n$ and using

$$\mathbf{S}_N(\lambda) = \frac{1}{2N+1} \sum_{n=-N}^{N} \mathbf{X}_n e^{-i\lambda n} \tag{7.48}$$

we found in Proposition 4.9 that $\mathbf{S}_N(0) \to \mathbf{m}$ if and only if the atom of $\boldsymbol{\xi}(\cdot)$ at $\{0\}$ is the vector of constants $\mathbf{m}$, or in terms of $\mathbf{F}$, $F_{jj}(\{0\}) = |m_j|^2$, $j = 0, 1, \ldots, T-1$. Relating the matrix spectral measures $\mathbf{F}$ and $\mathcal{F}$ via (6.41), along with $\mathcal{F}_{kk}(0) = F(\{2\pi k/T\}, \{2\pi k/T\}) = F_0(\{2\pi k/T\})$, we find that the preceding condition on the $F_{jj}(\{0\})$ are equivalent to (7.47).

7.4 PC SEQUENCES AS PROJECTIONS OF STATIONARY SEQUENCES

Since by Proposition 6.7 PC sequences are strongly harmonizable, they have a stationary dilation, meaning they can be expressed (Proposition 5.9) as the orthogonal projection

$$X_t = P_{\mathcal{H}_X} Y_t = (Y_t | \mathcal{H}_X) \tag{7.49}$$

of a stationary sequence Y_t in some larger space $\mathcal{H}_Y \supseteq \mathcal{H}_X$. See the problems for a very simple example of a PC sequence obtained from a projection.

More importantly, in the case of PC sequences we can always construct from X_t a stationary dilation Y_t solving (7.49) (see Miamee [160]) as follows.

Given a PC-T sequence X_t, let $\mathcal{H}$ denote the direct sum of T copies of $\mathcal{H}_X$, namely, $\mathcal{H} = \mathcal{H}_X^T$ on which the inner product between two vectors $\mathbf{X} = (x_1, x_2, \ldots, x_T)$ and $\mathbf{Y} = (y_1, y_2, \ldots, y_T)$ is defined to be

$$\langle\langle \mathbf{X}, \mathbf{Y} \rangle\rangle = \sum_{k=1}^{T} \langle x_k, y_k \rangle,$$

where $\langle \cdot, \cdot \rangle$ is the inner product on each copy of $\mathcal{H}_X$. The sequence $\mathbf{Y}_t = [X_t, X_{t+1}, \ldots, X_{t+T-1}]$ is stationary in $\mathcal{H}^T$ because

$$\langle\langle \mathbf{Y}_s, \mathbf{Y}_t \rangle\rangle = \sum_{k=0}^{T-1} \langle X_{s+k}, X_{t+k} \rangle = \sum_{k=1}^{T} \langle X_{s+k}, X_{t+k} \rangle = \langle\langle \mathbf{Y}_{s+1}, \mathbf{Y}_{t+1} \rangle\rangle. \quad (7.50)$$

Furthermore, it is clear that X_t is the projection of $\mathbf{Y}_t$ onto the space $\mathcal{H}_X \times \mathbf{0} \times \mathbf{0} \cdots \times \mathbf{0}$, which is essentially $\mathcal{H}_X$.

PROBLEMS AND SUPPLEMENTS

7.1 The representations following from (7.2) can also be obtained for PC processes defined on other index sets. For example, every PC field indexed on $\mathbb{Z}^2$ can be represented in similar forms, also with a finite sum (see Hurd et al. [118]). A similar representation is possible for continuous time but with possibly a countable sum (see Hurd and Kallianpur [111]). Groups of unitary operators may be used to form the almost periodically unitary processes, a class of almost PC processes [112].

7.2 For the case $T = 2$ show that if $f(\lambda) = e^{i\lambda/2}$ for $0 \leq \lambda < \pi$ and $f(\lambda) = e^{i\lambda/2}e^{i\pi}$ for $\pi \leq \lambda < 2\pi$, then $V = \int_0^{2\pi} f(\lambda)dE(\lambda)$ solves $V^2 = U_X = \int_0^{2\pi} e^{i\lambda/T} dE(\lambda)$.

7.3 Put the PC sequences given by the following examples into the form $X_t = V^t[P(t)]$, for V unitary:

(a) amplitude scale modulated stationary sequences (Section 2.1.3);

(b) time scale modulated stationary sequences (Section 2.1.4);

7.4 The TSR when X_t is stationary. If X_t is stationary the representation (7.43) holds for every $T \geq 1$. This motivates the question: Is there anything different in the proper PC case (meaning the sequence is PC but not stationary) relative to the stationary case? Here is one difference.

Note that if X_t is PC-T, then for every $N \geq 1$ the restricted correlations

$$\mathbf{R}_\tau^N = [R(s+\tau, t+\tau)]_{s,t=0}^{N-1} \quad (7.51)$$

are periodic in τ, $\mathbf{R}_\tau^N = \mathbf{R}_{\tau+T}^N$. For some N this periodicity need not be proper (meaning all the elements of $\mathbf{R}_\tau^N$ are constant with respect to τ), but in order for X_t to be PC-T there must be some finite N for which the periodicity is proper.

But if X_t is stationary, $\mathbf{R}_\tau^N = \mathbf{R}_{\tau+1}^N$ for every N. So by taking $N = kT$ a representation of the form (7.43) can be obtained for any PC-T sequence (including stationary sequences). But for sufficiently large N, the matrices Φ and Σ will depend on τ for a PC-T sequence, but they will never depend on τ for a stationary sequence.

To construct a proper PC-T sequence for which $\mathbf{R}_\tau^N$ is constant with respect to τ for small N, consider the PMA model $X_t = \theta_0(t)\xi_t + \theta_k(t)\xi_{t-k}$ for $k > T$ and choose $\theta_0(t), \theta_k(t)$ to be properly periodic but $\theta_0^2(t) + \theta_k^2(t) = \sigma^2$. Then $\mathbf{R}_\tau^N = \mathbf{R}_{\tau+1}^N$ for $1 \le N < k$.

7.5 Show the operator U defined on $\mathcal{L}_X = \text{sp}\{X_t, t \in \mathbb{Z}\}$ by $U(\sum_{j=1}^n a_j X_{t_j}) = \sum_{j=1}^n a_j X_{t_j+T}$ is well defined, linear, and inner product preserving. Hint: Denote $\mathbf{z}_1 = \sum_{j=1}^n a_j X_{t_j}$ and $\mathbf{z}_1 = \sum_{k=1}^m b_k X_{t_k}$. To show U is well defined, assume $\mathbf{z}_1 = \mathbf{z}_2$ and show $U\mathbf{z}_1 = U\mathbf{z}_2$. For linearity, show $U(a\mathbf{z}_1 + \mathbf{z}_2) = aU\mathbf{z}_1 + U\mathbf{z}_2$. To show that inner products are preserved, note that for any two vectors $\mathbf{z}_1, \mathbf{z}_2 \in \mathcal{L}_X$, (1.7) implies

$$\langle \mathbf{z}_1, \mathbf{z}_2 \rangle = \langle U\mathbf{z}_1, U\mathbf{z}_2 \rangle.$$

7.6 A simple example of a PC sequence obtained as a projection. Suppose the vectors Y_t are orthonormal, $\langle Y_s, Y_t \rangle = \delta_{s-t}$, so Y_t is stationary white noise. Let $\mathcal{H}_X = \overline{\text{sp}}\{Y_{2t} : t \in \mathbb{Z}\}$. Then

$$X_t = P_{\mathcal{H}_X} Y_t = \begin{cases} Y_t, & t \text{ even} \\ 0, & t \text{ odd} \end{cases}$$

So X_t is clearly properly PC with period 2.

7.7 Makagon and Salehi [142] show that every periodically stationary X_t (in the strict sense of Definition 1.2) can be represented in law by a time varying linear combination of strictly stationary sequences. This may be viewed as an extension of Gladyshev's theorem, which is a representation in $L^2(\Omega, \mathcal{F}, P)$.

CHAPTER 8

PREDICTION OF PC SEQUENCES

The primary objective of this chapter is to present, for PC sequences, some of the basic ideas that are fundamental to linear prediction.

We will first give some notation and then proceed to the Wold decomposition and linear prediction before presenting the solution for sequences that are *periodic autoregressions of order one* (PAR(1)) .

Although many aspects of the prediction problem can be solved via the connection to multivariate sequences, we believe that viewing PC sequences as nonstationary can facilitate our general understanding of nonstationary processes, and so we take this tack. Some of the following definitions, already given in Chapter 4, are repeated here due to the new context. We begin with the subspaces that are important for discussing the prediction problem. In the following discussion, the closure is with respect to $L^2(\Omega, \mathcal{F}, P)$.

Definition 8.1 *The time domain of a second order sequence X_t is the subspace generated by all the values of X_t, with $t \in \mathbb{Z}$,*

$$\mathcal{H} = \overline{\text{sp}}\{X_t : t \in \mathbb{Z}\}. \tag{8.1}$$

The Hilbert subspace generated by all the values of X_s for $s \le t$ is denoted by

$$\mathcal{H}(t) = \overline{\text{sp}}\{X_s : s \le t\} \tag{8.2}$$

and is called the past of the process up to and including time t. The remote past is defined by

$$\mathcal{H}(-\infty) = \bigcap_{t \in \mathbb{Z}} \mathcal{H}(t). \tag{8.3}$$

We note that $\mathcal{H}(t)$ is nondecreasing with respect to t, that is, $\mathcal{H}(t) \supseteq \mathcal{H}(s)$ for $t \ge s$ and for this reason we can also express

$$\mathcal{H}(-\infty) = \bigcap_{t<0} \mathcal{H}(t) = \bigcap_{t_k<0} \mathcal{H}(t_k) = \lim_{t_k \to -\infty} \mathcal{H}(t_k), \tag{8.4}$$

where $\{t_k\}$ is any sequence of integers that converges to $-\infty$.

Definition 8.2 *A second order random sequence is called purely nondeterministic or regular if*

$$\mathcal{H}(-\infty) = \{0\}$$

and is called (globally) deterministic or singular if

$$\mathcal{H}(-\infty) = \mathcal{H},$$

or equivalently, if

$$\mathcal{H}(s) = \mathcal{H}(t), \text{ for all } s, t \text{ in } \mathbb{Z}.$$

A sequence is called nondeterministic at time t if $X_t \notin \mathcal{H}(t-1)$ and deterministic at time t if $X_t \in \mathcal{H}(t-1)$. It is called nondeterministic if it is nondeterministic for some $t \in \mathbb{Z}$.

Note the negative of nondeterminism, $X_t \in \mathcal{H}(t-1)$ for all t, is equivalent to the condition $\mathcal{H}(-\infty) = \mathcal{H}$ given above as the definition of determinism.

Also note in the case of stationary sequences, if $X_t \notin \mathcal{H}(t-1)$ for some t then $X_t \notin \mathcal{H}(t-1)$ for all t, and so in that case it is convenient to state the condition for nondeterminism as $X_0 \notin \mathcal{H}(-1)$, as we did in Section 4.2. However, the following PC-2 sequence illustrates that for nonstationary sequences we need to permit the issue of determinism to be time dependent. Suppose $\{Y_t\}$ is a stationary purely nondeterministic sequence and a sequence X_t is constructed by "doubling" Y_t; that is, X_t is the sequence

$$\ldots Y_{-1}\; Y_{-1}\; Y_0\; Y_0\; Y_1\; Y_1 \ldots$$

or in symbols

$$X_t = \begin{cases} Y_{t/2} & t \text{ even} \\ Y_{(t-1)/2} & t \text{ odd} \end{cases}. \tag{8.5}$$

It is easily seen that X_t is purely nondeterministic because $\mathcal{H}_X(2t) = \mathcal{H}_Y(t)$ implies

$$\bigcap_{t\in\mathbb{Z}} \mathcal{H}_X(t) = \bigcap_{t\in\mathbb{Z}} \mathcal{H}_Y(t) = \{0\}.$$

But we find that $X_t \in \mathcal{H}_X(t-1)$ for t odd and $X_t \notin \mathcal{H}_X(t-1)$ for t even. So X_t is purely nondeterministic and hence nondeterministic although it is deterministic (locally) for t odd. This is not a contradiction because pure nondeterminism is a global quality whereas determinism for specific times is local.

Sometimes a sequence X_t can be expressed in terms of another (not necessarily unique) sequence ξ_t, for example, when $\{\xi_t\}$ is a basis (in some sense) for X_t. This motivates the concepts of causality, autonomy, and stability.

Causality, Autonomy, and Stability.

Definition 8.3 (Causal) *A sequence X_t is called* causal *with respect to the sequence ξ_t if $\mathcal{H}_X(t) \subset \mathcal{H}_\xi(t)$ for all $t \in \mathbb{Z}$.*

In summary, causality addresses the issue of time order. If X_t is causal with respect to ξ_t, then X_t cannot depend on future values of ξ_t, that is, on ξ_s for $s > t$.

If X_t is causal with respect to an orthogonal sequence ξ_t then the orthogonality of ξ_t implies

$$\bigcap_{t} \mathcal{H}_X(t) \subset \bigcap_{t} \mathcal{H}_\xi(t) = \{0\},$$

showing that X_t is regular or purely nondeterministic.

If X_t is causal with respect to the orthonormal sequence ξ_t, then for all $t \in \mathbb{Z}$

$$X_t = \sum_{k=-\infty}^{t} \alpha_k(t)\xi_k \tag{8.6}$$

and

$$\sum_{k=-\infty}^{t} |\alpha_k(t)|^2 < \infty.$$

Setting $j = t - k$ gives

$$X_t = \sum_{j=0}^{\infty} \psi_j(t)\xi_{t-j} \text{ and } \sum_{j=0}^{\infty} |\psi_j(t)|^2 < \infty \tag{8.7}$$

where $\psi_j(t) = \alpha_{t-j}(t)$.

Remark. See Lütkepohl [141] for some other notions of causality.

Definition 8.4 (Autonomous) *The sequence X_t has an autonomous*[1] *representation in terms of ξ_t if for all $t \in \mathbb{Z}$*

$$X_t = \sum_{j=-\infty}^{\infty} \psi_j \xi_{t-j}, \tag{8.8}$$

where the sum is in the mean-square sense. The condition $\sum_j |\psi_j|^2 < \infty$ along with the orthogonality of ξ_t and boundedness of $\|\xi_t\|$ are sufficient for the existence of X_t for each fixed t.

In summary, X_t is *autonomous* with respect to ξ_t if the coefficients of the representation do not depend on t.

Definition 8.5 (Stable) *The sequence X_t has a stable representation in terms of the sequence ξ_t if for all $t \in \mathbb{Z}$*

$$X_t = \sum_{j=-\infty}^{\infty} \psi_j(t) \xi_{t-j} \tag{8.9}$$

with $\sum_{j=-\infty}^{\infty} |\psi_j(t)| < \infty$.

In other words, X_t is *stable* with respect to ξ_t if the coefficients of the representation are summable.

8.1 WOLD DECOMPOSITION

As in the case for stationary sequences, the Wold decomposition is closely related to the relationship between the propagating unitary operator U_X defined in (7.2) and the subspaces we have just defined. We now give these relationships for PC sequences.

Lemma 8.1 *If X_t is PC-T with unitary T-shift U, then*

(a) $\mathcal{H}(t+T) = U\mathcal{H}(t)$;

(b) $\mathcal{H} = U\mathcal{H}$;

(c) $\mathcal{H}(-\infty) = U\mathcal{H}(-\infty)$.

Proof. For item (a) recall that if $A : \mathcal{H} \mapsto \mathcal{H}$ and M is any subset of $\mathcal{H}$, then $AM = \{\mathbf{y} \in \mathcal{H} : \mathbf{y} = A\mathbf{x}, \mathbf{x} \in M\}$. Thus taking M to be $\mathcal{L}(t) = \text{sp}\{X_s :$

[1]The word *autonomous* conveys the notion of time invariance and usually refers to the solution of an unforced deterministic system; see Elaydi [54, page 2.]

$s \le t\}$ then $\mathcal{L}(t+T) = U\mathcal{L}(t)$; for if $\mathbf{z} \in \mathcal{L}(t)$, then $\mathbf{z} = \sum_{j=1}^n \alpha_j X_{t_j} : t_j \le t$ so that $U\mathbf{z} = \sum_{j=1}^n \alpha_j X_{t_j+T} \in \mathcal{L}(t+T)$, which means $U\,\mathcal{L}(t) \subset \mathcal{L}(t+T)$.

Now let $\mathbf{z} \in U\mathcal{H}(t)$, so $\mathbf{z} = U\mathbf{w}$ for $\mathbf{w} \in \mathcal{H}(t)$. Then $\mathbf{w} = \lim \mathbf{w}_n$ for $\mathbf{w}_n \in \mathcal{L}(t)$. By continuity of U we can write $\mathbf{z} = U\mathbf{w} = U(\lim \mathbf{w}_n) = \lim U\mathbf{w}_n$. But since $U\mathbf{w}_n \in \mathcal{L}(t+T)$ for all n we conclude that $\mathbf{z} \in \overline{\mathcal{L}(t+T)} = \mathcal{H}(t+T)$. Thus $U\mathcal{H}(t) \subset \mathcal{H}(t+T)$. A similar argument using the continuity of U^{-1} produces $\mathcal{H}(t+T) \subset U\mathcal{H}(t)$.

For item (b), the proof is essentially the same. One first shows that $U\mathcal{L} \subset \mathcal{L}$, where $\mathcal{L} = \text{sp}\{X_s, s \in \mathbb{Z}\}$ and $U\mathcal{H} \subset \mathcal{H}$ follows from the continuity of U. Then $\mathcal{H} \subset U\mathcal{H}$ follows similarly.

To prove item (c), first suppose $\mathbf{z} \in \mathcal{H}(-\infty)$ which means $\mathbf{z} \in \mathcal{H}(t)$ for all t. But then $\mathbf{z} \in \mathcal{H}(t+T) = U\mathcal{H}(t)$ so there is a $\mathbf{y} \in \mathcal{H}(t)$ (for all $t \in \mathbb{Z}$) with $\mathbf{z} = U\mathbf{y}$. Hence $\mathbf{z} \in U\mathcal{H}(-\infty)$. ∎

The Wold decomposition now follows, in essence, from the facts given in the preceding lemma.

Proposition 8.1 (Wold Decomposition for PC Sequences) *Any PC-T sequence X_t has a unique decomposition*

$$X_t = Y_t + Z_t \tag{8.10}$$

with

(a) $\mathcal{H}_X(t) = \mathcal{H}_Y(t) \oplus \mathcal{H}_Z(t)$;

(b) *Y_t is deterministic and Z_t is purely nondeterministic;*

(c) *Y_t and Z_t are PC-T ;*

(d) *U_Y and U_Z are restrictions of U_X to $\mathcal{H}_X(-\infty)$ and $\mathcal{H}_X(-\infty)^\perp$, respectively.*

Proof. For each $t \in \mathbb{Z}$ set $Y_t = (X_t|\mathcal{H}(-\infty))$ and $Z_t = X_t - Y_t$. On the one hand, for each s, $Y_s \in \mathcal{H}_X(-\infty)$, and on the other hand, $Z_t = X_t - Y_t = X_t - (X_t|\mathcal{H}_X(-\infty)) \perp \mathcal{H}_X(-\infty)$ for each t. Hence $Y_s \perp Z_t$ for all $s, t \in \mathbb{Z}$, which means $\mathcal{H}_Y \perp \mathcal{H}_X$. For each $t \in \mathbb{Z}, Y_t \in \mathcal{H}_X(-\infty) \subseteq \mathcal{H}_X(t)$ and hence $Z_t = X_t - Y_t \in \mathcal{H}_X(t)$ and therefore $\mathcal{H}_Y(t) \oplus \mathcal{H}_Z(t) \subseteq \mathcal{H}_X(t)$. In order to complete the proof of item (a) it suffices to show that $\mathcal{H}_X(t) \subseteq \mathcal{H}_Y(t) \oplus \mathcal{H}_Z(t)$. Pick $\mathbf{w} \in \mathcal{H}_X(t)$; then $\mathbf{w} = \lim \mathbf{w}_n$, where each vector $\mathbf{w}_n \in \mathcal{M}_X(t)$ and hence $\mathbf{w}_n = \mathbf{u}_n + \mathbf{v}_n$, $\mathbf{u}_n \in \mathcal{L}_Y(t)$, $\mathbf{v}_n \in \mathcal{L}_Z(t)$. But since $\mathbf{w}_n$ is Cauchy and $\|\mathbf{u}_n - \mathbf{u}_m\| \le \|\mathbf{w}_n - \mathbf{w}_m\|$, $\mathbf{u}_n$ must be Cauchy. Hence $\mathbf{u}_n \to \mathbf{u} \in \mathcal{H}_Y(t)$. Similarly $\mathbf{v}_n \to \mathbf{v} \in \mathcal{H}_Z(t)$. Taking the limit of $\mathbf{w}_n = \mathbf{u}_n + \mathbf{v}_n$ we get $\mathbf{w} = \mathbf{u} + \mathbf{v}$, which shows $\mathcal{H}_X(t) \subset \mathcal{H}_Y(t) \oplus \mathcal{H}_Z(t)$.

For item (b), we first show that $\mathcal{H}_X(-\infty) = \mathcal{H}_Y(t)$ for all $t \in \mathbb{Z}$, which implies Y_t is deterministic. Since it is clear that $\mathcal{H}_Y(t) \subset \mathcal{H}_X(-\infty)$ for each t, we show that $\mathcal{H}_Y(t) \subset \mathcal{H}_X(-\infty)$. For contradiction, suppose for fixed t there is a nonzero $\mathbf{u} \in \mathcal{H}_X(-\infty) \ominus \mathcal{H}_Y(t)$, then for each $s \leq t$ we have $\mathbf{u} \perp Y_s$ and $\mathbf{u} \perp Z_s$ so that $\mathbf{u} \perp X_s = Y_s + Z_s$ and hence $\mathbf{u} \perp \mathcal{H}_X(t)$, and therefore $\mathbf{u} \perp \mathcal{H}_X(-\infty)$. This implies $\mathbf{u} = \mathbf{0}$, which is a contradiction.

For item (c), using Lemma 8.1, for any $t \in \mathbb{Z}$, we can write

$$Y_{t+T} = (X_{t+T}|H_X(-\infty)) = (U_X X_t|H_X(-\infty)) = U_X(X_t|H_X(-\infty)),$$

from which one can conclude that Y_t is PC-T.

The argument in the proof of item (c) also shows that U_X and U_Y act the same on H_Y, which together with the fact $H_X(-\infty) = H_Y$ proved above gives $U_Y = U_X|_{H_X(\infty)}$, which proves the statement in item (d) about Y_t. The statements about Z_t follow easily. ∎

.

8.2 INNOVATIONS

Suppose for the nondecreasing spaces $\mathcal{H}_X(t)$ there is some t_0 where the inclusion $\mathcal{H}_X(t_0) \supset \mathcal{H}_X(t_0 - 1)$ is proper. Then something is added to the history of the sequence at time t_0, and this motivates the following definition.

Definition 8.6 *The innovation space of the sequence X_t at time t is defined to be*

$$\mathcal{I}_X(t) = \mathcal{H}_X(t) \ominus \mathcal{H}_X(t-1) = \{\mathbf{x} \in \mathcal{H}_X(t) : \mathbf{x} \perp \mathcal{H}_X(t-1)\}.$$

The preceding can also be expressed as

$$\mathcal{H}_X(t) = \mathcal{I}_X(t) \oplus \mathcal{H}_X(t-1)$$

and this suggests that, continuing to iterate, one can express the entire history as

$$\mathcal{H}_X(t) = \mathcal{I}_X(t) \oplus \mathcal{I}_X(t-1) \oplus \ldots,$$

but this in not quite correct because any nonzero vector $\mathbf{x} \in \mathcal{H}_X(-\infty)$ will not necessarily be in $\oplus \sum_{j \leq t} \mathcal{I}_X(j)$. But such vectors are in $\mathcal{H}_X(-\infty)$. So our next guess is

$$\mathcal{H}_X(t) = \mathcal{I}_X(t) \oplus \mathcal{I}_X(t-1) \oplus \cdots \oplus \mathcal{H}_X(-\infty).$$

This is in fact true. We will now make it a bit more precise while presenting some basic facts about $\mathcal{I}_X(t)$ that are important for understanding its role in prediction. We do this for PC-T sequences and mention the connection to the univariate stationary case as well as the multivariate stationary case.

Lemma 8.2 *If X_t is a PC-T sequence with unitary operator U, then*

(a) $\mathcal{I}_X(t+T) = U\mathcal{I}_X(t)$;

(b) $\mathcal{I}_X(t) \perp \mathcal{H}_X(-\infty)$ *for every* $t \in \mathbb{Z}$, *which implies*

$$\mathcal{I}_X(t) = \mathcal{H}_Z(t) \ominus \mathcal{H}_Z(t-1);$$

(c) $d_X(t) = \dim \mathcal{I}_X(t) = 0$ *or* 1 *and* $d_X(t+T) = d_X(t)$, $t \in \mathbb{Z}$.

Proof. For item (a), first write

$$\begin{aligned}\mathcal{I}_X(t+T) &= \mathcal{H}_X(t+T) \ominus \mathcal{H}_X(t+T-1) \\ &= U\mathcal{H}_X(t) \ominus U\mathcal{H}_X(t-1) \\ &= U[\mathcal{H}_X(t) \ominus \mathcal{H}_X(t-1)]\end{aligned}$$

where the last line follows from the fact that U is unitary (see problem 3.4 in Chapter 3).

For item (b), take two vectors $\mathbf{u} \in \mathcal{I}_X(t)$ and $\mathbf{v} \in \mathcal{H}_X(-\infty)$. Thus $\mathbf{u} \in \mathcal{H}_X(t)$ and $\mathbf{u} \perp \mathcal{H}_X(t-1)$ but $\mathbf{v} \in \mathcal{H}_X(s)$ for all s and in particular for $s = t-1$ so $\langle \mathbf{u}, \mathbf{v} \rangle = 0$. To see the second part of item (b) write

$$\begin{aligned}\mathcal{I}_X(t) = \mathcal{H}_X(t) \ominus \mathcal{H}_X(t-1) &= [\mathcal{H}_Y(t) \oplus \mathcal{H}_Z(t)] \ominus [\mathcal{H}_Y(t-1) \oplus \mathcal{H}_Z(t-1)] \\ &= [\mathcal{H}_Y(t) \ominus \mathcal{H}_Y(t-1)] \oplus [\mathcal{H}_Z(t) \ominus \mathcal{H}_Z(t-1)] \\ &= [\mathcal{H}_Z(t) \ominus \mathcal{H}_Z(t-1)] = \mathcal{I}_Z(t),\end{aligned}$$

because for all s, t we have $\mathcal{H}_Z(s) \perp \mathcal{H}_Y(t)$ and $\mathcal{H}_Y(t) \ominus \mathcal{H}_Y(t-1) = \{\mathbf{0}\}$.

For item (c), we only need to observe that the vector X_t is either in $\mathcal{H}_X(t-1)$, in which case $\dim \mathcal{I}_X(t) = 0$, or if not, then $\dim \mathcal{I}_X(t) = 1$. The periodicity in the innovation dimension, $d_X(t) = d_X(t+T)$ follows immediately from item (a). ∎

If X_t is stationary then $T = 1$ and from this it follows that either $d_X(t) = 0$ for all $t \in \mathbb{Z}$ or $d_X(t) = 1$ for all $t \in \mathbb{Z}$. In the first case we can conclude that X_t is singular or globally deterministic; in the second case X_t has a regular or purely nondeterministic component. The PC-2 sequence X_t formed by "doubling" (see (8.5)) clearly has $d_X(t) = 1$ for t even and $d_X(t) = 0$ for t odd.

The preceding lemma also shows that the *block innovation* at time t, defined by

$$\begin{aligned}\mathcal{I}_X^T(t) &= \mathcal{H}_X(t) \ominus \mathcal{H}_X(t-T) \\ &= \mathcal{I}_X(t) \oplus \mathcal{I}_X(t-1) \oplus \cdots \oplus \mathcal{I}_X(t-T+1),\end{aligned} \tag{8.11}$$

is of dimension

$$d_X^T(t) = \sum_{s=0}^{T-1} d_X(t-s). \tag{8.12}$$

Furthermore, the periodicity $d_X(t) = d_X(t+T)$ implies $d_X^T(t)$ is constant with respect to t. This constant d_X^T gives a very natural way to define the *rank* of the PC-T sequences.

Definition 8.7 *The rank r_X of a PC-T sequence is the number of times in one period that the innovation is not trivial, that is, the number of times in one period that $d_X(t) = 1$.*

This is consistent with the rank (see Section 4.4.3) of the T-variate stationary process obtained by blocking. It is also clear that $\mathcal{I}_X^T(t) \perp \mathcal{H}_X(-\infty)$ for every $t \in \mathbb{Z}$.

The PC-2 sequence X_t formed by "doubling" a regular sequence Y_t is thus of rank 1.

Now we can state and prove our earlier suggestion, that $\mathcal{H}_Z(t)$ is precisely the orthogonal sum of the current and previous innovation spaces.

Proposition 8.2 *If X_t is PC-T then*

$$\mathcal{H}_Z(t) = \mathcal{I}_X(t) \oplus \mathcal{I}_X(t-1) \oplus \ldots = \oplus \sum_{j=0}^{\infty} \mathcal{I}_X(t-j).$$

Proof. The inclusion

$$\oplus \sum_{j=0}^{\infty} \mathcal{I}_X(t-j) \subset \mathcal{H}_Z(t)$$

follows from item (b) in lemma 8.2. To show the opposite inclusion, we use the truth of preceding inclusion to write

$$\mathcal{H}_Z(t) = \oplus \sum_{j=0}^{\infty} \mathcal{I}_X(t-j) \oplus \mathcal{M},$$

for some subspace $\mathcal{M}$. Now if $\mathbf{u} \in \mathcal{H}_Z(t)$ we can write $\mathbf{u} = \mathbf{v} + \mathbf{w}$, where $\mathbf{v} \in \sum_{j=0}^{\infty} \oplus \mathcal{I}_X(t-j)$ and $\mathbf{w} \in \mathcal{M}$, which implies $\mathbf{w} \perp \mathcal{I}_X(t-j)$ for $j = 0, 1, 2, \ldots$. Thus $\mathbf{w} \in \mathcal{H}_Z(t)$ and $\mathbf{w} \perp \mathcal{I}_X(t)$, which imply $\mathbf{u} \in \mathcal{H}_Z(t-1)$. This argument can be carried out inductively to reach the conclusion $\mathbf{w} \in \mathcal{H}_Z(t-j)$ for $j \in \mathbb{Z}$, which, because Z_t is purely nondeterministic, implies

$$\mathbf{w} \in \bigcap_{j=0}^{\infty} \mathcal{H}_Z(t-j) = \{\mathbf{0}\}.$$

Hence $\mathbf{w} = \mathbf{0}$ and consequently $\mathbf{u} \in \oplus \sum_{j=0}^{\infty} \mathcal{I}_X(t-j)$. ∎

The following corollary follows easily from (8.11).

Corollary 8.2.1 *If X_t is a PC-T sequence, then*

$$\mathcal{H}_Z(t) = \oplus \sum_{j=0}^{\infty} \mathcal{I}_X^T(t - jT).$$

The representation of $\mathcal{H}_Z(t)$ by an orthogonal sum of innovation spaces gives rise to the moving average representation of Z_t (or of X_t when X_t is purely nondeterministic). From the previous discussion about $d_X(t)$ we now denote

$$D^+ = \{t : d_X(t) > 0\} \tag{8.13}$$

to be the set of time indices where X_t has positive innovation dimension. We note that D^+ is a periodic set in the sense that $t \in D^+$ implies $t + kT \in D^+$ for every $k \in \mathbb{Z}$, and that $d_X^T = \text{card}\,(D^+ \cap \{0, 1, ..., T-1\})$. This notation permits us to give the moving average representation even when X_t is not of full rank, meaning $d_X^T < T$.

Proposition 8.3 (Moving Average Representation) *The second order process X_t is a purely nondeterministic PC-T sequence of rank d_X^T if and only if there exists a T-periodic set (of indices) D^+ with $d_X^T = card(\, D^+ \cap \{0, 1, ..., T-1\})$ and a set of orthonormal innovation vectors*

$$\mathcal{I} = \{\xi_m : m \in D^+\} \tag{8.14}$$

such that for every t

$$X_t = \sum_{j \geq 0 \,: t-j \in D^+} a_j(t)\xi_{t-j}, \tag{8.15}$$

where

$$\sum_{j \geq 0 \,: t-j \in D^+} |a_j(t)|^2 < \infty \tag{8.16}$$

and

$$a_j(t + kT) = a_j(t) \tag{8.17}$$

for every j, k, t with $t - j \in D^+$.

Remark. To clarify the notation, note that if $s \notin D^+$, then there is no corresponding innovation vector $\xi_s \in I$. An alternative representation having T innovations per period, some of which may be ignored by $a_j(t)$, is given in the Remark following the proof.

Proof. Suppose X_t is given by (8.15) where the $\{a_j(t)\}$ and ξ_m have the stated properties. The orthonormality of the ξ_m and the square summability (8.16) together ensure that X_t is a second order random process. Now to be specific we take $t \geq s$ and observe

$$\begin{aligned} R(t,s) &= \langle X_t, X_s \rangle \\ &= \sum_{j\geq 0: t-j\in D^+} \sum_{k\geq 0: s-k\in D^+} a_j(t)\overline{a_k(s)}\langle \xi_{t-j}, \xi_{s-k}\rangle \\ &= \sum_{k\geq 0: s-k\in D^+} a_{k+t-s}(t)\overline{a_k(s)} \\ &= \sum_{k\geq 0: s-k\in D^+} a_{k+t+T-s-T}(t+T)\overline{a_k(s+T)} \\ &= R(t+T, s+T), \end{aligned} \tag{8.18}$$

where we use the fact that for $t-j \in D^+$ and $s-k \in D^+$,

$$\langle \xi_{t-j}, \xi_{s-k}\rangle = \begin{cases} 1 & t-j = s-k \\ 0 & t-j \neq s-k \end{cases},$$

and so X_t is PC-T.

Since it is clear that

$$\mathcal{H}_X(t) \subset \mathcal{H}_\xi(t) = \overline{\text{sp}}\{\xi_s : s \leq t, s \in D^+\},$$

then

$$\bigcap_t \mathcal{H}_X(t) \subset \bigcap_t \mathcal{H}_\xi(t) = \{0\}$$

showing that X_t is purely nondeterministic (regular).

To see that X_t is of rank card$(D^+ \cap \{0,1,...,T-1\})$ we note from (8.15) that for $t \in \{0,1,...,T-1\}$ there are exactly card$(D^+ \cap \{0,1,....,T-1\})$ values of t for which X_t depends on ξ_t; for the others, X_t depends only on the past innovations ($j > 0$; i.e., $j = 0$ is not permitted). In other words, X_t has exactly card$(D^+ \cap \{0,1,...,T-1\})$ nonzero innovations for $t \in \{0,1,...,T-1\}$ (so $d_X(t) = \dim \mathcal{I}_X(t) = 1$ for precisely these t) and this implies rank $(X) =$ card$(D^+ \cap \{0,1,...,T-1\})$.

Conversely, suppose X_t is PC-T, purely nondeterministic, and of rank d_X^T. Since the innovation spaces $\mathcal{I}_X(p)$ appearing in (8.13) are of dimension at most one and $\mathcal{I}_X(p) \perp \mathcal{I}_X(q)$ for $p \neq q$, we may express any vector $Y \in \mathcal{H}_X(t)$ as

$$Y = \sum_{j\geq 0: t-j\in D^+} \langle Y, \xi_{t-j}\rangle \xi_{t-j},$$

where $D^+ = \{t : d_X(t) = \dim \mathcal{I}_X(t) = 1\}$ and ξ_p is the unit vector of $\mathcal{I}_X(p)$ when $p \in D^+$.

Finally, since $X_t \in \mathcal{H}_X(t)$ we may write

$$X_t = \sum_{j\geq 0 : t-j \in D^+} a_j(t)\xi_{t-j},$$

where for each t we must have

$$\sum_{j\geq 0 : t-j \in D^+} |a_j(t)|^2 < \infty,$$

which are (8.15) and (8.16). To obtain the periodicity of the $a_j(t)$ we first write

$$X_{t+T} = \sum_{j\geq 0 : t+T-j \in D^+} a_j(t+T)\xi_{t+T-j} = \sum_{j\geq 0 : t-j \in D^+} a_j(t+T)\xi_{t+T-j},$$

where the change of indexing in the last expression follows because D^+ is a periodic set (with period T, due to X_t being PC-T). But then we may also express

$$X_{t+T} = U_X X_t = \sum_{j\geq 0 : t-j \in D^+} a_j(t) U_X[\xi_{t-j}] = \sum_{j\geq 0 : t-j \in D^+} a_j(t)\xi_{t+T-j},$$

where U_X may be brought inside the sum due to mean-square convergence of the partial sums and continuity of U_X. The last equality follows from the fact that for $p \in D^+$ we have $\xi_{p+T} = U_X \xi_p$, a conclusion that may be drawn from $\mathcal{I}_X(t+T) = U_X \mathcal{I}_X(t)$, which was established in Lemma 8.2. Then from the uniqueness of the decomposition it follows that

$$a_j(t) = a_j(t+T)$$

whenever $t - j \in D^+$, as we have claimed in (8.17). ∎

Remark. The regular part has an alternative representation that may be useful. More precisely, a second order sequence X_t is PC-T, purely nondeterministic, and of rank d_X^T if and only if there exist a periodic set (of indices) D^+ of period T having $d_X^T = \text{card}(D^+ \cap \{0, 1, ..., T-1\})$ and a sequence of orthonormal vectors

$$\mathcal{I} = \{\xi'_m : m \in \mathbb{Z}\} \tag{8.19}$$

such that for every t

$$X_t = \sum_{j\geq 0} a'_j(t)\xi'_{t-j} \tag{8.20}$$

with

$$\sum_{j\geq 0} |a_j'(t)|^2 < \infty \tag{8.21}$$

and

$$a_j'(t+kT) = a_j'(t) \tag{8.22}$$

for every k, t and $j \geq 0$ but

$$a_j'(t) = 0 \tag{8.23}$$

whenever $t - j \notin D^+$. The representation (8.20) for X_t has T orthonormal vectors per period but ignores some of them via (8.23) if $d_X^T < T$.

8.3 PERIODIC AUTOREGRESSIONS OF ORDER 1

From Section 2.1.7, a second order sequence X_t is called a periodic autoregression of order 1, or PAR(1), if it satisfies

$$X_t = \phi(t)X_{t-1} + \sigma(t)\xi_t, \tag{8.24}$$

where $\{\xi_t : t \in \mathbb{Z}\}$ is a collection of orthonormal random variables and $\phi(t) = \phi(t+T)$ and $\sigma(t) = \sigma(t+T)$. In the introduction to this topic in Section 2.1.7 we set $\sigma(t) \equiv 1$ but here we will permit $\sigma(t), t = 0, 1, \ldots, T-1$ to be arbitrary real numbers, which can be taken to be nonnegative without any loss in generality. However, we assume that not all $\sigma(t)$ are zero.

There are two important related concepts that must be considered in addressing the solution X_t of this simple system, namely, boundedness of solutions and causality. Boundedness means that $\sup_t \|X_t\| < \infty$ and causality means $\mathcal{H}_X(t) \subset \mathcal{H}_\xi(t)$. If X_t is causal then X_t cannot depend on future values of ξ_t (i.e., on ξ_s for $s > t$). From Chapter 2 we already know that the number $A = \prod_{t=0}^{T-1} \phi(t)$ plays a crucial role in the nature of the solutions to (8.24). The following theorem gives the relationship between these notions.

Theorem 8.1 *Let X_t be a PAR(1) sequence given by (8.24). Any two of the following conditions implies the other one.*

(a) *X_t is bounded;*

(b) *$\mathcal{H}_X(t) \subset \mathcal{H}_\xi(t)$ (causality);*

(c) $|A| < 1$.

Proof. Assume (a) and (b) hold. Set

$$\begin{aligned} A_0(t) &= 1 \\ A_1(t) &= \phi(t) \\ A_2(t) &= \phi(t)\phi(t-1) \\ &\vdots \\ A_{T-1}(t) &= \phi(t)\phi(t-1)\ldots\phi(t-T+2) \\ A = A_T &= \phi(t)\phi(t-1)\ldots\phi(t-T+1) \end{aligned} \tag{8.25}$$

and then the first recursion of (8.24) produces

$$X_t = \phi(t)[\phi(t-1)X_{t-2} + \sigma(t-1)\xi_{t-1}] + \sigma(t)\xi_t$$

and repeating $N-2$ times gives

$$X_t = \sum_{j=0}^{N-1} A_j(t)\sigma(t-j)\xi_{t-j} + A_N(t)X_{t-N}, \tag{8.26}$$

where $A_j(t) = A^p A_r(t)$ for $j = pT + r$, $0 \leq r < T$. For $N = MT$ the causality condition implies that

$$\|X_t\|^2 = \left(\sum_{k=0}^{T-1} A_k^2(t)\sigma^2(t-k)\right)\sum_{j=0}^{M-1} A^{2j} + A^{2M}\|X_{t-MT}\|^2. \tag{8.27}$$

Since X_t is assumed to be bounded, we must have

$$\left(\sum_{k=0}^{T-1} A_k^2(t)\sigma^2(t-k)\right)\sum_{j=0}^{\infty} A^{2j} < \infty,$$

which implies $|A| < 1$.

Now assume (b) and (c) hold. Since we have assumed causality, we have (8.27). Taking the limit as $M \to \infty$ of both sides of (8.27), the limit of the second term on the right becomes zero (because $|A| < 1$) and we get

$$\|X_t\|^2 = \frac{1}{1-A^2}\sum_{k=0}^{T-1} A_k^2(t)\sigma^2(t-k) \text{ for every } t.$$

Therefore X_t is bounded.

Finally assume that (a) and (c) hold. Using (8.26) with $N = MT$ we get

$$X_t = \sum_{j=0}^{MT-1} A_j(t)\sigma(t-j)\xi_{t-j} + A^{2M}X_{t-2M+1}. \tag{8.28}$$

Letting M go to infinity, since X_t is bounded and $|A| < 1$, the second term on the right goes to zero, giving

$$X_t = \sum_{j=0}^{\infty} A_j(t)\sigma(t-j)\xi_{t-j}, \tag{8.29}$$

which shows X_t is causal. ∎

The preceding theorem gives sufficient conditions for X_t to be PC.

Proposition 8.4 *Any two of the conditions in the statement of Theorem 8.1 implies the solution to (8.24) is PC-T.*

Proof. Since any two of the conditions implies the third and hence (8.29) is alwasy true along with $\sum_j |A_j(t)|^2\sigma^2(t-j) < \infty$. Taking $s > t$ to be specific, the thus compute

$$R(s,t) = E\{X_s X_t\} = \sum_{j=0}^{\infty} A_j(t)A_{s-t+j}(s)\sigma^2(t-j),$$

which together with $A_j(t) = A_j(t+T)$ implies $R(s,t) = R(s+T, t+T)$ meaning X_t is PC-T. ∎

Note that $|\phi(t)| > 1$ is possible for some t and still $|A| < 1$. In other terms, the system described by (8.24) can be locally expanding for some t and locally contracting for other t and yet contracting on the whole, $|A| < 1$. It is also clear that under any two of the conditions of Theorem 8.1, the resulting causality $\mathcal{H}_X(t) \subset \mathcal{H}_\xi(t)$ implies X_t is purely nondeterministic.

So when a PAR(1) sequence is PC-T and either causal or $|A| < 1$, it must have an infinite moving average representation of the form (8.15) or (8.20). The next corollary explicitly gives the coefficients $a_j(t)$ and $a'_j(t)$ appearing in (8.15) and (8.20).

Corollary 8.4.1 *If X_t is a causal PC solution to the PAR(1) system (8.24), then we have the moving average representation*

$$X_t = \sum_{j=0}^{\infty} a_j(t)\xi_{t-j}, \tag{8.30}$$

where $a_j(t) = A^p A_r(t)\sigma(t-r) = a_j(t+T)$ for $j = pT + r$, $0 \le r < T$ with A and $A_p(t)$ defined by (8.25) and $\sum_{j=0}^{\infty} |a_j(t)|^2 < \infty$.

Proof. The proof follows from identifying the coefficients $a_j(t)$ with the coefficients in representation (8.29). It is clear that $a_j(t) = a_j(t+T)$ from the periodicity of $A_r(t)$ and $\sigma(t)$. We obtain representations of the type (8.15) by ignoring the extraneous ξ_{t-j}, where $\sigma(t-j) = 0$, and we obtain a representation of type (8.20) if we include them but then their coefficients must satisfy $a_j(t) = 0$. ∎

Before continuing further, we need to say something about the effects of $\sigma(t) = 0$ and $\phi(t) = 0$. First, if $\sigma(t_0) = 0$, then it is clear that $\mathcal{H}_X(t_0) = \mathcal{H}_X(t_0 - 1)$ and hence $d_X(t_0) = 0$. Thus rank $(X) = \text{card}\,\{t \in 0, 1, \ldots T-1 : \sigma(t) \neq 0\}$. Furthermore, X_t is *deterministic* since $X_t \in \mathcal{H}_X(t-1)$ at all t for which $\sigma(t) = 0$. Hence if X_t is a PAR(1) sequence and PC, then it is deterministic if and only if it is not full rank.

The occurrence of $\phi(t) = 0$ does not affect rank but only the memory. For if $\phi(t_0) = 0$, then clearly $A = 0$ and so the infinite moving average representation (8.29) terminates after some finite number J of terms, where $J < T$. If for some t we have both $\sigma(t) = 0$ and $\phi(t) = 0$, then $X_t = 0$, but still $X_t \in \mathcal{H}_X(t-1)$ so that also $d_X(t) = 0$, meaning there is no innovation at t that contributes to the rank.

8.4 SPECTRAL DENSITY OF REGULAR PC SEQUENCES

We begin with the spectral density

$$\mathbf{f}(\lambda) = \frac{d\mathbf{F}}{d\mu}(\lambda)$$

(μ is Lebesgue measure) of the T-variate stationary sequence $\mathbf{X}_n$ made from T-blocks of a PC-T process X_t.

For a regular PC-T sequence of rank $r = d_X^T$, it may be seen from the infinite moving average representation (8.15) that X_t depends only on r innovation vectors from the block of indices $\{t, t-1, \ldots, t-T+1\}$. In order to make a correspondence between t and a block number n we assume $t = nT$. We define $\boldsymbol{\xi}_n$ to be the column vector whose components are $\xi_j \in \mathcal{I} : j \in \{nT - T + 1 \leq j \leq nT\} \cap D^+$. Using this notation X_t may be expressed in terms of the vectors $\boldsymbol{\xi}_k$ for $k \leq n$ and the coefficients are taken from the $a_j(t)$, where $t - j \in D^+$. Similarly, X_{t-1} can be expressed in terms of the same $\boldsymbol{\xi}_k$ except the coefficient for the innovation occurring at t, which, if among the components of $\boldsymbol{\xi}_n$, must be zero. This can be continued to X_{t-T+1} to obtain

$$\mathbf{X}_n = \sum_{p \geq 0} \mathbf{A}_p \boldsymbol{\xi}_{n-p}, \tag{8.31}$$

where $\mathbf{A}_p$ is of dimension $T \times r$ and (8.16) implies

$$\sum_{p \geq 0} |\mathbf{A}_p|^2 < \infty, \tag{8.32}$$

which means

$$\sum_{p \geq 0} |A_p^{ij}|^2 < \infty$$

for all the entries A_p^{ij} of $\mathbf{A}_p$, whose dimension is $T \times r$. To obtain $\mathbf{A}_0$ explicitly, denote $\{j_1, j_2, \ldots, j_r\} = \{j : 0 \leq j \leq T-1 \text{ and } t-j \in D^+\}$, where the j_k are also ordered $j_1 < j_2 \cdots < j_r$. Then

$$\mathbf{A}_0 = \begin{bmatrix} a_{j_1}(t) & a_{j_2}(t) & \cdots & a_{j_r}(t) \\ a_{j_1}(t-1) & a_{j_2}(t-1) & \cdots & a_{j_r}(t-1) \\ \vdots & \vdots & \vdots & \vdots \\ a_{j_1}(t-T+1) & a_{j_2}(t-T+1) & \cdots & a_{j_r}(t-T+1) \end{bmatrix}, \tag{8.33}$$

where we note that for $j = j_1, j_2, \ldots, j_r$ the coefficient $a_j(t')$ is never present whenever $t' - j \notin D^+$. But, in addition, for row k we take $a_j(t-k) = 0$ if $j < k$.

The spectral density of $\mathbf{X}_n$ can be written (see Theorem 4.11)

$$\mathbf{f}(\lambda) = \mathbf{\Phi}(\lambda)\mathbf{\Phi}^*(\lambda), \tag{8.34}$$

where

$$\mathbf{\Phi}(\lambda) = \sum_{p=0}^{\infty} \mathbf{A}_p e^{i\lambda p}.$$

Recall that the matrix valued measures $\mathcal{F}$ and $\mathbf{F}$, linked in (6.41) by the continuous transformation $\mathbf{V}$, are both absolutely continuous or neither are. And since the collection measures $\{F_0, F_1, \ldots, F_{T-1}\}$ are formed by splicing of the elements of $\mathbf{F}$, then $\mathcal{F}$, $\mathbf{F}$, and the collection $\{F_0, F_1, \ldots, F_{T-1}\}$ all are absolutely continuous (meaning all elements of the matrix are absolutely continuous) or none are. Applied to the current case, since $\mathbf{F}$ is absolutely continuous with density $\mathbf{f}(\lambda)$, then $\mathcal{F}$ and the collection $\{F_0, F_1, \ldots, F_{T-1}\}$ are absolutely continuous, where the densities of the collection are $\{f_0, f_1, \ldots, f_{T-1}\}$.

We can also note that when the PC sequence has rank r, then from Theorem 4.11 and Proposition 4.14, $\mathbf{f}(\lambda)$ is of rank r for a.e. λ and from the invertibility and continuity of $\mathbf{V}(\lambda)$ we can conclude that rank $\mathbf{f}(\lambda) =$ rank $f_{\mathbf{Z}}(\lambda)$ for a.e. λ. The rank is reflected into the collection $\{f_0, f_1, \ldots, f_{T-1}\}$ through the way they are formed from the elements of $f_{\mathbf{Z}}$.

8.4.1 Spectral Densities for PAR(1)

There are two ways to compute the spectral densities for a PAR(1): a direct method that retains the nonstationary setup, and the lifting method that translates the problem to a vector autoregression or VAR. We will present them both.

8.4.1.1 The Direct Method This is a frequency domain approach in the sense discussed in Chapter 4. It was presented in a treatment of a first order autoregression with almost periodic coefficients [147].

We assume $|A| < 1$ so X_t given by (8.24) is PC-T. Let V denote the unitary operator, mapping $\mathcal{H}_X$ to itself, defined by $V\xi_t = \xi_{t+1}$.

Let Ψ be the unitary linear transformation defined by $\Psi : \xi_t \longrightarrow e^{it\cdot}$, which maps M_ξ onto $L^2 = L^2([0, 2\pi), d\mu)$. Here and in the sequel $d\mu$ denotes the normalized Lebesgue measure and $[0, 2\pi)$ is regarded as a group with addition mod 2π. It is easy to see that $\Psi V \Psi^{-1}$ is the operator of multiplication by $e^{i\cdot}$ and the process X_t is unitarily equivalent to the L^2 sequence h_t. In terms of the equivalent sequence h_t the system (8.24) takes the form

$$h_t(\cdot) = \phi(t)h_{t-1}(\cdot) + \sigma(t)e^{it\cdot}, \qquad t \in Z \tag{8.35}$$

and the moving average representation (8.30) takes the form

$$h_t = \sum_{j=0}^{\infty} a_j(t)e^{i(t-j)\lambda}, \tag{8.36}$$

where $a_j(t) = A^p A_r(t)\sigma(t-r) = a_j(t+T)$ for $j = pT + r$, $0 \le r < T$ with A and $A_p(t)$ defined by (8.25) and $\sum_{j=0}^{\infty} |a_j(t)|^2 < \infty$.

Now one can express h_t as $h_t = e^{it\cdot}g_t$, in terms of a periodic sequence g_t given by

$$g_t(\lambda) = \left[1 - Ae^{-iT\lambda}\right]^{-1} G_t(\lambda). \tag{8.37}$$

Indeed, if $|A| < 1$, then from (8.36) we can write

$$\begin{aligned} g_t(\lambda) &= e^{-it\lambda}h_t(\lambda) \\ &= e^{-it\lambda}\left(\sum_{j=0}^{\infty} A_j(t)\sigma(t-j)e^{-ij\lambda}\right)e^{i(t-j)\lambda} \\ &= \sum_{j=0}^{\infty} A_j(t)\sigma(t-j)e^{-ij\lambda} \\ &= \sum_{N=0}^{\infty} A^N e^{-iNT\lambda}\sum_{k=0}^{T-1} A_k(t)\sigma(t-k)e^{-ik\lambda} \\ &= \left[1 - Ae^{-iT\lambda}\right]^{-1}\left(\sum_{k=0}^{T-1} A_k(t)\sigma(t-k)e^{-ik\lambda}\right). \end{aligned}$$

We identify

$$G_t(\lambda) = \left(\sum_{k=0}^{T-1} A_k(t)\sigma(t-k)e^{-ik\lambda}\right)$$

and note that $G_t(\lambda)$ is the source of the periodicity $g_t(\lambda) = g_{t+T}(\lambda)$. Also, since rank deficiency of X_t corresponds to $\sigma(t) = 0$ for some values of t, then $G_t(\lambda)$ also carries the rank information. If $A_k(t) \neq 0, k = 0, 1, \ldots, T-1$ for some t, then rank deficiency means that, for at least one value of k, a term $e^{-ik\lambda}$ does not appear in the Fourier series of $G_t(\lambda)$.

Recall from (1.17) that the spectral measures F_k can be defined by

$$\int_0^{2\pi} e^{ip\lambda}F_k(d\lambda) = \frac{1}{T}\sum_{t=0}^{T-1} e^{(2\pi itk/T)}(X_{t+p}, X_t), \tag{8.38}$$

from which we obtain the following.

Proposition 8.5 *Let $(\phi(t))$ be a T-periodic sequence of nonzero complex numbers with $|A| = |\phi(1)\cdots\phi(T)| < 1$. Let (F_k), $k = 0, \ldots, T-1$ be the spectrum of the (PC) solution to the system (8.24). Then the measures γ_k are absolutely continuous with respect to the normalized Lebesgue measure $d\lambda$ and*

$$\frac{dF_k}{d\lambda}(\lambda) = \sum_{l=0}^{T-1} |1 - Ae^{iT(\lambda)}|^{-2}\hat{G}_l(\lambda + 2\pi l/T)\overline{\hat{G}_{l-k}(\lambda + 2\pi l/T)}, \tag{8.39}$$

where $G_t(\lambda) = \sum_{k=0}^{T-1} A_k(t)\sigma(t-k)e^{-ik\lambda}$ and $\hat{G}_j(\lambda) = \frac{1}{T}\sum_{t=0}^{T-1} G_t(\lambda)e^{2\pi ijt/T}$, $j \in \mathbb{Z}$.

Proof. Let $g_t(\lambda)$ be given by (8.37) and from the periodicity, $g_t(\lambda) = g_{t+T}(\lambda)$ we can write (dropping the λ) $g_t = \sum_{j=0}^{T-1} e^{-2\pi ij/T}\hat{g}_j$,

$$\begin{aligned}\frac{1}{T}\sum_{n=0}^{T-1} e^{2\pi itk/T} g_{t+p}\overline{g_t} &= \frac{1}{T}\sum_{j=0}^{T-1}\sum_{l=0}^{T-1} e^{-2\pi ipl/T}\hat{g}_l\overline{\hat{g}_j}\left(\sum_{t=0}^{T-1} e^{2\pi it(j-l+k)/T}\right)\\ &= \sum_{l=0}^{T-1} e^{-2\pi ipl/T}\hat{g}_l\overline{\hat{g}_{l-k}}.\end{aligned}$$

Hence, in view of (8.38) and the equivalence of (X_t) and $(e^{it\cdot}g_t)$,

$$\begin{aligned}\int_0^{2\pi} e^{ip\lambda}\gamma_k(d\lambda) &= \frac{1}{T}\sum_{t=0}^{T-1} e^{2\pi itk/T}\int_0^{2\pi} e^{i(t+p)\lambda}e^{-it\lambda}g_{t+p}(\lambda)\overline{g_t(\lambda)}d\lambda\\ &= \int_0^{2\pi} e^{ip\lambda}\sum_{l=0}^{T-1} e^{-2\pi ipl/T}\hat{g}_l(\lambda)\overline{\hat{g}_{l-k}(\lambda)}d\lambda\\ &= \sum_{l=0}^{T-1}\int_0^{2\pi} e^{ip(\lambda-2\pi l/T)}\hat{g}_l(\lambda)\overline{\hat{g}_{l-k}(\lambda)}d\lambda\\ &= \int_0^{2\pi} e^{ip\lambda}\left(\sum_{l=0}^{T-1}\hat{g}_l(\lambda+2\pi l/T)\overline{\hat{g}_{l-k}(\lambda+2\pi l/T)}\right)d\lambda.\end{aligned}$$

Since $\hat{g}_t(\lambda+2\pi l/T) = [1-Pe^{-iT(\lambda)}]^{-1}\hat{G}_t(\lambda+2\pi l/T)$ and the Fourier coefficients determine a measure, the proposition is proved. ∎

8.4.1.2 The Method of Lifting Although the following is easily proved for PARMA(p, q) systems, this generality is not needed here. We omit the proof.

Proposition 8.6 *The univariate $PAR(1)$ system (8.24) can be expressed as a T-variate $VAR(1)$*

$$\mathbf{X}_n = \mathbf{\Phi}_0^{-1}\mathbf{\Phi}_1\mathbf{X}_{n-1} + \mathbf{\Phi}_0^{-1}\mathbf{\Theta}_0\mathbf{\Xi}_n, \tag{8.40}$$

where

$$\mathbf{X}_n = \begin{bmatrix} X_{nT} \\ X_{nT-1} \\ \vdots \\ X_{nT-T+1}\end{bmatrix}, \tag{8.41}$$

$$\mathbf{\Xi}_n = \begin{bmatrix} \xi_{nT} \\ \xi_{nT-1} \\ \vdots \\ \xi_{nT-T+1}\end{bmatrix}, \tag{8.42}$$

with $Cov(\Xi_m, \Xi_n) = \delta_{m-n} I_T$,

$$\mathbf{\Phi}_0 = \begin{bmatrix} 1 & -\phi(T) & \dots & \dots & \dots \\ 0 & 1 & -\phi(T-1) & \dots & \dots \\ 0 & 0 & 1 & \phi(T-2) & \dots \\ \vdots & \vdots & \vdots & \vdots & \vdots \\ 0 & 0 & 0 & \dots & 1 \end{bmatrix}, \tag{8.43}$$

$$\mathbf{\Phi}_1 = \begin{bmatrix} 0 & 0 & \dots & 0 \\ 0 & 0 & \dots & 0 \\ 0 & 0 & \dots & 0 \\ \vdots & \vdots & \vdots & \vdots \\ \phi(1) & \dots & \dots & \dots \end{bmatrix}. \tag{8.44}$$

By premultiplying each side of (8.40) by the invertible transformation $\mathbf{\Phi}_0^{-1}$ we can obtain the usual form

$$\mathbf{\Phi}(B)\mathbf{X}_n = \mathbf{\Phi}_0^{-1}\mathbf{\Theta}_0\Xi_n,$$

where $\mathbf{\Phi}(z) = \mathbf{I_T} - \mathbf{\Phi_0^{-1}\Phi_1} z$. Note the elements of $\mathbf{X}_n$ are ordered from later to earlier and do not overlap with the elements of $\mathbf{X}_{n-1}$. If the elements of $\mathbf{X}_n$ are ordered from earlier to later, the matrix $\mathbf{\Phi}_0$ will be lower triangular rather than upper triangular.

Using well known results [28, Theorem 11.3.1] from multivariate sequences, if det $\mathbf{\Phi}(z) \neq 0$ for $|z| \leq 1$, then $\mathbf{X}_n$ is causal in the sense that

$$\mathbf{X}_n = \sum_{j=0}^{\infty} \mathbf{\Psi}_j \Xi_{n-j}, \tag{8.45}$$

where $\mathbf{\Psi}(z) = \mathbf{\Phi}^{-1}(z)\mathbf{\Phi}_0^{-1}\mathbf{\Theta}_0$ with $\sum_{j=0}^{\infty} \|\mathbf{\Psi}_j\| < \infty$. Then the spectral density is of the form (4.82),

$$\mathbf{f}(\lambda) = \mathbf{\Phi}(e^{-i\lambda})^{-1}\mathbf{\Phi}_0^{-1}\mathbf{\Theta}_0\mathbf{\Theta}_0^*[\mathbf{\Phi}_0^{-1}]^*[\mathbf{\Phi}(e^{-i\lambda})^{-1}]^*, \tag{8.46}$$

which exists and is continuous (hence bounded) for all $\lambda \in [0, 2\pi)$. Since by assumption $\mathbf{\Phi}(e^{-i\lambda})$ is invertible for all $\lambda \in [0, 2\pi)$, and $\mathbf{\Phi}_0$ is always invertible, then clearly rank $\mathbf{f}(\lambda)$ = rank $\mathbf{\Theta}_0 = r$. As long as the condition for causality is met, the rank of $\mathbf{f}(\lambda)$, which is the (innovation) rank of the sequence, is governed by $\mathbf{\Theta}_0$.

An expression of the form (8.46) was introduced for PARMA sequences by Sakai [205].

8.5 LEAST MEAN-SQUARE PREDICTION

The problem of linear mean-square prediction or interpolation addresses the formation of a least mean-square estimator $\widehat{X}_t$ for $t \in S_p$ (the indices of prediction) based on observed values of X_s for $s \in S_o$ (the indices of observations). Here the set S_o can be an infinite past, $S_o = \{s \in \mathbb{Z} : s \le t\}$, or a finite past $S_o = \{s \in \mathbb{Z} : t - N \le s \le t\}$. Sometimes the observations are also called predictors, but $\widehat{X}_t$ is always an optimal predictor comprised from the predictors in S_o.

8.5.1 Prediction Based on Infinite Past

Here we will consider only the case $S_p = \{t + \delta\}$; that is, we predict $X_{t+\delta}$. And we will focus here on regular PC-T sequences because the prediction error for the singular component is zero; the prediction error comes only from the regular component.

As discussed in Chapter 4, the optimal least squares predictor $\widehat{X}_{t+\delta}$ is that vector in the subspace of observations $\mathcal{H}_X(t)$ that minimizes

$$\|\widehat{X}_{t+\delta} - X_{t+\delta}\|;$$

the solution (the sought vector) is the orthogonal projection $(X_{t+\delta}|\mathcal{H}_X(t))$ of $X_{t+\delta}$ onto $\mathcal{H}_X(t)$. Now since X_t is assumed regular so it has representation (8.15), then

$$\begin{aligned} X_{t+\delta} &= \sum_{j\ge 0\,:t+\delta-j\in D^+} a_j(t+\delta)\xi_{t+\delta-j} \\ &= \sum_{j=0\,:t+\delta-j\in D^+}^{\delta-1} a_j(t+\delta)\xi_{t+\delta-j} + \sum_{j=\delta\,:t+\delta-j\in D^+}^{\infty} a_j(t+\delta)\xi_{t+\delta-j} \end{aligned} \tag{8.47}$$

makes it easy to see that

$$\widehat{X}_{t+\delta} = (X_{t+\delta}|\mathcal{H}_X(t)) = \sum_{j=\delta\,:t+\delta-j\in D^+}^{\infty} a_j(t+\delta)\xi_{t+\delta-j} \tag{8.48}$$

because of the orthogonality of the ξ_t. The prediction error

$$e_{t+\delta} = X_{t+\delta} - (X_{t+\delta}|\mathcal{H}_X(t))$$

has the variance

$$\sigma^2_{\widehat{X},t}(t+\delta) = \|e_{t+\delta}\|^2 = \sum_{j=0\,:t+\delta-j\in D^+}^{\delta-1} |a_j(t+\delta)|^2.$$

Since we also have

$$\|X_{t+\delta}\|^2 = \sum_{j\geq 0\,:t+\delta-j\in D^+} |a_j(t+\delta)|^2,$$

it follows that if $|a_j(t)|$ decreases slowly with respect to j, then $\|e_{t+\delta}\|^2$ will be a small fraction of $\|X_{t+\delta}\|^2$. If $|a_j(t)|$ decreases quickly with respect to j, $\|e_{t+\delta}\|^2$ will be a larger fraction of $\|X_{t+\delta}\|^2$. To get a real example we examine the PAR(1) PC sequences.

8.5.2 Prediction for a PAR(1) Sequence

Now we can address the linear prediction of a PAR(1) that is PC. We wish to express $P_{\mathcal{H}_X(t)}X(t+\delta)$ explicitly in terms of the coefficients $\phi(t)$ and $\sigma(t)$.

Proposition 8.7 *If X_t is a PC solution to (8.24) where X_t is causal (or $|A|<1$), then using the notation of Corollary 8.4.1,*

$$\widehat{X}_{t+\delta} = (X_{t+\delta}|\mathcal{H}_X(t)) = A^p A_r(t+\delta)X_t, \tag{8.49}$$

where $\delta = pT + r$ for $0\leq r<T$.

Proof. First we observe that

$$\widehat{X}_{t+\delta} = (X_{t+\delta}|\mathcal{H}_X(t)) = (X_{t+\delta}|\mathcal{H}'_\xi(t)),$$

where $\mathcal{H}'_\xi(t) = \overline{\text{sp}}\{\xi_s : s\leq t \text{ and } s\in D^+\}$. Here, D^+ turns out to be the set $\{t\in\mathbb{Z} : \sigma(t)\neq 0\}$.

The desired projection can be written explicitly from the infinite moving average representation (8.29)

$$\begin{aligned}
\widehat{X}_{t+\delta} &= \sum_{j=\delta:t+\delta-j\in D^+}^{\infty} A_j(t+\delta)\sigma(t+\delta-j)\xi_{t+\delta-j} \\
&= \sum_{k=0:t-k\in D^+}^{\infty} A_{k+\delta}(t+\delta)\sigma(t-k)\xi_{t-k} \\
&= A^p A_r(t+\delta)\sum_{k=0}^{\infty} A_k(t)\sigma(t-k)\xi_{t-k} = A^p A_r(t+\delta)X_t,
\end{aligned} \tag{8.50}$$

where the third line follows from the definition of $A_j(t)$ and the relation

$$A_{j+\delta}(t+\delta) = A_\delta(t+\delta)A_j(t),$$

which can be derived from (8.25). This finally yields $A_\delta(t+\delta) = A^p A_r(t+\delta)$, where $\delta = pT + r$ with $0\leq r<T$. ∎

It is now quite easy to obtain the prediction error.

Corollary 8.7.1 *Under the conditions of Proposition 8.7*

$$\sigma^2_{\widehat{X},t}(t+\delta) = \|X_{t+\delta} - \widehat{X}_{t+\delta}\|^2 = \sum_{j=0}^{\delta-1} |a_j(t+\delta)|^2 = \sum_{j=0}^{\delta-1} [A^{p(j)}A_{r(j)}(t)]^2 \sigma^2(t-r) \tag{8.51}$$

where $j = p(j)T + r(j)$, $0 \le r(j) < T$.

In the case of predicting to $t+1$ ($\delta = 1$) based on $\mathcal{H}_X(t)$, the *prediction error* is zero if $\sigma(t) = 0$, and on the other hand, the *prediction* is 0 if $A^0A_1(t+1) = \phi(t+1) = 0$, and then the prediction error has variance

$$\sigma^2_{\widehat{X},t}(t+\delta) = \|X_{t+1} - \widehat{X}_{t+1}\|^2 = \|X_{t+1}\|^2.$$

The prediction error can be zero for other t, δ provided $\delta < T-1$ because we assume that $\sigma(t) > 0$ for at least one t in $\{0, 1, \ldots, T-1\}$. Aside from this constraint for general δ, any occurrences of $\sigma^2(t-r) = 0$ clearly diminish the prediction error. On the other hand, the prediction will be 0 whenever there is a $s \in \{t, t+1, \ldots t+\delta\}$ with $\phi(s) = 0$ because then $A^pA_r(t+\delta) = 0$. This of course again leads to

$$\sigma^2_{\widehat{X},t}(t+\delta) = \|X_{t+\delta} - \widehat{X}_{t+\delta}\|^2 = \|X_{t+\delta}\|^2.$$

As in the stationary case, the general solution to the prediction of PC sequences based on the past infinite sequence of observations is considered beyond our current goals. But the path to it, via spectral theory for multivariate sequences, seems clear. From a practical viewpoint, its need for practical computing has been diminished due to ever-increasing computing capacity that makes finite past computation practical for very large finite pasts. Thus it becomes very important to solve the problem of prediction based on finite pasts.

8.5.3 Finite Past Prediction

The problem addressed here is the prediction of $X(t+\delta)$ based on the n observations $\{X_{t-n+1}, \ldots, X_t\}$. The best linear predictor is, of course, the orthogonal projection of $X_{t+\delta}$ onto $\mathcal{M}(t;n) = \text{sp}\{X_s, t-n+1 \le s \le t\})$ and we will thus denote

$$\widehat{X}_{t+\delta,t;n} = (X_{t+\delta}|\mathcal{M}(t;n)). \tag{8.52}$$

We shall now consider only real sequences and $\delta = 1$, and seek the coefficients in the linear expression

$$\widehat{X}_{t+1,t;n} = \sum_{j=1}^{n} \alpha_{nj}^{(t+1)} X_{t-j+1}. \tag{8.53}$$

The normal equations arising from the properties of projection are

$$E\{[X_{t+1} - \widehat{X}_{t+1,t;n}]X_s\} = 0, \quad s = t-n+1, \dots, t,$$

or

$$\sum_{j=1}^{n} \alpha_{nj}^{(t+1)} R(t-j+1, s) = R(t+1, s), \quad s = t-n+1, \dots, t, \tag{8.54}$$

which can be expressed in matrix form as

$$\begin{bmatrix} R_{t+1,t} \\ \vdots \\ R_{t+1,t-n+1} \end{bmatrix} = \begin{bmatrix} R_{t,t} & \cdots & R_{t-n+1,t} \\ R_{t,t-1} & \cdots & R_{t-n+1,t-1} \\ \vdots & \vdots & \vdots \\ R_{t,t-n+1} & \cdots & R_{t-n+1,t-n+1} \end{bmatrix} \begin{bmatrix} \alpha_{n1}^{(t+1)} \\ \alpha_{n2}^{(t+1)} \\ \vdots \\ \alpha_{nn}^{(t+1)} \end{bmatrix}, \tag{8.55}$$

where we use $R_{u,v}$ for $R(u, v)$ as needed. In a shorter notation (8.55) becomes

$$\mathbf{r}_{t+1,t:t-n+1} = \mathbf{R}(t, n)\boldsymbol{\alpha}_n^{(t+1)}. \tag{8.56}$$

For any $\boldsymbol{\alpha}_n^{(t+1)} = [\alpha_{n1}^{(t+1)} \alpha_{n2}^{(t+1)} \dots \alpha_{nn}^{(t+1)}]'$ that solves (8.55), the prediction error

$$\epsilon_{t+1,n} = X_{t+1} - \widehat{X}_{t+1,t;n} \tag{8.57}$$

has variance

$$\begin{aligned} \sigma_n^2(t+1) &= \|X_{t+1} - \widehat{X}_{t+1,t;n}\|^2 \\ &= E\{[X_{t+1} - \widehat{X}_{t+1,t;n}]X_{t+1}\} \\ &= R(t+1, t+1) - \sum_{j=1}^{n} \alpha_{nj}^{(t+1)} R(t+1, t-j+1) \\ &= R(t+1, t+1) - \mathbf{r}'_{t+1,t:t-n+1}\boldsymbol{\alpha}_n^{(t+1)}. \end{aligned} \tag{8.58}$$

In the stationary case the prediction of X_{t-n} based on $\mathcal{M}(t; n) = \text{sp}\{X_s : t-n+1 \le s \le t\}$ is the same problem as prediction of X_{t+1} due to the symmetry arising from stationarity (Section 4.2.5). But in the nonstationary case it must be addressed separately. But still from the arguments above, the coefficients $\beta_{nj}^{(t-n)}$ in the best linear estimator

$$\widehat{X}_{t-n,t;n} = \sum_{j=1}^{n} \beta_{nj}^{(t-n)} X_{t-j+1} \tag{8.59}$$

are determined by

$$\mathbf{r}_{t-n,t:t-n+1} = \mathbf{R}(t, n)\boldsymbol{\beta}_n^{(t-n)} \tag{8.60}$$

and the prediction error

$$\epsilon_{t-n,n} = X_{t-n} - \widehat{X}_{t-n,t;n} \tag{8.61}$$

has variance

$$\begin{aligned}
\sigma_n^2(t-n) &= \|X_{t-n} - (X_{t-n}|\mathcal{M}(t;n))\|^2 \\
&= E\{[X_{t-n} - \widehat{X}_{t-n,t;n}]X_{t-n}\} \\
&= R(t-n,t-n) - \sum_{j=1}^{n} \beta_{nj}^{(t-n)} R(t-n,t-j+1) \\
&= R(t-n,t-n) - \mathbf{r}'_{t-n,t:t-n+1}\boldsymbol{\beta}_n^{(t-n)}. \qquad (8.62)
\end{aligned}$$

Now we present for PC sequences, some important properties of $\widehat{X}_{t+1,t;n}$ and $\widehat{X}_{t-n,t;n}$ corresponding to Propositions 4.5 and 4.15,and again in a manner motivated by Pourahmadi [183, Chapter 7].

Proposition 8.8 *If X_t is PC-T and $\widehat{X}_{t+1,t;n}$ and $\widehat{X}_{t-n,t;n}$ are the best linear predictors of X_{t+1} and X_{t-n} based on $\{X_{t-n+1},\dots,X_t\}$, then*

(a) *if $\boldsymbol{\alpha}_n^{(t+1)}$ solves (8.55) and $\boldsymbol{\beta}_n^{(t-n)}$ solves (8.60), then irrespective of invertibility of $\mathbf{R}(t,n)$, they are also solutions when t is replaced with $t+T$;*

(b) *for n fixed, irrespective of the invertibility of $\mathbf{R}(t,n)$,*

$$\sigma_n^2(t+1) = \sigma_n^2(t+T+1) \quad \textit{and} \quad \sigma_n^2(t-n) = \sigma_n^2(t-n+T); \tag{8.63}$$

(c) *for t fixed, $\sigma_n^2(t+1)$ is bounded and nonincreasing with respect to n, and $\sigma_n^2(t+1) \to \sigma^2_{\widehat{X},t}(t+1)$, which is 0 if $\{X_t\}$ is deterministic at $t+1$;*

(d) *$\mathbf{R}(t+1,n+1)$ is invertible if and only if $\sigma_n^2(t+1) > 0$ and $\mathbf{R}(t,n)$ is invertible; whenever $\mathbf{R}(t,n)$ is invertible*

$$\begin{aligned}
\sigma_n^2(t+1) &= R(t+1,t+1) - \mathbf{r}'_{t+1,t:t-n+1}\mathbf{R}(t,n)^{-1}\mathbf{r}_{t+1,t:t-n+1} \\
&= \frac{|\mathbf{R}(t+1,n+1)|}{|\mathbf{R}(t,n)|}. \qquad (8.64)
\end{aligned}$$

If $\sigma_n^2(t+1) > 0$, then $\operatorname{rank}\mathbf{R}(t+1,n+1) = \operatorname{rank}\mathbf{R}(t,n) + 1$.

(e) *$\mathbf{R}(t,n+1)$ is invertible if and only if $\sigma_n^2(t-n) > 0$ and $\mathbf{R}(t,n)$ is invertible; whenever $\mathbf{R}(t,n)$ is invertible*

$$\begin{aligned}
\sigma_n^2(t-n) &= R(t-n,t-n) - \mathbf{r}'_{t-n,t:t-n+1}\mathbf{R}(t,n)^{-1}\mathbf{r}_{t-n,t:t-n+1} \\
&= \frac{|\mathbf{R}(t,n+1)|}{|\mathbf{R}(t,n)|}. \qquad (8.65)
\end{aligned}$$

If $\sigma_n^2(t-n) > 0$, *then* $\text{rank}\,\mathbf{R}(t,n+1) = \text{rank}\,\mathbf{R}(t,n) + 1$.

(f) *if* X_t *is nondeterministic at* t_0+1, *then* $\sigma_n^2(t_0+1) > 0$ *for all* $n \geq 1$; *if in addition,* $|\mathbf{R}(t_0,n)| \neq 0$ *for* $n \geq 1$, *then*

$$\sigma^2_{\widehat{X},t}(t+1) = \exp\left(\lim \frac{1}{n} \log |\mathbf{R}(t,n)|\right) > 0.$$

Proof. For (a), the vectors $\mathbf{r}_{t+1,t:t-n+1}, \mathbf{r}_{t-n,t:t-n+1}$ and matrix $\mathbf{R}(t,n)$ in (8.55) and (8.60) are invariant when t is replaced with $t+T$.

For (b), the result is clear due to (a) if $\boldsymbol{\alpha}_n^{(t+1)}$ is unique. But even if not a unique solution of (8.55), $\boldsymbol{\alpha}_n^{(t+1)}$ still represents the projection $P_{\mathcal{M}(t;n)}X_{t+1}$ because it solves the normal equations (8.54) and the claim follows by using the top line of (8.58). The same argument gives the result for $\sigma_n^2(t-n)$.

For (c), $\sigma_n^2(t+1)$ is bounded and nonincreasing because of the top line of (8.58); the limit argument follows first because $\lim_{n\to\infty}\mathcal{M}(t;n) = \text{sp}\{X_s, s \leq t\}$ and then because the predictor $\widehat{X}_{t+1}$ that achieves error variance $\sigma^2_{\widehat{X},t}(t+1)$ can be approximated arbitrarily closely by elements of $\text{sp}\{X_s, s \leq t\}$.

For (d), note that

$$\mathbf{R}(t+1,n+1) = \begin{bmatrix} R_{t+1,t+1} & \mathbf{r}'_{t+1,t:t-n+1} \\ \mathbf{r}_{t+1,t:t-n+1} & \mathbf{R}(t,n) \end{bmatrix}, \tag{8.66}$$

from which it follows (see [183, Chapter 7] or [148, Appendix A]) that

$$\begin{aligned} |\mathbf{R}(t+1,n+1)| &= [R_{t+1,t+1} - \mathbf{r}'_{t+1,t:t-n+1}\mathbf{R}(t,n)^{-1}\mathbf{r}_{t+1,t:t-n+1}] \times |\mathbf{R}(t,n)| \\ &= \sigma_n^2(t+1)|\mathbf{R}(t,n)|. \end{aligned} \tag{8.67}$$

First, (8.67) implies we can write (8.64) when $|\mathbf{R}(t,n)| \neq 0$. To finish the proof, note that $|\mathbf{R}(t+1,n+1)| \neq 0$ if and only if the random variables $\{X_{t+1},\ldots,X_{t-n+1}\}$ are LI, and this occurs if and only if $\sigma_n^2(t+1) \neq 0$. From (8.67), $|\mathbf{R}(t+1,n+1)| \neq 0$ then implies $|\mathbf{R}(t,n)| \neq 0$.

The proof of (e) follows in the same manner by using

$$\mathbf{R}(t,n+1) = \begin{bmatrix} \mathbf{R}(t,n) & \mathbf{r}_{t-n,t:t-n+1} \\ \mathbf{r}'_{t-n,t:t-n+1} & R_{t-n,t-n} \end{bmatrix} \tag{8.68}$$

and

$$\begin{aligned} |\mathbf{R}(t,n+1)| &= [R_{t-n,t-n} - \mathbf{r}'_{t-n,t:t-n+1}\mathbf{R}(t,n)^{-1}\mathbf{r}_{t-n,t:t-n+1}] \times |\mathbf{R}(t,n)| \\ &= \sigma_n^2(t-n)|\mathbf{R}(t,n)|. \end{aligned} \tag{8.69}$$

For (f), if X_t is nondeterministic at t_0+1, then $X_{t_0+1} \notin \mathcal{H}_X(t_0)$ and so certainly $\sigma_n^2(t+1) = \|X_{t+1} - P_{\mathcal{M}(t;n)}X_{t+1}\|^2 > 0$ for all $n \geq 1$. If $|\mathbf{R}(t_0,n)| \neq 0$

for $n \geq 1$, then $\sigma_n^2(t+1) > 0$ implies $|\mathbf{R}(t_0+1, n+1)| \neq 0$ for all $n \geq 1$. Since $\sigma_n^2(t+1) \to \sigma_{\widehat{X},t}^2(t+1)$ we finally obtain

$$\begin{aligned}
\log \sigma_{\widehat{X},t}^2(t+1) &= \lim_{n\to\infty} \frac{1}{n} \sum_{k=1}^{n} \sigma_n(t+1) \\
&= \lim_{n\to\infty} \frac{1}{n} \sum_{k=1}^{n} \log \frac{|\mathbf{R}(t_0+1, k+1)|}{|\mathbf{R}(t_0, k)|} \\
&= \lim_{n\to\infty} \log |\mathbf{R}(t_0, n)|.
\end{aligned}$$

This completes the proof. ∎

8.5.3.1 Partial Autocorrelations For a second order random sequence X_t, the partial autocorrelation is defined, for $n \geq 1$, as

$$\begin{aligned}
\pi(t, n+1) &= \text{Corr}\,\{[X_{t+1} - \widehat{X}_{t+1,t;n}][X_{t-n} - \widehat{X}_{t-n,t;n}]\} \\
&= \frac{E\{\epsilon_{t+1,n}\epsilon_{t-n,n}\}}{\sigma_n(t+1)\sigma_n(t-n)},
\end{aligned} \tag{8.70}$$

which gives the immediate interpretation that $\pi(t, n+1)$ is the correlation of the prediction errors $\epsilon_{t+1,n}$ with $\epsilon_{t-n,n}$. Another interpretation is that $\pi(t, n+1)$ is the correlation between X_{t+1} and X_{t-n} when the effects on the variables $\{X_{t-n+1}, \ldots X_t\}$ are removed. Note that when $n = 0$, we obtain $\pi(t, 1) = \text{Corr}\{X_{t+1}X_t\}$, there are none between. In the nonstationary case, it is possible for either or both $\epsilon_{t+1,n-1}$ and $\epsilon_{t-n+1,n-1}$ to be zero, and this may also be true of the random variables X_{t+1} and X_{t-n} and of the predictors $\widehat{X}_{t+1,n-1}$ and $\widehat{X}_{t-n+1,n-1}$. Since $\pi(t, n+1)$ is defined as a correlation, we define it to be zero when either of the random variables $\epsilon_{t+1,n}, \epsilon_{t-n,n}$ is zero. Subsequently we will give some examples of simple PC sequences that exhibit some of these situations.

For nonstationary processes we hardly expect $\pi(t, n+1)$ to be constant with respect to t for fixed n, as in the stationary case.

Lemma 8.3 *If X_t is a PC-T process, then*

$$\pi(t, n+1) = \pi(t+T, n+1) \text{ for all } n \geq 1.$$

Proof. Using the periodicity $\boldsymbol{\alpha}_{n-1}^{(t+1)} = \boldsymbol{\alpha}_{n-1}^{(t+T+1)}$ (similarly for $\boldsymbol{\beta}_{n-1}^{(t-n)}$ and $\sigma_{n-1}(t+1)$, see Proposition 8.8) and $R(s,t) = R(s+T, t+T)$, the periodicity of $\pi(t, n+1)$ follows from

$$
\begin{aligned}
&E\{\epsilon_{t+1,n}\epsilon_{t-n,n}\} && (8.71)\\
&= E\Big\{\Big[X_{t+1} - \sum_{j=1}^{n} \alpha_{nj}^{(t+1)} X_{t-j+1}\Big] \times \Big[X_{t-n} - \sum_{k=1}^{n} \beta_{nk}^{(t-n)} X_{t-k+1}\Big]\Big\}\\
&= R(t+1, t-n) - \sum_{j=1}^{n} \alpha_{nj}^{(t+1)} R(t-j+1, t-n)\\
&\quad - \sum_{k=1}^{n} \beta_{nk}^{(t-n)} R(t+1, t-k+1)\\
&\quad + \sum_{j=1}^{n}\sum_{k=1}^{n} \alpha_{nj}^{(t+1)} \beta_{nk}^{(t-n)} R(t-j+1, t-k+1). \quad \blacksquare && (8.72)
\end{aligned}
$$

(8.73)

Remark. This result has obvious implications on the estimation of $\pi(t, n+1)$ for PC sequences.

Again we note that the vectors $\boldsymbol{\alpha}_n^{(t+1)}$ and $\boldsymbol{\beta}_n^{(t-n)}$ need not be unique solutions to the forward and backward Yule–Walker equations because as long as they are solutions they represent the projections. The expression for $\pi(t, n+1)$ can be shortened, since for each k in the last line of (8.73),

$$
\sum_{j=1}^{n-1} \alpha_{n(n-1)}^{(t+1)} R(t-j+1, t-k+1) = R(t+1, t-k+1),
$$

causing the cancellation of the last two lines and producing

$$
\pi(t, n+1) = \frac{R(t+1, t-n) - \mathbf{r}'_{t-n,t:t-n+1} \boldsymbol{\alpha}_n^{(t+1)}}{\sigma_n(t+1)\sigma_n(t-n)}. \tag{8.74}
$$

If in the last line of (8.73) we sum first on k, then we additionally obtain

$$
\pi(t, n+1) = \frac{R(t+1, t-n) - \mathbf{r}'_{t+1,t:t-n+1} \boldsymbol{\beta}_n^{(t-n)}}{\sigma_n(t+1)\sigma_n(t-n)}. \tag{8.75}
$$

Next, we give some simple PC sequences that demonstrate the t dependence of the various quantities comprising $\pi(t, n)$ in (8.70).

For the first simple example, consider again the sequence X_t given by doubling Y_t,

$$
\begin{array}{ccccccccc}
\cdots & Y_{-1} & Y_{-1} & Y_0 & Y_0 & Y_1 & Y_1 & \cdots & \\
\updownarrow & \updownarrow & \updownarrow & \updownarrow & \updownarrow & \updownarrow & \updownarrow & \updownarrow & , \\
\cdots & X_{-2} & X_{-1} & X_0 & X_1 & X_2 & X_3 & \cdots &
\end{array}
$$

where Y_t is an orthonormal sequence. Then using (8.5) and $\|Y_t\| = 1$, we obtain ($n = 0$)

$$\pi(t,1) = E\{X_{t+1}X_t\} = \begin{cases} E\{Y_{t/2}Y_{t/2}\} = 1 & t \text{ even} \\ E\{Y_{(t+1)/2}Y_{(t-1)/2}\} = 0 & t \text{ odd} \end{cases}.$$

Next, $\pi(t,2) = 0$ because

$$\begin{aligned} \epsilon_{t-1,1} &= X_{t-1} - \widehat{X}_{t-1,1} = 0 \quad t \text{ odd}, \\ \epsilon_{t+1,1} &= X_{t+1} - \widehat{X}_{t+1,1} = 0 \quad t \text{ even}, \end{aligned}$$

where these follow from $X_{t-1} = X_t$ for t odd and $X_{t+1} = X_t$ for t even. From here it is easy to see that for $n \geq 2$, $\pi(t, n+1) = 0$ because $\epsilon_{t+1,n-1}$ is always 0 or X_{t+1} and in the latter case, $X_{t+1} \perp X_{t-n+1}$.

If X_t is a causal PAR(1) sequence (8.24), then $\pi(t,1) = E\{X_{t+1}X_t\} = \phi(t+1)\sigma_X^2(t)$, which shows that $\pi(t,1)$ can be zero. Note that in order for $\sigma_X^2(t+1) = |\phi(t+1)|^2\sigma_X^2(t) + \sigma^2(t+1) = 0$, it is necessary for either $\phi(t+1) = 0$ or $\sigma_X^2(t) = 0$. Either of these produces $\pi(t,1) = 0$. For $n \geq 2$, observe that

$$\begin{aligned} \epsilon_{t+1,n-1} &= X_{t+1} - \widehat{X}_{t+1,n-1} = \sigma(t+1)\xi_{t+1}, \\ \epsilon_{t-n+1,n-1} &= X_{t-n+1} - (X_{t-n+1}|\mathcal{M}(t;n-1)), \end{aligned}$$

so $\xi_{t+1} \perp [X_{t-n+1} - \widehat{X}_{t-n+1,n-1}]$ implies $\pi(t, n+1) = 0$.

Note that $\sigma_X(t) = 0$ means a rank deficiency occurs for any collection that includes X_t. But $\pi(t,1) = 0$ only means $E\{X_{t+1}X_t\} = 0$, which occurs, for example, if X_t is an orthogonal sequence. But for PC sequences, the above examples show the possibility of $\pi(t,1) \neq 0$ for some t while $\pi(t,1) = 0$ for other t.

8.5.3.2 Durbin–Levinson Algorithm The idea of the Durbin–Levinson algorithm is to find a computationally economical way to compute $\boldsymbol{\alpha}_{n+1}^{(t+1)}$ given the vector of predictor coefficients $\boldsymbol{\alpha}_n^{(t+1)}$ (a solution of (8.56)). We follow the general presentation of Pourahmadi [183, Chapter 7] (which is similar to that of Sharf [207]). Write the matrix equation (8.56) with $n+1$ replacing n, as

$$\begin{bmatrix} \mathbf{r}_{t+1,t:t-n+1} \\ R_{t+1,t-n} \end{bmatrix} = \begin{bmatrix} \mathbf{R}(t,n) & \mathbf{r}_{t-n,t:t-n+1} \\ \mathbf{r}'_{t-n,t:t-n+1} & R_{t-n,t-n} \end{bmatrix} \begin{bmatrix} \boldsymbol{\alpha}_u \\ \alpha_l \end{bmatrix}. \tag{8.76}$$

We seek the vector of coefficients $\boldsymbol{\alpha}_{n+1}^{(t+1)\prime} = [\boldsymbol{\alpha}'_u \alpha_l]'$. Writing the two equations separately produces

$$\begin{aligned} \mathbf{r}_{t+1,t:t-n+1} &= \mathbf{R}(t,n)\boldsymbol{\alpha}_u + \mathbf{r}_{t-n,t:t-n+1}\alpha_l, \\ R_{t+1,t-n} &= \mathbf{r}'_{t-n,t:t-n+1}\boldsymbol{\alpha}_u + \alpha_l R_{t-n,t-n}. \end{aligned} \tag{8.77}$$

Since $\mathbf{r}_{t+1,t:t-n+1} = \mathbf{R}(t,n)\boldsymbol{\alpha}_n^{(t+1)}$ it is natural to try $\boldsymbol{\alpha}_u = \boldsymbol{\alpha}_n^{(t+1)} + \mathbf{w}$, which transforms the preceding into

$$\begin{aligned} \mathbf{0} &= \mathbf{R}(t,n)\mathbf{w} + \mathbf{r}_{t-n,t:t-n+1}\alpha_l \\ R_{t+1,t-n} &= \mathbf{r}'_{t-n,t:t-n+1}(\boldsymbol{\alpha}_n^{(t+1)} + \mathbf{w}) + \alpha_l R_{t-n,t-n}. \end{aligned} \tag{8.78}$$

But the top line is solved by $\mathbf{w} = -\alpha_l \boldsymbol{\beta}_n^{(t-n)}$ and the bottom line gives

$$\alpha_l = \frac{R_{t+1,t-n} - \mathbf{r}'_{t-n,t:t-n+1}\boldsymbol{\alpha}_n^{(t+1)}}{R_{t-n,t-n} - \mathbf{r}'_{t-n,t:t-n+1}\boldsymbol{\beta}_n^{(t-n)}} = \frac{\pi(t,n+1)\sigma_n(t+1)}{\sigma_n(t-n)}. \tag{8.79}$$

So given $\boldsymbol{\alpha}_n^{(t+1)}$ and $\boldsymbol{\beta}_n^{(t-n)}$, if $\pi(t,n+1) \neq 0$ (meaning both X_{t+1} and X_{t-n} are LI of $\mathcal{M}(t,n-1)$), determine α_l from the preceding and then $\boldsymbol{\alpha}_u = \boldsymbol{\alpha}_n^{(t+1)} - \alpha_l \boldsymbol{\beta}_n^{(t-n)}$. If $\pi(t,n+1) = 0$ then (8.77) is solved by $\alpha_l = 0$ and $\boldsymbol{\alpha}_u = \boldsymbol{\alpha}_n^{(t+1)}$, which makes perfect sense because X_{t-n} does not add anything new.

For the backward coefficients, we solve for $\boldsymbol{\beta}_{n+1}^{t-n}$ in terms of $\boldsymbol{\beta}_n^{t-n}$ (predicting to the same time $t-n$ based on a longer sample into the future). Beginning with (8.60) we obtain

$$\begin{bmatrix} R_{t-n,t+1} \\ \mathbf{r}_{t-n,t:t-n+1} \end{bmatrix} = \begin{bmatrix} R_{t+1,t+1} & \mathbf{r}'_{t+1,t:t-n+1} \\ \mathbf{r}_{t+1,t:t-n+1} & \mathbf{R}(t,n) \end{bmatrix} \begin{bmatrix} \beta_u \\ \boldsymbol{\beta}_l \end{bmatrix}, \tag{8.80}$$

which leads, as above, to

$$\beta_u = \frac{R_{t-n,t+1} - \mathbf{r}'_{t+1,t:t-n+1}\boldsymbol{\beta}_n^{(t-n)}}{R_{t+1,t+1} - \mathbf{r}'_{t+1,t:t-n+1}\boldsymbol{\alpha}_n^{(t+1)}} = \frac{\pi(t,n+1)\sigma_n(t-n)}{\sigma_n(t+1)} \tag{8.81}$$

and $\boldsymbol{\beta}_{n+1}^{t-n} = [\beta_u \ \boldsymbol{\beta}_l']'$, where $\boldsymbol{\beta}_l = \boldsymbol{\beta}_n^{t-n} - \beta_u \boldsymbol{\alpha}_n^{t+1}$. Given that we wish to compute the coefficients up through some $n = n_0$, we begin with $n = 0$ and directly obtain

$$\boldsymbol{\alpha}_0^{t+1} = \{R_{t+1,t}/R_{t,t}\}, \qquad \boldsymbol{\beta}_0^{t-1} = \{R_{t-1,t}/R_{t,t}\}$$

for each $t = 0, 1, \ldots, T-1$. All the coefficients needed for $\{\boldsymbol{\alpha}_1^{t+1}, \boldsymbol{\beta}_1^{t-2} : t = 0, 1, \ldots, T-1\}$ are present in the first set $\{\boldsymbol{\alpha}_0^{t+1}, \boldsymbol{\beta}_0^{t-1} : t = 0, 1, \ldots, T-1\}$ and coefficients for any other values of t can be obtained by periodicity. The process is continued recursively to $n = n_0$.

The expressions (8.79) and (8.81) give, for PC sequences, the connection between the last regression coefficient and the partial autocorrelation (see the discussion of Durbin–Levinson algorithm in Chapter 4, or in [28, Section 3.4] for the stationary case. Another, and perhaps more practical, solution along similar lines is the innovations algorithm.

8.5.3.3 Innovations Algorithm Section 4.2.5.3 showed the close connection between the Cholesky decomposition and the innovations algorithm whenever $\mathbf{R}$ is positive definite. The treatment given there was for general positive definite $\mathbf{R}$, not just those arising from covariances of stationary sequences. However, the covariance of a nonstationary sequence need not be positive definite, although it certainly must be nonnegative definite. Here we modify the discussion of Section 4.2.5.3 to accommodate covariances $\mathbf{R}$ that are not necessarily positive definite, thus not necessarily of full rank.

Proposition 8.9 (Cholesky Decomposition for a NND matrix) *If the $n \times n$ matrix $\mathbf{R}$ is nonnegative definite and of rank r, then*

(a) *there exists a $n \times n$ lower triangular matrix $\boldsymbol{\Theta}$ of rank r for which*

$$\mathbf{R} = \boldsymbol{\Theta}\, \boldsymbol{\Theta}'; \tag{8.82}$$

(b) *there exists a lower semitriangular $n \times r$ matrix $\tilde{\boldsymbol{\Theta}}$ for which*

$$\mathbf{R} = \widetilde{\boldsymbol{\Theta}}\, \widetilde{\boldsymbol{\Theta}}'. \tag{8.83}$$

Proof. We use the same notation as in the proof of Proposition 4.6 and rely on the fact (see [56, Theorem 2]) that $\mathbf{R}$ is nonnegative definite and of rank r if and only if there exist random variables $\{X_1, X_2, \ldots, X_n\}$ of finite variance with $\mathbf{R} = \mathrm{Cov}\mathbf{X}$, where $\mathbf{X} = (X_1, X_2, \ldots, X_n)'$, and r is the maximum number of LI vectors that can be found in $\{X_1, X_2, \ldots, X_n\}$.

The proof is a simple modification of the proof of Proposition 4.6. In the Gram–Schmidt orthogonalization process, whenever we find that $X_k \in \mathcal{M}_{k-1} = \mathrm{sp}\{X_1, X_2, \ldots, X_{k-1}\}$, meaning that $Y_k = X_k - P_{\mathcal{M}_{k-1}} X_k = 0$ (the prediction error is null), we can proceed in two ways. One way gives us (a) and the other gives us (b). In the first case, we set the new basis (or innovation) vector η_k to be a dummy unit vector, call it η_k', orthogonal to $\mathcal{M}_n$ and to all previous dummy vectors. The resulting $\boldsymbol{\Theta}$ will be lower triangular and still $\mathbf{X} = \boldsymbol{\Theta}\boldsymbol{\eta}$, with $\mathbf{R} = \boldsymbol{\Theta} E\{\boldsymbol{\eta}\boldsymbol{\eta}'\}\boldsymbol{\Theta}' = \boldsymbol{\Theta}\boldsymbol{\Theta}'$, giving (8.82) as required. However, since no X_j depends on any of the $n - r$ dummy vectors, there will be $n - r$ columns of $\boldsymbol{\Theta}$ that will be zero.

Alternatively, we do not introduce a dummy η_k when $Y_k = X_k - P_{\mathcal{M}_{k-1}} X_k = 0$ and only retain the η_j required to represent $\{X_1, X_2, \ldots, X_k\}$. Hence $\boldsymbol{\eta}$ will contain only r elements when $k = n$, and the matrix of coefficients $\widetilde{\boldsymbol{\Theta}}$ will be $n \times r$. But $\widetilde{\boldsymbol{\Theta}}$ will have a semitriangular property: denoting $c_i = \max_j\{\widetilde{\theta}_{ij} \neq 0\}$, then c_i is nondecreasing and $c_n = r$. ∎

See the problems at the end of the chapter for the connection between the Cholesky decomposition and rank revealing factorizations (see Gu and Miaranian [83]).

The recursive computing of the matrices $\boldsymbol{\Theta}$ and $\widetilde{\boldsymbol{\Theta}}$ requires only a simple modification to Proposition 4.7.

Proposition 8.10 (The Innovation Algorithm for a NND Matrix) *If the $n \times n$ matrix $\mathbf{R}$ is nonnegative definite, then the lower triangular matrix $\boldsymbol{\Theta}$ in (8.82) can be computed recursively as follows. First set $\theta_{11} = [R(1,1)]^{1/2}$. The remainder of the coefficients $\theta_{k+1,j}$ are computed left to right beginning with $k = 1$ (row 2) as follows. For $j = 1, 2, \ldots, k$ set*

$$\theta_{k+1,j} = \begin{cases} \theta_{jj}^{-1}[R(k+1,j) - \sum_{m=1}^{j-1} \theta_{jm}\theta_{k+1,m}] & \text{if} \quad \theta_{jj} \neq 0 \\ 0 & \text{if} \quad \theta_{jj} = 0 \end{cases} . \tag{8.84}$$

For the diagonal term, set

$$\theta_{k+1,k+1} = \left[R(k+1,k+1) - \sum_{j=1}^{k} \theta_{k+1,j}^2 \right]^{1/2} . \tag{8.85}$$

Subsequent rows ($k = 2, \ldots, n-1$) are computed in increasing order. The matrix $\widetilde{\boldsymbol{\Theta}}$ is just $\boldsymbol{\Theta}$ with the null columns removed.

In the context of PC sequences, the innovations algorithm for full rank PC sequences was presented by Anderson, Meerschaert, and Vecchia [10] and Lund and Basawa [140].

PROBLEMS AND SUPPLEMENTS

8.1 How much of the claim of the Wold decomposition of X_t (Proposition 8.1) can be obtained without using the result that $\mathcal{H}_X(-\infty)$ is invariant under U_X? That is, for an arbitrary subspace $\mathcal{M}$ of $\mathcal{H}_X$ we can always write

$$X_t = Y_t + Z_t,$$

where $Y_t = P_{\mathcal{M}} X_t$ and $Z_t = X_t - Y_t = P_{\mathcal{M}^\perp} X_t$ and it is clear that $Y_t \perp Z_s$ for any $s, t \in \mathbb{Z}$ or equivalently

$$H_X(t) = H_Y(t) \oplus H_Z(t)$$

because if $\mathbf{x} \in H_X(t)$ then $\mathbf{x} = \lim \mathbf{x}_n$, where $\mathbf{x}_n$ is a linear combination of the vectors $\{X_s, s \leq t\}$. But then $\mathbf{x}_n = \mathbf{y}_n + \mathbf{z}_n$, where $\mathbf{y}_n = P_{\mathcal{M}}\mathbf{x}_n$ and $\mathbf{z}_n = \mathbf{x}_n - \mathbf{y}_n$, and from the continuity of projection and the closedness of $\mathcal{M}$ it follows that there is a $\mathbf{y} \in \mathcal{H}_Y(t)$ and $\mathbf{z} \in \mathcal{H}_Z(t)$ such that $\mathbf{x} = \mathbf{y} + \mathbf{z}$. It also follows immediately that $\mathcal{H}_Y \perp \mathcal{H}_Z$.

Since X_t is PC-T, the linearity of U_X permits us to write

$$X_{t+T} = U_X X_t = U_X Y_t + U_X Z_t$$

but at the same time

$$X_{t+T} = Y_{t+T} + Z_{t+T}$$

so can we say that $Y_{t+T} = U_X Y_t$ for all t? Yes, we can provided that $U_X Y_t$ never escapes from $\mathcal{M}$; that is, $\mathcal{M}$ must be closed under U_X. Thus we see how the invariance of $H_X(-\infty)$ under U_X is used.

8.2 A subspace $\mathcal{M} \subset \mathcal{H}$ is called *reducing* for operator A if $A\mathcal{M} \subset \mathcal{M}$ and $A\mathcal{M}^\perp \subset \mathcal{M}^\perp$. Show that $\mathcal{H}_X(-\infty)$ is a reducing subspace for U_X.

8.3 Suppose X_t is PC-T and there are orthogonal processes Y_t and Z_t with Y_t deterministic and Z_t purely nondeterministic such that $X_t = Y_t + Z_t$. Show that $Y_t = (X_t|\mathcal{H}_X(-\infty))$.

8.4 Show by direct calculation that if a PAR(1) solution is causal, then for any $s > t$,

$$E\{X_s X_t\} = \phi(s)\phi(s-1)\cdots\phi(t+1)\sigma^2_{X,t},$$

where $\sigma^2_{X,t} = E\{X_t X_t\}$.

8.5 What statements corresponding to Proposition 8.8, part (c), can be made about $\sigma^2_n(t-n)$ as $n \to \infty$?

8.6 Cholesky decomposition and rank revealing factorizations. The issue of a Cholesky decomposition for NND (not necessarily positive definite) matrices has been examined in the computing literature in the broader context of rank revealing factorizations. In the case of a Cholesky decomposition of a $n \times n$ NND matrix A, the resuting rank revealing decomposition produces (see Gu and Miaranian [83])

$$\Pi A \Pi' = LDL', \tag{8.86}$$

where Π is a permutation matrix,

$$L = \begin{bmatrix} A_{n'} & 0 \\ B_{n'} & I_{n-n'} \end{bmatrix} \quad D = \begin{bmatrix} I_{n'} & 0 \\ 0 & C_{n-n'} \end{bmatrix}.$$

The $n' \times n'$ matrix $A_{n'}$ is the Cholesky factor of the full rank part of A and the rest describes the rank deficient part. The essence of this factorization is easily seen in the discussion in the text.

CHAPTER 9

ESTIMATION OF MEAN AND COVARIANCE

The main topic of this chapter is the estimation of the mean

$$m_t = E\{X_t\} = m_{t+T},$$

the covariance

$$R_{t+\tau,} = E\{[X_{t+\tau} - m_{t+\tau}][X_t - m_t]\} = R_{t+T+\tau,t+T},$$

and their Fourier coefficients

$$\widetilde{m}_k = \frac{1}{T}\sum_{t=0}^{T-1} m_t e^{-i2\pi kt/T}, \quad \text{and} \quad B_k(\tau) = T^{-1}\sum_{t=0}^{T-1} R_{t+\tau,t}e^{-i2\pi kt/T}.$$

Throughout this chapter it is assumed that X_t is a real PC-T sequence. In this discussion we shall mainly treat consistency in mean square (i.e. convergence in $L^2(\Omega, \mathcal{F}, P)$), in order to show what can be expected to be true and to make the connections between the harmonizable X_t and the lifted T-variate

Periodically Correlated Random Sequences:Spectral Theory and Practice. By H.L. Hurd and A.G. Miamee

stationary sequence $\mathbf{X}_n$. Strong (almost sure) consistency and asymptotic normality will also be discussed.

Although results for all the consistency issues can be established via the lifted stationary sequence $\mathbf{X}_n$, some of the direct methods are discussed because they can also be applied to the almost periodic case, where the bijective mapping to finite dimensional vector stationary sequences is not possible [109, 135].

9.1 ESTIMATION OF m_t: THEORY

We assume a finite sample $X_{0,1}, \ldots, X_{NT-1}$ and recall that the sample periodic mean $\hat{m}_{t,N}$ introduced in Chapter 1 is

$$\widehat{m}_{t,N} = \frac{1}{N} \sum_{n=0}^{N-1} X_{t+nT}, \quad t = 0, ..., T-1, \tag{9.1}$$

from which it is clear that $\widehat{m}_{t,N}$ is unbiased, $E\{\widehat{m}_{t,N}\} = m_t$. Results on the limiting behavior of $\widehat{m}_{t,N}$ can be obtained from the stationarity of the lifted T-variate sequence $\mathbf{X}_n$, or from estimating the Fourier coefficients $\widetilde{m}_k$ of m_t:

$$\widetilde{m}_k = \frac{1}{T} \sum_{t=0}^{T-1} m_t e^{-i2\pi kt/T}. \tag{9.2}$$

We begin with consistency of $\widehat{m}_{t,N}$.

In order to discuss the mean-square consistency of $\widehat{m}_{t,N}$ from a spectral viewpoint, recall that the matrix valued cross spectral measure for the T-variate stationary sequence $\{[\mathbf{X}_n]_p = X_{nT+p}, p = 0, 1, ..., T-1, \ n \in \mathbb{Z}\}$ is denoted as $\mathbf{F}$ and its density as $\mathbf{f}$. Recall we take $\mathbf{F}$ and $\mathbf{f}$ to be associated with the *covariance* of $\mathbf{X}_n$.

Proposition 9.1 (Mean-Square Consistency of $\hat{m}_{t,N}$) *If X_t is PC-T, then*

(a) $\lim_{N\to\infty} E\{[\widehat{m}_{t,N} - m_t]^2\} = 0$ *if and only if*

$$F_{tt}(\{0\}) = 0,$$

where $E\{X_{t+jT} X_t\} = \int_0^{2\pi} e^{i\lambda j} F_{tt}(d\lambda)$.

(b) $\sum_{k=-\infty}^{\infty} |R_{t+kT,t}| < \infty$ *is sufficient for* (a) *and then*

$$\lim_{N\to\infty} NE\{[\widehat{m}_{t,N} - m_t]^2\} = f_{tt}(0), \tag{9.3}$$

where $f_{tt}(\lambda)$ is continuous.

Proof. The first claim is just a direct application of Section 4.3.3, but we indicate the proof using the notation of the PC context. Indeed, since

$$\widehat{m}_{t,N} - m_t = \frac{1}{N}\sum_{p=0}^{N-1}[X_{t+pT} - EX_t],$$

we have

$$\begin{aligned} E|\widehat{m}_{t,N} - m_t|^2 &= \frac{1}{N^2}\sum_{p=0}^{N-1}\sum_{q=0}^{N-1}\int e^{i2\pi\lambda(p-q)/N}F_{tt}(d\lambda) \\ &= \int\left|\frac{e^{i2\pi\lambda}-1}{N(e^{i2\pi\lambda/N}-1)}\right|^2 F_{tt}(d\lambda), \end{aligned}$$

which converges to $F_{tt}(\{0\})$ as $N \to \infty$.

For the second part, use $R(t+pT, t+qT) = R(t+(p-q)T, t)$ (the PC structure) to write

$$\begin{aligned} NE|\widehat{m}_{t,N} - m_t|^2 &= \frac{1}{N}\sum_{p=0}^{N-1}\sum_{q=0}^{N-1}R(t+pT, t+qT) \\ &= \frac{1}{N}\sum_{p=0}^{N-1}\sum_{q=0}^{N-1}R(t+(p-q)T, t) \\ &= \frac{1}{N}\sum_{r=-N+1}^{N-1}\sum_{q\in I(N,r)}R(t+rT, t), \quad r = p-q \\ &= \sum_{r=-N+1}^{N-1}\frac{N-r}{N}R(t+rT, t) \longrightarrow \sum_{r=-\infty}^{\infty}R(t+rT, t) = f_{tt}(0) \end{aligned}$$

as $N \to \infty$. ∎

If $\sum_{k=-\infty}^{\infty}|R_{s+kT,t}| < \infty$ for some s, t, then the same argument gives

$$NE[\widehat{m}_{s,N} - m_s][\widehat{m}_{t,N} - t_s] \to f_{st}(0).$$

The slightly weaker condition, $\sum_{k=-\infty}^{\infty}|R_{u+kT,u}| < \infty$ for $u = s, t$, gives only that for arbitrary $\epsilon > 0$,

$$\begin{aligned} &N\,|E\{[\widehat{m}_{s,N} - m_s][\widehat{m}_{t,N} - m_t]\}| \\ &\leq N\left[E\{\widehat{m}_{s,N} - m_s\}^2\right]^{1/2}\left[E\{\widehat{m}_{t,N} - m_t\}^2\right]^{1/2} \\ &\leq f_{ss}^{1/2}(0)f_{tt}^{1/2}(0) + \epsilon \end{aligned}$$

for $N > N_0$. See the problems at the end of the chapter for some related issues.

It will be useful later (in spectral estimation) to have some more conditions that imply mean-square convergence of the averages of $[X_t - m_t]e^{-i\lambda t}$. Thus denote

$$J_N(\lambda) = \frac{1}{N}\sum_{t=0}^{N-1}[X_t - m_t]e^{-i\lambda t}, \tag{9.4}$$

$$\sigma_N^2(\lambda) = E|J_N(\lambda)|^2 = \frac{1}{N^2}\sum_{t=0}^{N-1}\sum_{s=0}^{N-1} R_{t,s}e^{-i\lambda(t-s)}. \tag{9.5}$$

Lemma 9.1 *Suppose X_t is a L^2 random sequence whose covariance $R_{s,t}$ satisfies $R_{t,t} \le M$ for all t; then either of the conditions (a),(b) below is sufficient for*

$$\lim_{N\to\infty} J_N(\lambda) = 0 \quad \textit{in mean square} \tag{9.6}$$

uniformly in λ.

(a) $\lim_{\tau\to\infty} R_{t+\tau,t} = 0$ *uniformly in t;*

(b) *for X_t PC-T, $\sum_{s=0}^{\infty}\sum_{t=0}^{T-1}|R_{s,t}| < \infty$.*

Proof. Clearly, for any fixed λ, (9.6) is equivalent to $\lim_{N\to\infty}\sigma_N^2(\lambda) = 0$. To see that (a) is sufficient, for arbitrary $\epsilon > 0$ choose N_0 such that $|R_{t+\tau,t}| < \epsilon/2$ for $|\tau| > N_0$. Defining sets $A = [0, N-1] \times [0, N-1]$ and

$$B = A\bigcap\{(s,t) : |s-t| < N_0\},$$

then

$$\begin{aligned} E|J_N(\lambda)|^2 &\le \frac{1}{N^2}\sum_{s=0}^{N-1}\sum_{t=0}^{N-1}|R_{s,t}| \\ &\le \frac{1}{N^2}\sum\sum_{(s,t)\in\, A-B}|R_{s,t}| + \frac{1}{N^2}\sum\sum_{(s,t)\in\, B}|R_{s,t}| \\ &\le \frac{\epsilon}{2} + \frac{4MN_0N}{N^2} < \epsilon \end{aligned}$$

if $N > 8N_0M/\epsilon$, where $M = \max|R_{s,t}| = \max|R_{t,t}|_{t=0}^{T-1}$.

Condition (b) follows easily from

$$E\{|J_N(\lambda)|^2\} \le \frac{1+[N/T]}{N^2}\sum_{s=0}^{\infty}\sum_{t=0}^{T-1}|R_{s,t}|,$$

or Proposition 5.12. In both of these cases the uniformity with respect to λ is clear. ∎

Remark. In the stationary case, condition (a) is $\lim_{u\to\infty} R_u = 0$ and condition (b) is $\sum_0^\infty |R_u| < \infty$.

Now we examine $\widehat{\widetilde{m}}_{k,N}$, where the harmonizability of X_t helps express the result spectrally.

Proposition 9.2 *If X_t is PC-T and $\xi(\lambda)$ is the random spectral measure associated with X_t, then in the mean-square sense we have*

$$\lim_{N\to\infty} \widehat{\widetilde{m}}_{k,N} = \xi(\{2\pi k/T\}), \quad \lim_{N\to\infty} \widehat{m}_{t,N} = \sum_{k=0}^{T-1} \xi(\{2\pi k/T\}) e^{i2\pi kt/T}. \tag{9.7}$$

Proof. Note that here N refers to the number of whole periods in the sample.

$$\begin{aligned} \widehat{\widetilde{m}}_{k,N} &= \frac{1}{NT}\sum_{j=0}^{NT-1} X_j e^{-i2\pi kj/T} \\ &= \frac{1}{N}\frac{1}{T}\sum_{p=0}^{N-1}\sum_{t=0}^{T-1} X_{t+pT} e^{-i2\pi k(t+pT)} \\ &= \frac{1}{T}\sum_{t=0}^{T-1} \widehat{m}_{t,N} e^{-i2\pi kt/T}. \end{aligned} \tag{9.8}$$

The first claim follows from Proposition 5.12. The second claim follows from the inversion of (9.2),

$$\widehat{m}_{t,N} = \sum_{k=0}^{T-1} \widehat{\widetilde{m}}_{k,N} e^{i2\pi kt/T}$$

and the first claim. ∎

Proposition 9.3 (Mean-Square Consistency of $\widehat{\widetilde{m}}_{k,N}$) *If X_t is PC-T, F is its spectral measure, and F_0 is the diagonal measure (associated with $B_0(\tau)$), then*

(a) $\lim_{N\to\infty} E\{[\widehat{\widetilde{m}}_{k,N} - \widetilde{m}_k]^2\} = 0$ *if and only if*

$$F(\{2\pi k/T\}, \{2\pi k/T\}) = F_0(\{2\pi k/T\}) = |\widetilde{m}_k|^2; \tag{9.9}$$

(b) *the condition* $\sum_{u=0}^{T-1}\sum_{k=0}^{\infty}|R_{u+kT,u}| < \infty$ *is sufficient for* (a) *and then*

$$\lim_{N\to\infty} NE\{[\widehat{\widetilde{m}}_{k,N} - \widetilde{m}_k]^2 = \frac{1}{T} f_0(2\pi k/T), \tag{9.10}$$

where $f_0(\lambda)$ *is the (continuous) density defined by*

$$B_0(\tau) = \int_0^{2\pi} e^{i\lambda\tau} f_0(\lambda) d\lambda. \tag{9.11}$$

Proof. For claim (a), since

$$\widehat{\widetilde{m}}_{k,N} - \widetilde{m}_k = \frac{1}{NT}\sum_{j=0}^{NT-1}[X_j - m_j]e^{-i2\pi kj/T}, \tag{9.12}$$

the convergence of $E\{[\widehat{\widetilde{m}}_{k,N} - \widetilde{m}_k]^2\}$ to zero is equivalent to the average of $[X_t - m_t]e^{-i2\pi kt/T}$ converging in L_2 to zero. A necessary and sufficient condition for this convergence is given in Proposition 5.12 and is expressed by (9.9). The interpretation of (9.9) is that the point mass $F_0(\{2\pi k/T\})$ of the diagonal measure is only large enough to account for $|\widetilde{m}_k|^2$; there is no *random component* of positive variance at frequency $\lambda = 2\pi k/T$.

For (b), $\sum_{u=0}^{T-1}\sum_{k=0}^{\infty}|R_{u+kT,u}| < \infty$ implies $\sum_{\tau=-\infty}^{\infty}|B_k(\tau)| < \infty$ and hence

$$f_k(\lambda) = \sum_{\tau=-\infty}^{\infty} B_k(\tau)e^{-i\lambda\tau}$$

is clearly continuous and solves (9.11). See Proposition 4.3.2 of Brockwell and Davis [28] for the stationary case. ∎

See the problems for another proof of (9.10).

From (9.10), if $\sum_{u=0}^{T-1}\sum_{k=0}^{\infty}|R_{u+kT,u}| < \infty$, for $u = s, t$, then it is easy to obtain

$$\begin{aligned} & N\left|E\{[\widehat{\widetilde{m}}_{j,N} - \widetilde{m}_j][\widehat{\widetilde{m}}_{k,N} - \widetilde{m}_k]\}\right| \\ \leq\ & N\left[E\{\widehat{\widetilde{m}}_{j,N} - \widetilde{m}_j\}^2\right]^{1/2}\left[E\{\widehat{\widetilde{m}}_{k,N} - \widetilde{m}_k\}^2\right]^{1/2} \\ \leq\ & f_0^{1/2}(j2\pi/T)f_0^{1/2}(k2\pi/T) + \epsilon \end{aligned}$$

for all N sufficiently large.

Almost Sure Consistency. Although almost sure consistency can be addressed directly, we now show how results from stationary univariate sequences can be adapted for PC sequences through the fact that the sequence $Y_t = X_{t+\Theta}$ is stationary if Θ is independent of X and uniformly distributed on $0, 1, \ldots, T-1$ (see Section 6.8.3). In particular, we use this device to study the almost sure limits of

$$J_{X,N}(\lambda) = \frac{1}{N}\sum_{t=0}^{N-1}[X_t - m_t]e^{-i\lambda t},$$

that is, to the strong law for the sequence $[X_t - m_t]e^{-i\lambda t}$. Note that $J_{X,N}(\lambda) \to 0$ a.s. implies $\widehat{\widetilde{m}}_{k,N} \to \widetilde{m}_k$ a.s.

Proposition 9.4 *If X_t is PC-T with mean m_t and Θ is independent of X_t, and uniformly distributed on $0, 1, \ldots, T-1$, then if either $J_{X,N}(\lambda)$ or $J_{Y,N}(\lambda)$ converges a.s., the other does also and*

$$\lim_{N\to\infty} J_{X,N}(\lambda) = e^{-i\lambda\Theta} \lim_{N\to\infty} J_{Y,N}(\lambda) \quad a.s. \tag{9.13}$$

Remark. This result is adapted from the case in which X_t is a PC process indexed on the reals; see Cambanis et al. [30]. A result for APC processes is given there also.

Proof. Set $Z_t = X_t - m_t$ and consider the difference (ω is sometimes suppressed)

$$\begin{aligned}
e^{-i\lambda\Theta}J_{Y,N}(\lambda) - J_{X,N}(\lambda) &= \frac{1}{N}\sum_{t=0}^{N-1}\left[Z_{t+\Theta}e^{-i\lambda t - i\lambda\Theta} - Z_t e^{-i\lambda t}\right] \\
&= \frac{1}{N}\sum_{t=N}^{N-1+\Theta} Z_t e^{-i\lambda t} - \frac{1}{N}\sum_{t=0}^{\Theta-1} Z_t e^{-i\lambda t} \\
&= f_N(\lambda,\omega) - h_N(\lambda,\omega),
\end{aligned}$$

where it is assumed $N > T > \Theta(\omega)$ for all ω. The quantity $h_N(\lambda,\omega)$ is $O(N^{-1})$ a.s. because $\sum_0^{T-1}|Z_t(\omega)| < \infty$ a.s. due to $E\{|Z_t|\} < \infty$, $t = 0, 1, \ldots, T-1$. The convergence of $f_N(\lambda,\omega)$ is a little more troublesome because the interval changes with N. But since

$$\begin{aligned}
E\{|f_N(\lambda,\omega)|^2\} &\le \frac{1}{N^2}\sum_{s=N}^{N-1+T}\sum_{t=N}^{N-1+T} E\{|Z_s Z_t|\} \\
&\le \frac{1}{N^2}\left[\sum_{t=N}^{N-1+T}\left(E\{|Z_t|^2\}\right)^{1/2}\right]^2 \\
&\le \frac{1}{N^2}\left[\sum_{t=N}^{N-1+T} E\{|Z_t|^2\}\sum_{t=N}^{N-1+T} 1\right] = \frac{T^2}{N^2}B_{0,0}
\end{aligned}$$

the Borel–Cantelli lemma provides the result $f_N(\lambda, \omega) \to 0$ a.s. for every fixed λ. ∎

Many of the known sufficient conditions for almost sure convergence of $J_{X,N}(\lambda)$ to zero appear in the article by Gaposhkin [62], where more subtle conditions, such as item (c) below are also given. The following conditions result from the application of the preceding proposition to some of these known conditions for stationary sequences.

Proposition 9.5 *If X_t is PC-T with mean m_t, then each of the following conditions is sufficient for $J_{X,N}(\lambda) \to 0$ a.s.*

(a) *for fixed λ, there exists $\alpha > 0$ and $N_0 > 0$ for which*

$$\begin{aligned}
\sigma_N^2(\lambda) &= \frac{1}{N^2} \sum_{s=0}^{N-1} \sum_{t=0}^{N-1} B_0(s-t) e^{-i2\pi(s-t)\lambda} \\
&= \frac{1}{N} \sum_{j=-N+1}^{N-1} \left(1 - \frac{|j|}{N}\right) B_0(j) e^{-i2\pi j\lambda} \\
&= \int_0^{2\pi} \frac{\sin^2 \pi N(\eta - \lambda)}{N^2 \sin^2 \pi(\eta - \lambda)} F_0(d\eta) \le \frac{K}{N^\alpha},
\end{aligned}$$

whenever $N > N_0$;

(b) *each of the conditions $B_0(\tau) = O(\tau^{-\alpha})$ for $\alpha > 0$ or $\sum_0^\infty |B_0(\tau)| < \infty$ is sufficient for condition* (a) *uniformly in λ;*

(c) *for fixed λ, convergence of the sequence*

$$\sum_{k=3}^{\infty} \frac{\sigma_k^2(\lambda)}{k \log k} \log \log k.$$

Proof. For the stationary sequence $Y_t = X_{t+\Theta} - m_{t+\Theta}$, conditions (a) and (b) imply $J_{Y,N}(\lambda) \to 0$ a.s. by application of Theorem 6.1 of Doob [49] to the covariance $B_0(\tau)$ of Y_t; condition (c) implies $J_{Y,N}(\lambda) \to 0$ a.s. by application of Theorem 4A of Gaposhkin [62]. The claim $J_{X,N}(\lambda) \to 0$ a.s. follows from Proposition 9.4. ∎

Remark. Note that $\max_t |R_{t+\tau,t}| = O(\tau^{-\alpha})$ and $\sum_{\tau=-\infty}^{\infty} \sum_{t=0}^{T-1} |R_{t+\tau,t}| < \infty$ are sufficient for the conditions given in (b). This latter condition is equivalent to condition (b) of Lemma 9.1.

In the stationary case (see Doob [49, page 493]) condition (a) is established for $\lambda = 0$ and then argued for $\lambda \neq 0$ by noting that $X_t' = X_t e^{-i\lambda t}$ is also stationary with spectral distribution function $F_{X'}(u) = F_X(u + \lambda)$ modulo $[-\pi, \pi)$ (or $[0, 2\pi)$). If X_t is PC-T, then $X_t' = X_t e^{-i\lambda t}$ also remains PC as

$$\begin{aligned} E\{X_s' X_t'\} &= E\{X_s X_t\} e^{-i\lambda(s-t)} \\ &= E\{X_{s+T} X_{t+T}\} e^{-i\lambda(s+T-t-T)} \\ &= E\{X_{s+T}' X_{t+T}'\} \end{aligned}$$

and its spectral measure $F_{X'}$ is that of X_t shifted down the main diagonal by λ (see Section 6.8.5), that is, $F_{X'}(\lambda_1, \lambda_2) = F_X(\lambda_1 + \lambda, \lambda_2 + \lambda)$ and so the main diagonal measure F_0 shifts down by λ modulo 2π.

Gaposhkin [62] also gives conditions for almost sure convergence (a) in spectral terms and (b) for a class of *quasistationary* processes. The application of these is left to the reader. Finally, I. Honda [98] applies existing results to the lifted $\mathbf{X}_n$ to show that if X_t is PC-T and Gaussian, then $J_{X,N}(\lambda) \to 0$ a.s. for all λ if and only if the diagonal spectral measure F_0 is absolutely continuous with respect to Lebesgue measure.

Asymptotic Normality. The two main approaches to asymptotic normality are via the assumptions (1) that the process is linear (an infinite moving average) and (2) that the process is mixing in some sense. We shall give results using both approaches. But first we see how the random shift can be used for this problem. Recall that both $J_{X,N}(\lambda)$ and $J_{Y,N}(\lambda)$ have mean-square limits. The direct proof that $E\{|e^{-i\lambda\Theta} J_{Y,N}(\lambda) - J_{X,N}(\lambda)|^2\} \to 0$ is even simpler than the a.s. result (Proposition 9.4) and is left as an exercise. This is enough to give the following, where $\Rightarrow$ denotes convergence in distribution.

Proposition 9.6 *If X_t is PC-T with mean m_t, Θ is independent of X_t, and uniformly distributed on $0, 1, \ldots, T-1$, then $J_{Y,N}(\lambda) \Rightarrow F_Y$ implies $J_{X,N}(\lambda) e^{i\lambda\Theta} \Rightarrow F_Y$.*

Proof. The proof is a consequence of $e^{-i\lambda\Theta} J_{Y,N}(\lambda) \xrightarrow{P} J_{X,N}(\lambda)$ and elementary facts about convergence in distribution. See Loève [139, page 278] or Brockwell and Davis [28, Proposition 6.3.3]. ∎

This result can give a way to obtain asymptotic normality for $\widehat{\widetilde{m}}_{k,N}$ for a single fixed k.

Proposition 9.7 *Suppose X_t is PC-T with mean m_t, Θ is independent of X_t, and uniformly distributed on $0, 1, \ldots, T-1$. Any condition on X that gives $\widehat{\widetilde{m}}_{k,N} \to \widetilde{m}_k$ in probability and asymptotic normality for the mean estimator of* *stationary sequence* $Y_t^{(k)} = X_{t+\Theta} e^{-i2\pi k(t+\Theta)/T}$ *will give asymptotic normality for* $\widehat{\widetilde{m}}_{k,N}$.

Proof. Denote $X_t^{(k)} = X_t e^{-i2\pi kt/T}$ (a PC-T sequence) and $Y_t^{(k)} = X_{t+\Theta}^{(k)}$. Observe that

$$J_{Y^{(k)},N}(0) = \frac{1}{NT} \sum_{t=0}^{NT-1} Y_t^{(k)} = e^{-i2\pi k\Theta/T} \frac{1}{NT} \sum_{jt=0}^{NT-1} X_{t+\Theta} e^{-i2\pi kt/T}.$$

The sum on the left $J_{Y^{(k)},N}(0) \Rightarrow N(\mu, \sigma^2)$ by hypothesis. But the rightmost quantity converges in probability to $e^{-i2\pi k\Theta/T} e^{i2\pi k\Theta/T} \widetilde{m}_k = \widetilde{m}_k$, so evidently $\mu = \widetilde{m}_k$. Taking $\lambda = 0$ in the preceding proposition gives $\widehat{\widetilde{m}}_{k,N} = J_{X^{(k)},N}(0) \Rightarrow N(\widetilde{m}_k, \sigma^2)$. ∎

Proposition 9.8 *Suppose* X_t *is a real linear PC-T sequence*

$$X_t = m_t + \sum_{j=-\infty}^{\infty} \psi_j(t) \xi_{t-j}, \tag{9.14}$$

where $m_t = m_{t+T}$ *and* $\psi_j(t) = \psi_j(t+T)$, $j \in \mathbb{Z}$ *are real,* $\sum_j |\psi_j(t)| < \infty$, $t = 0, 1, \ldots, T-1$, *and* $\{\xi_j\}$ *is a real zero mean i.i.d. sequence. Then if* $\mathbf{\Psi}$ *is of rank* T,

$$\sqrt{N} \left(\widehat{\mathbf{m}}_N - \mathbf{m} \right) \Rightarrow \mathbb{N} \left(0, \mathbf{\Psi \Sigma \Psi}' \right),$$

where $\widehat{\mathbf{m}}_N = (\widehat{m}_{T-1,N}, \widehat{m}_{T-2,N}, \ldots, \widehat{m}_{0,N})'$, $\mathbf{m} = (m_{T-1}, m_{T-1}, \ldots, m_0)'$, *and*

$$\mathbf{\Psi} = \sum_{j=-\infty}^{\infty} \mathbf{\Psi}_j$$

with

$$\mathbf{\Psi}_j = \begin{bmatrix} \psi_{jT-1}(T-1) & \psi_{jT}(T-1) & \ldots & \psi_{jT+T-2}(T-1) \\ \psi_{jT-2}(T-2) & \psi_{jT-1}(T-2) & \ldots & \psi_{jT+T-3}(T-2) \\ \vdots & \vdots & \vdots & \vdots \\ \psi_{jT-T}(0) & \psi_{jT-T+1}(0) & \ldots & \psi_{jT-1}(0) \end{bmatrix}. \tag{9.15}$$

Proof. The linear PC-T sequence X_t, lifted to $\mathbf{X}_n = (X_{nT-1}, X_{nT-2}, \ldots X_{nT-T})'$, has the representation

$$\mathbf{X}_n = \mathbf{m} + \sum_{j=-\infty}^{\infty} \mathbf{\Psi}_j \mathbf{\Xi}_{n-j} \tag{9.16}$$

with $\mathbf{\Psi}_j$ given by (9.15) and where $\mathbf{\Xi}_n = (\xi_{nT-1}, \xi_{nT-2}, \ldots \xi_{nT-T})'$ are i.i.d. with $\mathrm{Cov}(\mathbf{\Xi}_n, \mathbf{\Xi}_m) = \mathbf{I}_T \delta_{m-n}$. Since $\sum_{j=-\infty}^{\infty} |[\mathbf{\Psi}_j]_{p,q=0}^{T-1}| = \sum_{t=0}^{T-1} \sum_k |\psi_k(t)| < \infty$, Proposition 11.2.2 of Brockwell and Davis [28] may be applied to conclude the result. ∎

Since the Fourier coefficients $\widetilde{\mathbf{m}} = (\widetilde{m}_0, \widetilde{m}_1, \ldots \widetilde{m}_{T-1})'$, are linearly related to m_t via $\widetilde{\mathbf{m}} = T^{-1/2}\mathbf{V}^*(0)\mathbf{m}$, where $\mathbf{V}(\lambda)$ is the unitary matrix defined in Proposition (6.9), we obtain the following.

Corollary 9.8.1 *Under the conditions of Proposition 9.8,*

(a) *if* rank $\mathbf{\Psi} = T$, *then* $\sqrt{N}\left(\widehat{\widetilde{\mathbf{m}}}_N - \widetilde{\mathbf{m}}\right) \Rightarrow \mathbb{N}\left(0, T^{-1}\mathbf{V}\mathbf{\Psi}\mathbf{\Sigma}\mathbf{\Psi}'\mathbf{V}^*\right)$;

(b) *if* rank $\mathbf{\Psi} \le T$, *then for any* $\boldsymbol{\beta}$ *with* $\boldsymbol{\beta}'\mathbf{\Psi}\mathbf{\Sigma}\mathbf{\Psi}'\boldsymbol{\beta} > 0$, *we have*

$$\sqrt{N}\sum_{t=0}^{T-1} \beta_t(\widehat{m}_{t,N} - m_t) \Rightarrow \mathbb{N}\left(0, 2\pi\boldsymbol{\beta}'\mathbf{\Psi}\mathbf{\Sigma}\mathbf{\Psi}'\boldsymbol{\beta}\right);$$

(c) *if* rank $\mathbf{\Psi} \le T$, *then for any* $\boldsymbol{\beta}$ *with* $\boldsymbol{\beta}'V(0)\mathbf{\Psi}\mathbf{\Sigma}\mathbf{\Psi}'V^*(0)\boldsymbol{\beta} > 0$, *we have*

$$\sqrt{N}\sum_{k=0}^{T-1} \beta_k(\widehat{\widetilde{m}}_{k,N} - \widetilde{m}_k) \Rightarrow \mathbb{N}\left(0, T^{-1}\boldsymbol{\beta}'\mathbf{V}(0)\mathbf{\Psi}\mathbf{\Sigma}\mathbf{\Psi}'\mathbf{V}^*(0)\boldsymbol{\beta}\right).$$

This is a common way to address the convergence of $\sqrt{N}\,(\widehat{\mathbf{m}}_N - \mathbf{m})$ to a possibly degenerate normal, meaning its covariance $\mathbf{\Psi}\Sigma\mathbf{\Psi}'$ is possibly not of full rank. If $\mathbf{\Psi}\Sigma\mathbf{\Psi}'$ is of full rank, then $\boldsymbol{\beta}'\mathbf{\Psi}\Sigma\mathbf{\Psi}'\boldsymbol{\beta} > 0$ for all nonzero $\boldsymbol{\beta}$. The notation is still correct if we consider $\boldsymbol{\beta}$ to be a projection onto a subspace and positivity to mean positive definite.

Asymptotic Normality Via Mixing. Asymptotic normality can also be obtained from certain mixing conditions that govern the *memory* of the process. Various notions of mixing exist and here we shall utilize the concepts of *strong or α-mixing* and ϕ-mixing. For a sequence X_t, denote the Borel sigma-fields $\mathcal{F}_t = \mathcal{B}(X_s,\ s \le t)$ and $\mathcal{G}_t = \mathcal{B}(X_s,\ s \ge t)$. Note that we can ignore the presence of a nonrandom mean m_t in X_t because $\mathcal{F}_t = \mathcal{B}(X_s - m_s,\ s \le t)$ and similarly for $\mathcal{G}_t$. We define the α-mixing and ϕ-mixing functions to be

$$\alpha_{t,n} = \sup\{|P(A \cap B) - P(A)P(B)| : A \in \mathcal{F}_t,\ B \in \mathcal{G}_{t+n}\}, \quad (9.17)$$
$$\phi_{t,n} = \sup\{|P(B|A) - P(B)| : A \in \mathcal{F}_t, P(A) > 0,\ B \in \mathcal{G}_{t+n}\}, \quad (9.18)$$

and see that if X_t is stationary (strictly), $\alpha_{t,n}$ and $\phi_{t,n}$ are independent of t. If

$$\lim_{n\to\infty} \sup_t \alpha_{t,n} = 0 \quad \text{or} \quad \lim_{n\to\infty} \sup_t \phi_{t,n} = 0,$$

then X_t is correspondingly called *uniformly strongly mixing* or *uniformly ϕ-mixing.*

If X_t is periodically stationary (see Definition 1.2) with period T, then $\alpha_{t,n} = \alpha_{t+T,n}$ and $\phi_{t,n} = \phi_{t+T,n}$ for every n; furthermore, $\sup_t$ in the preceding displays can be replaced with $\max_{t=0,1,\ldots,T-1}$. If $\lim_{n\to\infty} \alpha_{t,n} = 0$ or $\lim_{n\to\infty} \phi_{t,n} = 0$ then we say that a sequence is strongly mixing or ϕ-mixing for the reference time t; but since $\phi_{t-k_1,n+k_1} \leq \phi_{t,n} \leq \phi_{t+k_2,n-k_2}$ for any $k_1, k_2 \geq 0$ it follows from the left inequality that ϕ-mixing at any reference time t implies it for all t (similarly for α-mixing). Furthermore, $\phi_{t_2,n} = O(\phi_{t_1,n})$ for any t_1, t_2 provided $\lim_{n\to\infty} \phi_{t_1,n}$ exists. To see this, in the inequality above set $t = t_2$ and $t - k_1 = t_1 - T$, $t + k_2 = t_1$, where we can take $T > t_1 - t_2 > 0$. Then

$$\frac{\phi_{t_1,n+k_1}}{\phi_{t_1,n}} \leq \frac{\phi_{t_2,n}}{\phi_{t_1,n}} \leq \frac{\phi_{t_1+T,n-k_2}}{\phi_{t_1,n}}$$

and the limits of the rightmost and leftmost quantities are both unity. The mixing sequences $\phi_{t_1,n}$ and $\phi_{t_2,n}$ may be a little different but are asymptotically the same.

Note if a periodically stationary sequence is constructed by interleaving stationary processes having different mixing rates, say, that for each t the sequence X_{t+jT} is ϕ-mixing with mixing function

$$\phi_n^{(t)} = \sup\{|P(B|A) - P(B)| : A \in \mathcal{F}_m^{(t)}, P(A) > 0,\ B \in \mathcal{G}_{m+n}^{(t)}\},$$

where $\mathcal{F}_m^{(t)} = \mathcal{B}(X_{t+jT},\ j \leq m)$ and $\mathcal{G}_{m+n}^{(t)} = \mathcal{B}(X_{t+jT},\ j \geq m+n)$. We may conclude from $\mathcal{F}_m^{(t)} \subset \mathcal{F}_{t+mT}$ and $\mathcal{G}_{m+n}^{(t)} \subset \mathcal{G}_{t+mT+nT}$ that $\phi_n^{(t)} \leq \phi_{t+mT,nT} = \phi_{t,nT}$ and since $\phi_{t_2,nT} = O(\phi_{t_1,nT})$ then the periodic mixing functions $\phi_{t,nT}$ are bounded below by $\max_t \phi_n^{(t)}$. In other words, the slowest individual mixing rate governs the periodic mixing rate.

Finally, setting $\bar{\alpha}_n = \max_t \alpha_{t,n}$, we see that X_t is uniformly strongly mixing $\lim_{n\to\infty} \bar{\alpha}_n = 0$ if and only if it is strongly mixing for some t. And then also $\bar{\alpha}_n = O(\alpha_{t,n})$ for any t. The same statements hold for ϕ-mixing.

Rosenblatt [197] showed that strong (i.e., α) mixing along with a moment condition gave asymptotic normality for the sample mean of a stationary process. Using mixing hypotheses, Rozanov [201, Section 11] gives a central limit result for the sample mean of a stationary multivariate process but requires that the spectral density matrix be of full rank at $\lambda = 0$. The result for the rank deficient case is only a slight elaboration of the Rozanov result. The ϕ-mixing facilitates CLT results for covariance and spectral estimation because of the following facts. If in (9.18), ζ_1 and ζ_2 are $\mathcal{F}_t$ and $\mathcal{G}_{t+n}$ measurable, respectively, then

$$|E\{\zeta_1\zeta_2\} - E\{\zeta_1\}E\{\zeta_2\}| \leq 2\phi_n^{1/r} E^{1/r}|\zeta_1|^r E^{1/s}|\zeta_2|^s, \tag{9.19}$$

where $r, s > 1$ and $1/r + 1/s = 1$. (See Ibragimov [120, Lemma 1.1] or Billinsgley [13, page 170].)

Proposition 9.9 *If X_t is PC-T and* (a) $\bar{\alpha}_n = O(n^{-1-\epsilon})$ *for some* $\epsilon > 0$, (b) $E|X_t|^{2+\delta} < \infty$, $t \in \mathbb{Z}$ *for* $\delta > 4/\epsilon$ *and* (c) *the spectral density matrix* $\mathbf{f}$ *of the blocked sequence* $\mathbf{X}_n$ *is bounded and continuous at* $\lambda = 0$, *then*

$$\sqrt{N}\sum_{t=0}^{T-1}\beta_t(\widehat{m}_{t,N} - m_t) \Rightarrow \mathbb{N}\left(0, 2\pi\boldsymbol{\beta}'\mathbf{f}(0)\boldsymbol{\beta}\right)$$

whenever $\boldsymbol{\beta}'\mathbf{f}(0)\boldsymbol{\beta} > 0$. *If* $\det[\mathbf{f}(0)] \neq 0$, *then*

$$\sqrt{N}(\widehat{\mathbf{m}}_N - \mathbf{m}) \Rightarrow \mathbb{N}\left(0, 2\pi\mathbf{f}(0)\right).$$

Proof. When $\mathbf{f}(0)$ is of full rank, this is Theorem 11.2 of Rozanov [201]. If $\boldsymbol{\beta}'\mathbf{f}(0)\boldsymbol{\beta} > 0$, then the sequence $Y_n = \sum_{j=0}^{T-1}\beta_j X_{nT+j}$ is α-mixing with mixing function $\alpha_{Y,n} = O(\bar{\alpha}_n)$. Furthermore, the asymptotic variance for the estimator $\widehat{m}_{Y,N} = N^{-1}\sum_{n=0}^{N-1}Y_n$ of $m_Y = \sum_{j=0}^{T-1}\beta_j m_j$ is clearly $\sigma^2_{\boldsymbol{\beta}} = \boldsymbol{\beta}'\mathbf{f}(0)\boldsymbol{\beta} > 0$.

Also, $E|Y_n|^{2+\delta} \leq C\sum_{j=0}^{T-1}|\beta_j|^{2+\delta}E|X_{nT+j}|^{2+\delta} < \infty$, $t \in \mathbb{Z}$ and thus $\sqrt{N}(\widehat{m}_{Y,N} - m_Y) \Rightarrow N(0, \boldsymbol{\beta}'\mathbf{f}(0)\boldsymbol{\beta})$. ∎

The asymptotic normality of $\widehat{\widetilde{m}}_{k,N}$, $k = 0, 1, \ldots, T-1$ can be established using $\widehat{\widetilde{\mathbf{m}}} - \widetilde{\mathbf{m}} = V(0)(\widehat{\mathbf{m}}_N - \mathbf{m})$, as in Proposition 9.8. If $\det[\mathbf{f}(0)] \neq 0$, then we easily see that $\bar{\alpha}_n = O(n^{-1-\epsilon})$ implies that the mixing function for $\mathbf{X}_n$ satisfies $\alpha_{\mathbf{X},n} = O((Tn)^{-1-\epsilon}) = O((n)^{-1-\epsilon})$ and so asymptotic normality for $\widehat{\mathbf{m}}_N$ follows directly from Rozanov [201, Section 11].

Corollary 9.9.1 *Under the conditions of Proposition 9.9, if* $\operatorname{rank}\mathbf{f}(0) = T$

$$\sqrt{N}\left(\widehat{\widetilde{\mathbf{m}}}_N - \widetilde{\mathbf{m}}\right) \Rightarrow \mathbb{N}\left(0, 2\pi\mathbf{V}\mathbf{f}(0)\mathbf{V}^*\right);$$

if $\operatorname{rank}\mathbf{f}(0) \leq T$, *then for any* $\boldsymbol{\beta}$ *with* $\boldsymbol{\beta}'V\mathbf{f}(0)V'\boldsymbol{\beta} > 0$, *we have*

$$\sqrt{N}\sum_{k=0}^{T-1}\beta_k(\widehat{\widetilde{m}}_{k,N} - \widetilde{m}_k) \Rightarrow \mathbb{N}\left(0, \boldsymbol{\beta}'\mathbf{V}\mathbf{f}(0)\mathbf{V}^*\boldsymbol{\beta}\right).$$

9.2 ESTIMATION OF m_t: PRACTICE

Since we can describe the mean of a PC-T sequence X_t by m_t or by the Fourier coefficients, $\widetilde{m}_k$, we wish to estimate either of these quantities. Here we describe corresponding estimators denoted by $\widehat{m}_{t,N}$ (9.1) and $\widehat{\widetilde{m}}_{k,N}$. Although we can estimate $\widetilde{m}_k$ directly from $\widehat{m}_{t,N}$, as in (9.8), here we estimate it directly from the series by the first line in (9.8), namely,

$$\widehat{\widetilde{m}}_{k,N} = \frac{1}{NT} \sum_{t=0}^{NT-1} X_t e^{-i2\pi kj/T} \tag{9.20}$$

because it can be implemented as a sample Fourier transform of $\{X_t, t = 0, 1, \ldots, NT-1\}$ evaluated at frequency $2\pi k/T$. This permits the use of a frequency-based method for assessing significance. In the previous sections we have already examined the consistency of these estimators under the assumption that X_t is PC with period T. To summarize, under mild and unsurprising hypotheses, these estimators are consistent in several senses. In the following paragraphs we describe the computation of $\widehat{m}_{t,N}$ and $\widehat{\widetilde{m}}_{k,N}$, and their implementation by programs `permest.m` and `permcoeff.m`.

9.2.1 Computation of $\widehat{m}_{t,N}$

Since the mean sequence m_t can be an arbitrary periodic real valued sequence, it is also of interest to know whether or not m_t is properly periodic or whether it is a constant. In order to help the perception of $m_t \equiv m$, our realization of the estimator $\widehat{m}_{t,N}$ (to be described more completely later) includes $1-\alpha$ confidence intervals around each point estimate; these confidence intervals are based on the assumption that for the tth season, the random variables $\{Y_t^{(p)} = X_{t+pT} - m_t, p = 0, 1, \ldots, N-1\}$ are Normal$(0, \sigma_t^2)$. Since σ_t^2 must be estimated, the confidence intervals are determined by a t distribution with $N-1$ degrees of freedom. The existence of nonoverlapping confidence intervals, as in Figure 1.1, gives a preliminary clue that $m_t \not\equiv m$. One-way analysis of variance can be used to test for $m_t \equiv m$ under the assumption that the random variables $\{Y_t^{(p)}\}$ are Normal$(0, \sigma^2)$ (homogeneous variances in t) and that the collections $\{Y_s^{(p)}\}$ and $\{Y_t^{(p)}\}$ are independent whenever $s \neq t$.

Note that the condition $m_t \equiv m$ conveys nothing conclusive about the presence of PC structure in the covariance. Indeed, the true mean of a PC sequence may be constant and a sequence may have a periodic mean and stationary covariance structure. So we cannot use the outcome of a test for $m_t \equiv m$ to avoid the further testing for covariance structure.

Program `permest.m`. The program `permest.m` implements the estimator $\widehat{m}_{t,N}$, but slightly more generally because the series may contain missing values and the length of the series may not be an integral number of periods. Given an input time series vector, and a specified period T, the program computes and returns the periodic mean based on all the values that are present (not missing) in the series. That is, for each t in the base period,

$$\widehat{m}_{t,N} = \frac{1}{N_t} \sum_{p=0}^{N_t-1} X_{t+pT}, \tag{9.21}$$

where $N_t = \text{card}\{p \in \{0,1,\ldots,N-1\} : X_{t+pT} \text{ not missing}\}$. Using a specified α, the $1-\alpha$ confidence intervals are computed at each $t = 1,2,\ldots,T$. The original series is plotted with missing values replaced by the periodic mean and marked by "x"; the periodic mean is also plotted along with $1-\alpha$ confidence intervals based on the normality assumption. The p-value for a one-way ANOVA test for equality of means is also computed and is present on the plots. The demeaned series $X_t - \widehat{m}_{t,N}$ is computed and returned to the calling program.

Figure 1.1 presented an application of `permest` to 40 periods ($T = 24$) of a solar radiation series from station DELTA of the Taconite Inlet Project. The p-value from the one-way ANOVA for $m(t) \equiv m$ was computed by MATLAB to be zero. This is not too surprising considering the clarity of the periodicity and the size of the dataset. But even after shortening the dataset to 2 periods we found that the p-value is $\sim 10^{-13}$ and for 4 periods it again returns a p-value of zero.

9.2.2 Computation of $\widehat{\widetilde{m}}_{k,N}$

Although we could construct frequency based tests when the length is not an integral number of periods, the payoff seems hardly worth the effort, so we assume here that the series is of length NT. In this case we have already noted in (9.8) that

$$\widehat{\widetilde{m}}_{k,N} = \frac{1}{NT} \widetilde{X}_{NT}(2\pi k/T),$$

where

$$\widetilde{X}_{NT}(\lambda) = \sum_{t=0}^{NT-1} X_t e^{-i\lambda t}. \tag{9.22}$$

Note that the FFT algorithm provides the Fourier transform (9.22) evaluated at the *Fourier frequencies* $\lambda_j = 2\pi j/NT$, $j = 0,1,\ldots NT-1$ so that $\widehat{\widetilde{m}}_{k,N}$ is the FFT coefficient with index $j = kN$.

Taking this view helps us to construct tests for specific $\widetilde{m}_k = 0$ and also for $\{\widetilde{m}_k = 0,\ k \neq 0\}$, which exactly corresponds to $m_t \equiv m$. These tests are based on *variance contrast*, which at some frequency index j_0 is the value of $|\widetilde{X}_{NT}(2\pi j_0/NT)|^2$ in contrast to the average of $|\widetilde{X}_{NT}(2\pi j/NT)|^2$ values in a neighborhood. More details will be given in Chapter 10.

Program `permcoeff.m`. For a real series and specified period T, the program `permcoeff.m` implements the estimator $\widehat{\widetilde{m}}_{k,N}$, but slightly more generally because the series may contain missing values and the series may not have a length of an integral number of periods. Missing values are set to the sample mean of the nonmissing values and the series is cut to the largest number N of whole periods in the series.

Table 9.1 Result from program `permcoeff.m` applied to 40 periods of solar radiation data of Figure 1.1. The overall test $\{\widetilde{m}_k = 0, k = 1, 2, \ldots 11\}$ gives a p-value of 0.

k	$\lvert\widehat{\widetilde{m}}_{k,N}\rvert$	$n1$	$n2$	Variance ratio	p-value
0	2.94e+002	1	16	1.12e+003	3.33e-016
1	6.98e+001	2	32	2.30e+003	0.00e+000
2	4.26e+000	2	32	9.26e+000	6.69e-004
3	9.42e-001	2	32	7.90e-001	4.62e-001
4	1.40e+000	2	32	1.37e+000	2.68e-001
5	1.82e-001	2	32	5.09e+000	1.20e-002
6	5.49e-001	2	32	5.24e-001	5.97e-001
7	9.20e-001	2	32	1.76e+000	1.88e-001
8	2.14e-001	2	32	9.41e-002	9.10e-001
9	1.76e-001	2	32	1.07e-001	8.98e-001
10	7.68e-001	2	32	1.57e+000	2.24e-001
11	6.61e-001	2	32	3.61e+000	3.87e-002

The values of $\widehat{\widetilde{m}}_{k,N}$ are computed only for $k = 0, 1, \ldots, \lfloor((T-1)/2)\rfloor$ because for real series, $\widetilde{m}_k = \overline{\widetilde{m}_{T-k}}$. In addition, the p-values for the test $\widetilde{m}_k = 0$, discussed in the previous paragraphs, are returned.

The results of `permcoeff.m` when applied to 40 periods (with $T = 24$) of the solar radiation data of Figure 1.1 are presented in Table 9.1. Note the $k = 0$ term corresponds to the sample mean. The value of $\lvert\widehat{\widetilde{m}}_{k,N}\rvert$ for $k = 1$ is incomputably significant (i.e., 0) and so a p-value correction is meaningless. However, when shortening the series to 4 periods, the value of n_A is just 2 and the p-value for $k = 1$ is $1.39e-005$, still significant, even when corrected by a factor of 11. The overall test $\{\widetilde{m}_k = 0,\ k \neq 0\}$ gives a p-value of zero.

9.3 ESTIMATION OF $R(t+\tau, t)$: THEORY

Now we address the estimation of $R(t+\tau, t)$ and its Fourier coefficients $B_k(\tau)$. For motivation, we recall that for a stationary sequence X_t having correlation function $R(\tau) = E\{X_{t+\tau}X_t\}$, the natural estimator for $R(\tau)$ based on a finite sample of length N is the well known

$$\widehat{R}_N(\tau) = \frac{1}{N}\sum_{t=0}^{N-\tau-1}[X_{t+\tau} - \widehat{m}_N][X_t - \widehat{m}_N], \tag{9.23}$$

which, under conditions such as $\lim_{\tau\to\infty} R(\tau) = 0$, or $\sum_{-\infty}^{\infty}|R(\tau)| < \infty$, is mean-square consistent. If X is Gaussian, then $\widehat{R}_N(\tau)$ is consistent if and only if the spectral d.f. (measure) for X has no discrete component (see Doob [49, Theorem 7.1]). The estimator for the autocorrelation function is

$$\widehat{\rho}_N(\tau) = \frac{\widehat{R}_N(\tau)}{\widehat{R}_N(0)}. \tag{9.24}$$

9.3.1 Estimation of $R(t+\tau,t)$

Corresponding results for PC sequences are easily obtained for estimation of the covariance $R(t+\tau,t)$ based on a finite sample $X_0, X_1, \ldots X_{NT-1}$. The periodicity $R(t+\tau,t) = R(t+\tau+T,t+T)$ suggests the estimator

$$\widehat{R}_N(t+\tau,t) = \frac{1}{N}\sum_{k=0}^{N-1}[X_{t+kT+\tau} - \widehat{m}_{t+\tau,N}][X_{t+kT} - \widehat{m}_{t,N}] \tag{9.25}$$

for $t = 0, 1, \ldots, T-1$ and for all τ possible, and then use $\widehat{R}_N(s,t) = \widehat{R}_N(s+T, t+T)$ as necessary. Denoting $\widehat{\sigma}^2_N(t) = \widehat{R}_N(t,t)$, the autocorrelation estimator is

$$\widehat{\rho}_N(t+\tau,t) = \frac{\widehat{R}_N(t+\tau,t)}{\widehat{\sigma}_N(t+\tau)\widehat{\sigma}_N(t)}. \tag{9.26}$$

Recall in the stationary case, that the sum in (9.23) may be divided by $N-\tau-1$ rather than N to obtain the maximum likelihood estimator. But N is often preferred because $\widehat{R}_N(\tau)$ will be a NND function of τ. Later we will have need for an estimator for $R(s,t)$ that is a NND function of s,t. Generally, $\widehat{R}_N(s,t)$ given by (9.25) will not be NND unless $N = KT$ for integer K. This can be ensured by either truncating the series to KT observations, where $K = \lfloor N/T \rfloor$, or by filling the series with zeros from $N+1$ to KT, where $K = \lfloor N/T \rfloor + 1$. This process produces estimates that may be interpreted as the components of the matrix autocovariance and autocorrelation functions of the lifted sequence $\mathbf{X}_n$.

That is, if

$$\widehat{R}^{\mathbf{X}}_K(h) = \frac{1}{K}\sum_{k=0}^{K-1}[\mathbf{X}_{k+h} - \widehat{\mathbf{m}}_K][\mathbf{X}_k - \widehat{\mathbf{m}}_K]^* \tag{9.27}$$

and

$$[\widehat{\rho}_K^{\mathbf{X}}(h)]_{ij} = \frac{[\widehat{R}_K^{\mathbf{X}}(h)]_{ij}}{([\widehat{R}_K^{\mathbf{X}}(0)]_{ii}[\widehat{R}_K^{\mathbf{X}}(0)]_{jj})^{1/2}}, \tag{9.28}$$

the mapping (1.10) implies that

$$[\mathbf{X}_{k+h}]_i = X_{i+(k+h)T}$$

and so the correspondence between $\widehat{R}_K^{\mathbf{X}}(h)$ and $\widehat{R}_N(t+\tau, t)$ for $N = KT$ is

$$[\widehat{R}_N^{\mathbf{X}}(h)]_{ij} = \widehat{R}_N(i + hT, j) = \widehat{R}_N(j + (i-j) + hT, j).$$

Sufficient conditions for consistency can be given for linear PC sequences or directly in terms of $R(t+\tau, t)$. For the former, Theorem 11.2.1 of Brockwell and Davis [28] gives the following.

Proposition 9.10 *If X_t is a linear PC-T sequence and satisfies the conditions of Proposition 9.8, then*

$$\lim_{N\to\infty} [\widehat{R}_N^{\mathbf{X}}(h)]_{ij} = [R^{\mathbf{X}}(h)]_{ij} \quad \textit{in probability.} \tag{9.29}$$

Next, we obtain consistency results for an estimator expressed more clearly in terms of X_t; specifically, consider

$$\widehat{R}_N^{\dagger}(t+\tau, t) = \frac{1}{N}\sum_{k=0}^{N-1}[X_{t+kT+\tau} - m_{t+\tau}][X_{t+kT} - m_t] \tag{9.30}$$

in which the mean sequence m_t is assumed to be known. The following lemma shows that under broad conditions, $|\widehat{R}_N^{\dagger}(t+\tau,t) - \widehat{R}_N(t+\tau,t)| \to 0$ in probability.

Lemma 9.2 *If X_t is PC-T and has fourth moments, and if $\widehat{m}_{s,N} \xrightarrow{P} m_s$ for $s = t$ or $s = t+\tau$, then*

$$\Delta R_N = \widehat{R}_N(t+\tau, t) - \widehat{R}_N^{\dagger}(t+\tau, t) \xrightarrow{P} 0.$$

Proof. Denote

$$Z_{t,\tau} = [X_{t+\tau} - \widehat{m}_{t+\tau,N}][X_t - \widehat{m}_{t,N}] - R(t+\tau, t) \tag{9.31}$$

and

$$Z_{t,\tau}^{\dagger} = [X_{t+\tau} - m_{t+\tau}][X_t - m_t] - R(t+\tau, t). \tag{9.32}$$

Then by a simple direct computation

$$\begin{aligned} \Delta R_N = \widehat{R}_N(t+\tau,t) - \widehat{R}_N^{\dagger}(t+\tau,t) &= \frac{1}{N}\sum_{j=0}^{N-1}\left[Z_{t+jT,\tau} - Z_{t+jT,\tau}^{\dagger}\right] \\ &= (m_{t+\tau} - \widehat{m}_{t+\tau,N})(\widehat{m}_{t,N} - m_t). \end{aligned}$$

Since $R(s,t)$ must be bounded due to $|R_{s,t}| \leq \max_t E\{X_t^2\} = M_2$, then

$$E\{|\widehat{m}_{t,N} - m_t|^2\} \leq \frac{1}{N^2} \sum_{j=0}^{N-1} \sum_{j'=0}^{N-1} R(t+jT, t+j'T) \leq M_2,$$

showing that the error $\widehat{m}_{t,N} - m_t$ is bounded in probability for each $t = 0, 1, \ldots, T-1$. Hence if either $\widehat{m}_{t,N} \xrightarrow{P} m_t$ or $\widehat{m}_{t+\tau,N} \xrightarrow{P} m_{t+\tau}$, the result follows by straightforward convergence results. (See Proposition 6.1.1 of Brockwell and Davis [28].) ∎

The convergence $\Delta R_N \xrightarrow{P} 0$ is sufficient to ensure the equality of the limits in the various modes of convergence of interest here.

The following proposition gives conditions for consistency in terms of the zero mean sequence $Z_{t,\tau}^{\dagger}$ whose covariance is $R_{Z^{\dagger}}(t_1, t_2, \tau) = E\{Z_{t_1,\tau}^{\dagger} Z_{t_2,\tau}^{\dagger}\}$.

Proposition 9.11 *If X is PC-T with bounded fourth moments, then each of the following is sufficient for mean-square consistency of $\widehat{R}_N^{\dagger}(t+\tau, t)$ (for t, τ fixed):*

(a) $\lim_{N\to\infty} \frac{1}{N^2} \sum_{j=0}^{N-1} \sum_{k=0}^{N-1} R_{Z^{\dagger}}(t+jT, t+kT, \tau) = 0$;

(b) $\lim_{k\to\infty} R_{Z^{\dagger}}(t+jT, t+jT+kT, \tau) = 0$ *uniformly in j.*

If $Z_{t+jT,\tau}^{\dagger}$ is stationary in j, these simplify to

(a′) $\lim_{N\to\infty} \frac{1}{N} \sum_{j=0}^{N-1} \left(1 - \frac{|j|}{N}\right) R_{Z^{\dagger}}(t, t+jT, \tau) = 0$;

(b′) $\lim_{k\to\infty} R_{Z^{\dagger}}(t, t+kT, \tau) = 0$;

(c′) $\sum_{j=0}^{\infty} |R_{Z^{\dagger}}(t, t+jT, \tau)|^2 < \infty$.

If X_t is Gaussian, the following conditions suffice for (b′) and (c′):

(b″) $R(u+kT, v) \xrightarrow{k\to\infty} 0$ *for* $(u,v) = (t+\tau, t), (t,t), (t+\tau, t+\tau), (t, t+\tau)$;

(c″) $\sum_{j=0}^{\infty} |R(u+jT, v)|^2 < \infty$ *for* $(u,v) = (t+\tau, t), (t,t), (t+\tau, t+\tau), (t, t+\tau)$.

Proof. Item (a) follows from the computation

$$\text{Var}\,[\widehat{R}_N^{\dagger}(t+\tau, t)] = \frac{1}{N^2} \sum_{j=0}^{N-1} \sum_{k=0}^{N-1} R_{Z^{\dagger}}(t+jT, t+kT, \tau).$$

Item (b) is an application of (a) from Lemma 9.1 to the sequence $Z^\dagger_{t,\tau}$. Item (a′) follows from item (a) using the Toeplitz structure of stationary covariances. Item (b′) can be seen directly from item (b), and item (c′) is sufficient for (b′). Item (b″) is sufficient for item (b′) by use of Isserlis' formula,

$$E\{X_{t_1}X_{t_2}X_{t_3}X_{t_4}\} = E\{X_{t_1}X_{t_2}\}E\{X_{t_3}X_{t_4}\}+ \\ E\{X_{t_1}X_{t_3}\}E\{X_{t_2}X_{t_4}\} + E\{X_{t_1}X_{t_4}\}E\{X_{t_2}X_{t_3}\}, \tag{9.33}$$

which holds when X_t is real and Gaussian. For then

$$\begin{aligned} R_{Z^\dagger}(t_1,t_2,\tau) &= E\{Z^\dagger_{t_1,\tau}Z^\dagger_{t_2,\tau}\} \\ &= R(t_1+\tau,t_2+\tau)R(t_1,t_2) + R(t_1+\tau,t_2)R(t_1,t_2+\tau). \end{aligned} \tag{9.34}$$

Again by using (9.35) and the Schwarz inequality, item (c″) suffices for $\sum_{j=0}^{\infty} |R_{Z^\dagger}(t,t+jT,\tau)|^2 < \infty$ which implies (c′), or (b′) directly. ∎

It is important to observe that the presence of a discrete component in the spectrum of $Z^\dagger_{t,\tau}$ is in contradiction to the conditions of this proposition. In the Gaussian case, the presence of any discrete components in the spectrum of X_t is in contradiction to the conditions of this proposition.

Almost Sure Consistency. Conditions for almost sure consistency may be obtained from the stationary case when $\{Z^\dagger_{t,\tau}\}$ is PC in t for fixed τ, for then $\{Z^\dagger_{t+jT,\tau}\}$ is stationary in j for fixed t, τ.

Proposition 9.12 *If X is PC-T with bounded fourth moments and $\{Z^\dagger_{t+jT,\tau}\}$ is stationary in j for fixed t,τ, then the following are sufficient for a.s. consistency of $\widehat{R}^\dagger_N(t+\tau,t)$ (for t,τ fixed):*

(a) *for some $\alpha > 0$*

$$\frac{1}{N}\sum_{j=-N+1}^{N-1}\left(1-\frac{|j|}{N}\right)R_{Z^\dagger}(t+jT,t,\tau) \le \frac{K}{N^\alpha},$$

where $R_{Z^\dagger}(t+jT,t,\tau) = E\{Z^\dagger_{t+jT,\tau}Z^\dagger_{t,\tau}\}$;

(b) *either $\sum_{j=-\infty}^{\infty} |R_{Z^\dagger}(t+jT,t,\tau)| < \infty$ or $R_{Z^\dagger}(t+jT,t,\tau) = O(j^{-\alpha})$ is sufficient for (a);*

(c) *if X_t is Gaussian, the conditions $\sum_{j=0}^{\infty} |R(u+jT,v)|^2 < \infty$ or $R(u+jT,v) = O(j^{-\alpha})$ for $(u,v) = (t+\tau,t),(t,t),(t+\tau,t+\tau),(t,t+\tau)$ are sufficient for those in (2).*

Proof. Item (a) results from the application of Theorem 6.2 of Doob [49] to the stationary sequence $\{Z^\dagger_{t+jT,\tau}\}$. Both claims of item (b) are an application of Doob Theorem 7.1.1 to the stationary sequence $\{Z^\dagger_{t+jT,\tau}\}$. For item (c), application of the Isserlis formula (9.34) yields

$$\sum_{j=-\infty}^{\infty} |R_{Z^\dagger}(t+jT,t,\tau)| \le \sum_{j=-\infty}^{\infty} |R(t+jT+\tau,t+\tau)R(t+jT,t)| + |R(t+jT+\tau,t)R(t+jT,t+\tau)|$$

and, for example, the first of the sums on the right-hand side is bounded by

$$\left[\sum_{j=-\infty}^{\infty} R(t+jT+\tau,t+\tau)^2\right]^{1/2} \left[\sum_{j=-\infty}^{\infty} R(t+jT,t)^2\right]^{1/2}$$

and the other similarly. Thus the first condition in item (2) is satisfied. If $R(u+jT,v) = O(j^{-\alpha})$ for $(u,v) = (t+\tau,t),(t,t),(t+\tau,t+\tau),(t,t+\tau)$, then again by (9.35) we conclude the second condition, $R_{Z^\dagger}(t+jT,t,\tau) = O(j^{-\alpha})$, in item (2) is satisfied. ∎

Asymptotic Normality. The first results on asymptotic normality for estimators of covariances of PC sequences were due to Pagano [175] for periodic autoregressions, and to Vecchia and Ballerini [219], who obtained asymptotic normality results for one sided (causal) infinite periodic moving averages (8.15) under the additional conditions $\sum_{j\ge 0} |a_j(t)| < \infty$ and the orthonormal sequence ξ_k has fourth moments, $E\{\xi_k^4\} < \infty$. We will mainly concentrate here on the results based on ϕ-mixing (introduced by Ibragimov [120]), but an outline of the linear model approach is given in the supplements. The mixing approach has been used to obtain consistency and asymptotic normality for covariance and spectral estimators for almost PC processes in continuous time [108, 109, 135]. Although the results are easier here for discrete time PC processes, the main ideas are present. The following lemma relates the mixing function for $Z_{t,\tau} = X_{t+\tau}X_t$ in terms of the mixing function for X_t.

Lemma 9.3 *If X_t is periodically stationary with period T and uniformly ϕ-mixing with mixing function $\bar{\phi}_n$, then for arbitrary real numbers β_1, β_2 and arbitrary integers t_1, τ_1, t_2, τ_2, the sequence $\zeta_j = \beta_1 X_{t_1+\tau_1+jT}X_{t_1+jT} + \beta_2 X_{t_2+\tau_2+jT}X_{t_2+jT}$ is ϕ-mixing with mixing function $\bar{\phi}_{(n-n_0)T}$, where $n_0 \ge 0$ is a fixed integer.*

Proof. Set $n_{\min} = \lfloor \min\{t_1, \tau_1, t_2, \tau_2\}/T \rfloor$, $n_{\max} = \lfloor \max\{t_1, \tau_1, t_2, \tau_2\}/T \rfloor + 1$. Since

$$\begin{aligned} \mathcal{F}_{\zeta,m} &= \mathcal{B}(\zeta_j,\ j \le m) \subset \mathcal{B}(X_s,\ s \le mT + n_{\max}T), \\ \mathcal{G}_{\zeta,m+n} &= \mathcal{B}(\zeta_j,\ j \ge m+n) \subset \mathcal{B}(X_s,\ s \ge mT + nT + n_{\min}T), \end{aligned}$$

we have

$$\phi_{\zeta,n} = \sup\{|P(B|A) - P(B)| : A \in \mathcal{F}_{\zeta,m}, P(A) > 0,\ B \in \mathcal{G}_{\zeta,m+n}\} \le \bar{\phi}_{(n-n_0)T},$$

where $n_0 = n_{\max} - n_{\min}$. Since we are only interested in $\bar{\phi}_{(n-n_0)T}$ as $n \to \infty$, we can take $\bar{\phi}_n = 0$ for $n \le 0$. ∎

Proposition 9.13 *If X_t is periodically stationary with period T, $EX_t^4 < \infty$, $t \in \mathbb{Z}$ and uniformly ϕ-mixing with $\sum_{n=-\infty}^{\infty} (\bar{\phi}_n)^{1/2} < \infty$, then for any $t = 0, 1, ..., T-1$, $\tau \in \mathbb{Z}$ for which $\sigma^2(t,\tau) = \sum_{j=-\infty}^{\infty} R_{Z^\dagger}(t + jT, t, \tau) > 0$,*

$$\sqrt{N}[\widehat{R}_N^\dagger(t+\tau, t) - R(t+\tau, t)] \Rightarrow \mathbb{N}\left(0, \sigma^2(t,\tau)\right).$$

Proof. The hypotheses imply that $Z_{t+jT,\tau}^\dagger$ is a zero mean stationary sequence in j with $E\{(Z_{t+jT,\tau}^\dagger)^2\} < \infty$, $j \in \mathbb{Z}$ and the estimator for the mean of $Z_{t+jT,\tau}^\dagger$ is $\widehat{R}_N^\dagger(t+\tau, t) - R(t+\tau, t)$. The preceding lemma with $\beta_1 = 1$, $\beta_2 = 0$ implies for every $t = 0, 1, \ldots, T-1$ and $\tau \in \mathbb{Z}$, that $Z_{t+jT,\tau}^\dagger$ is ϕ-mixing with mixing function $\bar{\phi}_{(n-n_0)T}$ and hence the results of Ibragimov [120, Theorem 1.5] can be applied directly. The condition $\sum_{n=0}^{\infty} (\bar{\phi}_{(n-n_0)T})^{1/2} < \infty$ along with the relation (9.19) for $r = s = 2$ ensures the existence of the sum $\sigma^2(t,\tau) = \sum_{j=-\infty}^{\infty} E\{Z_{t+jT,\tau}^\dagger Z_{t,\tau}^\dagger\} = \sum_{j=-\infty}^{\infty} R_{Z^\dagger}(t + jT, t, \tau)$ even though the limit may be zero, an example of which is given in the supplements. ∎

Corollary 9.13.1 *Under the conditions of Proposition 9.13, for any vector $\boldsymbol{\beta}$ such that $\boldsymbol{\beta}'\boldsymbol{\Sigma}\boldsymbol{\beta} > 0$, we have*

$$\sqrt{N} \sum_{k=1}^{K} \beta_k (\widehat{R}_{t_k+\tau_k, t_k, N}^\dagger - R_{t_k+\tau_k, t_k}) \Rightarrow \mathbb{N}\left(0, \boldsymbol{\beta}'\boldsymbol{\Sigma}\boldsymbol{\beta}\right),$$

where

$$[\boldsymbol{\Sigma}]_{p,q} = \sum_{j=-\infty}^{\infty} E\{Z_{t_p+jT,\tau_p}^\dagger Z_{t_q,\tau_q}^\dagger\}$$

and $\beta_k,\ k=1,2,\ldots,K$ *are arbitrary real numbers and* $(t_k,\tau_k),\ k=1,2,\ldots,K$ *are arbitrary unique pairs of integers.*

Proof. Consider now for arbitrary real β_k and integers t_k,τ_k the sequence

$$\zeta_j=\sum_{k=1}^{K}\beta_k Z^{\dagger}_{t_k+jT,\tau_k},$$

which is clearly stationary in j and of mean zero, $E\{\zeta_j\}=\sum_{k=1}^{K}\beta_k E\{Z^{\dagger}_{t_k+jT,\tau_k}\}=0$. From the preceding lemma ζ_j is ϕ-mixing with mixing function $\bar{\phi}_{(n-n_0)T}$ for which $\sum_{n=0}^{\infty}(\bar{\phi}_{(n-n_0)T})^{1/2}<\infty$ and where n_0 depends on the collection $\{(t_k,\tau_k):k=1,2,\ldots,K\}$. Setting

$$J_N=\frac{1}{N}\sum_{j=0}^{N-1}\zeta_j=\sum_{k=1}^{K}\beta_k(\widehat{R}^{\dagger}_{t_k+\tau_k,t_k,N}-R_{t_k+\tau_k,t_k})$$

produces

$$\begin{aligned}
N\,\mathrm{Var}\,(J_N) &= NE\{J_N^2\}\\
&= \frac{1}{N}\sum_{j=0}^{N-1}\sum_{j'=0}^{N-1}\sum_{k=1}^{K}\sum_{k'=1}^{K}\beta_k\beta_{k'}E\{Z^{\dagger}_{t_k+jT,\tau_k}Z^{\dagger}_{t_{k'}+j'T,\tau_{k'}}\}\\
&= \sum_{k=1}^{K}\sum_{k'=1}^{K}\beta_k\beta_{k'}\sum_{j=-N+1}^{N-1}\left(1-\frac{|j|}{N}\right)E\{Z^{\dagger}_{t_k+jT,\tau_k}Z^{\dagger}_{t_{k'},\tau_{k'}}\},
\end{aligned}$$

which shows the asymptotic variance of $\sqrt{N}J_N$ is $\boldsymbol{\beta}'\boldsymbol{\Sigma}\boldsymbol{\beta}$. The sums defining $[\boldsymbol{\Sigma}]$ exist due to $\sum_{n=0}^{\infty}(\bar{\phi}_{(n-n_0)T})^{1/2}<\infty$ and the inequality (9.19). The preceding proposition (or Ibragimov [120, Theorem 1.5]) can be applied provided the asymptotic variance is positive, $\boldsymbol{\beta}'\boldsymbol{\Sigma}\boldsymbol{\beta}>0$. ∎

If X_t is stationary and Gaussian, several facts combine to make a result that depends only on a summability of covariances.

Proposition 9.14 *If* X_t *is periodically stationary with period* T, *Gaussian, and* $\sum_{t=0}^{T-1}\sum_{\tau=-\infty}^{\infty}|R(t+\tau,t)|^2<\infty$, *then*

$$\sqrt{N}[(\widehat{R}^{\dagger}_N(t+\tau,t)-R(t+\tau,t)]\Rightarrow\mathbb{N}\left(0,\sigma^2(t,\tau)\right),\ t=0,1,...,T-1,$$

where $\sigma^2(t,\tau)=\sum_{j=-\infty}^{\infty}R_{Z^{\dagger}}(t+jT,t,\tau)$ *and*

$$R_{Z^{\dagger}}(t+jT,t,\tau)=R(t+jT+\tau,t+\tau)R(t+jT,t)+R(t+jT+\tau,t)R(t+jT,t+\tau).$$

9.3.2 Estimation of $B_k(\tau)$

Since in the stationary case the natural estimator for the covariance is (9.23), examination of (6.2) suggests the following form for the estimator of $B_k(\tau)$ based on a sample of length NT:

$$\widehat{B}_{k,NT}(\tau) = \frac{1}{NT} \sum_{t \in I_{NT,\tau}} [X_{t+\tau} - \widehat{m}_{t+\tau,N}][X_t - \widehat{m}_{t,N}]e^{-i2\pi kt/T}, \tag{9.35}$$

where the interval over which the sum is performed must be modified for $\tau < 0$ because X_t is no longer stationary.

$$I_{NT,\tau} = \begin{cases} [0, NT - \tau - 1], & \tau \geq 0 \\ [-\tau, NT - 1], & \tau < 0 \end{cases}. \tag{9.36}$$

An alternative approach is to base the estimation of $B_k(\tau)$ on its Fourier relationship to $R(t + \tau, t)$. For this approach we would seek conditions for which the estimator $\widehat{R}_N^\dagger(t+\tau, t)$ is consistent for each $t = 0, 1, \ldots, T-1$, and then use the fact that each element of the vector $\mathbf{B}(\tau) = (B_0(\tau), B_1(\tau), \ldots, B_{T-1}(\tau))'$ is given by (6.2), which in vector form is

$$\mathbf{B}(\tau) = T^{-1/2} V * (0)[R(0 + \tau, 0), R(1 + \tau, 1), \ldots, R(T - 1 + \tau, T - 1)]'.$$

In this manner many of the preceding results adapt easily to give conditions for consistency. We will concentrate on the first approach because it can be applied to the case of almost PC processes for which the covariance $R_{t+\tau,t}$ cannot be directly estimated. See the supplements for a list of references that address the estimation of the Fourier coefficients of $B(t, \tau)$ in the almost periodic case.

As in the estimation of $R(t + \tau, t)$, we actually study

$$\widehat{B}^\dagger_{k,NT}(\tau) = \frac{1}{NT} \sum_{t \in I_{NT,\tau}} [X_{t+\tau} - m_{t+\tau}][X_t - m_t]e^{-i2\pi kt/T} \tag{9.37}$$

and use the following lemma.

Lemma 9.4 *If X_t is PC-T and has fourth moments, and if $\widehat{m}_{t,N} \xrightarrow{P} m_t$ for $t = 0, 1, \ldots, T - 1$, then*

$$\Delta B_N = \widehat{B}_{k,NT}(\tau) - \widehat{B}^\dagger_{k,NT}(\tau) \xrightarrow{P} 0.$$

Proof. Computing ΔB_N directly gives

$$
\begin{aligned}
|\Delta B_N| &= \frac{1}{N}\left| \sum_{t\in I_{NT,\tau}} \left[X_{t+\tau}(m_t - \widehat{m}_t) + X_t(\widehat{m}_{t+\tau} - m_{t+\tau}) \right.\right. \\
&\quad \left.\left. -(\widehat{m}_{t+\tau}\widehat{m}_t - m_{t+\tau}m_t)\right] e^{-i2\pi kt/T} \right| \\
&\le \frac{1}{NT}\sum_{t\in I_{NT,\tau}} \Big[|X_{t+\tau}(m_t - \widehat{m}_t)| + |X_t(\widehat{m}_{t+\tau} - m_{t+\tau})| \\
&\quad + |\widehat{m}_{t+\tau}\widehat{m}_t - m_{t+\tau}m_t|\Big] \\
&\le \left[\frac{1}{NT}\sum_{t\in I_{NT,\tau}} X_{t+\tau}^2\right]^{1/2}\left[\frac{1}{NT}\sum_{t\in I_{NT,\tau}} (m_t - \widehat{m}_t)^2\right]^{1/2} \\
&+ \left[\frac{1}{NT}\sum_{t\in I_{NT,\tau}} X_t^2\right]^{1/2}\left[\frac{1}{NT}\sum_{t\in I_{NT,\tau}} (\widehat{m}_{t+\tau} - m_{t+\tau})^2\right]^{1/2} \\
&+ \left[\frac{1}{NT}\sum_{t\in I_{NT,\tau}} \widehat{m}_{t+\tau}^2\right]^{1/2}\left[\frac{1}{NT}\sum_{t\in I_{NT,\tau}} (m_t - \widehat{m}_t)^2\right]^{1/2} \\
&+ \left[\frac{1}{NT}\sum_{t\in I_{NT,\tau}} \widehat{m}_t^2\right]^{1/2}\left[\frac{1}{NT}\sum_{t\in I_{NT,\tau}} (\widehat{m}_{t+\tau} - m_{t+\tau})^2\right]^{1/2}, \qquad (9.38)
\end{aligned}
$$

where the latter bounds are due to the Schwarz inequality for finite sums. Each of the quantities on the left in (9.38) are bounded in probability, due to the Chebychev inequality and the fact that they have finite expectation, for example,

$$
E\left[\frac{1}{NT}\sum_{t\in I_{NT,\tau}} X_t^2\right]^{1/2} \le \left[E\frac{1}{NT}\sum_{t\in I_{NT,\tau}} X_t^2\right]^{1/2} \le \left[\max_t E\{X_t^2\}\right]^{1/2}.
$$

By hypothesis, the quantity in each of the right brackets in (9.38) converges to zero in probability, giving the claimed result using the same argument as in Lemma 9.2. ∎

For the next lemma, set $M_2 = \max_t E\{X_t^2\}$.

Lemma 9.5 *If X_t is PC-T, then*

$$
E\{\widehat{B}_{k,NT}^{\dagger}(\tau)\} = B_k(\tau)\left[1 - \frac{\tau}{NT}\right] + \frac{\epsilon_{NT}(\tau,)}{NT} \qquad (9.39)
$$

where $\epsilon_{NT}(\tau,)$ is bounded for all N and τ. If $\sum_{\tau=-\infty}^{\infty}|R(t+\tau,t)|<\infty$ for $t=0,1,\ldots T-1$, then there is a constant K for which

$$\sum_{\tau=-\infty}^{\infty}|\epsilon_{NT}(\tau,)|\leq K. \tag{9.40}$$

Proof. We shall show the result for $\tau\geq 0$. For any N,

$$\begin{aligned} E\{\widehat{B}_{k,NT}^{\dagger}(\tau)\} &= \frac{1}{NT}\sum_{t=0}^{NT-\tau-1}R(t+\tau,t)e^{-i2\pi kt/T} \\ &= \frac{1}{NT}\sum_{t=0}^{mT-1}R(t+\tau,t)e^{-i2\pi kt/T} \\ &\quad +\frac{1}{NT}\sum_{t=mT}^{NT-\tau-1}R(t+\tau,t)e^{-i2\pi kt/T} \\ &= \frac{mT}{NT}B_k(\tau)+\frac{1}{NT}\sum_{t=mT}^{NT-\tau-1}R(t+\tau,t)e^{-i2\pi kt/T} \end{aligned} \tag{9.41}$$

where m is the largest integer in $(NT-\tau-1)/T$. Since $NT-\tau-1=mT+a$, where $0\leq a<T$, then the required $\epsilon_{NT}(\tau,)$ is given by

$$\epsilon_{NT}(\tau)=-aB_k(\tau)+\frac{1}{NT}\sum_{mT}^{NT-\tau-1}R(t+\tau,t)e^{-i2\pi kt/T}$$

and

$$|\epsilon_{NT}(\tau)|\leq 2\sum_{t=0}^{T-1}|R(t+\tau,t)|\leq 2TM_2, \tag{9.42}$$

where M_2 is the bound on the second moments of X_t. The second part of the claim follows from (9.42). ∎

Proposition 9.15 *If X is PC-T with $E\{X_t^4\}\leq M_4$, then each of the following is sufficient for the mean-square consistency of $\widehat{B}_{k,NT}^{\dagger}(\tau)$ for $k=0,1,\ldots,T-1$.*

(a) $\lim_{u\to\infty}R_{Z^{\dagger}}(t+u,t,\tau)=0$ *uniformly in t;*

(b) *if $Z_{t,\tau}^{\dagger}$ is PC in t and $\sum_{t_1=-\infty}^{\infty}\sum_{t_2=0}^{T-1}|R_{Z^{\dagger}}(t_1,t_2,\tau)|<\infty$;*

(c) *if X is Gaussian and $\sum_{\tau=-\infty}^{\infty}\sum_{t=0}^{T-1}R^2(t+\tau,t)<\infty$;*

(d) *if* X *is Gaussian and* $\lim_{\tau\to\infty} R(t+\tau,t) = 0$ *uniformly in* t.

Proof. We shall show the result for $\tau \geq 0$. First note that if

$$\begin{aligned} J_{k,NT}(\tau) &= \frac{1}{NT}\sum_{t=0}^{NT-\tau} Z^{\dagger}_{t,\tau}e^{-i2\pi kt/T} \\ &= \widehat{B}^{\dagger}_{k,NT}(\tau) - E\{\widehat{B}^{\dagger}_{k,NT}(\tau)\} \end{aligned} \tag{9.43}$$

converges in mean square to zero as $N\to\infty$, then the triangle inequality and Lemma 9.5 produce the desired result $\widehat{B}^{*}_{N}(t,\tau) \to B(t,\tau)$ in the mean-square sense.

That condition (a) is sufficient follows from

$$E\{J^2_{k,NT}(\tau)\} \leq \frac{1}{(NT)^2}\sum_{u=0}^{NT-1}\sum_{v=0}^{NT-1} |R_{Z^{\dagger}}(u,v,\tau)| \tag{9.44}$$

and performing the sum in two disjoint regions B and $A-B$, where A and B are defined in the proof of Lemma 9.1. As in that proof, N_0 may be chosen large enough that $|R_{Z^{\dagger}}(u,v,\tau)| < \epsilon/2$ everywhere in $A-B$ and so the part of the sum in (9.44) over $A-B$ satisfies the same inequality. Having fixed N_0, the value of N may be chosen to make the sum over B arbitrarily small. Note this is precisely condition (b$'$) of Proposition 9.11 holding for $t = 0, 1, \ldots, T-1$ (uniformly in t).

The sufficiency of (b) follows from the application of condition (b) of Lemma 9.1 to the sequence $Z^{\dagger}_{t,\tau}$. This is condition (c$'$) of Proposition 9.11 holding uniformly in t.

For (c) we use the fact that for any $\epsilon > 0$ there is an N_0 for which $|u-v| > N_0$ implies $|R(u,v)| < \epsilon^{1/2}$ for all v. Thus using the Isserlis formula,

$$|R_{Z^{\dagger}}(u,v,\tau)| \leq |R(u+\tau,v+\tau)R(u,v)| + |R(u+\tau,v)R(u,v+\tau)| \leq \epsilon/2$$

for all v if $|u-v| > N_0 + |\tau|$. The argument of condition (a) may then be applied to obtain the conclusion. This condition is sufficient for condition (b$'$) of Proposition 9.11 to hold for all t,τ.

To establish (d), we use the Isserlis formula again to obtain

$$\sum_{u,v=0}^{N-1} |R_{Z^{\dagger}}(u,v,\tau)| \leq \sum_{u,v=0}^{N-1} [|R(u+\tau,v+\tau)R(u,v)| + |R(u+\tau,v)R(u,v+\tau)|]$$

and then the Schwarz inequality, applied, for example, to the first term on the right, to obtain

$$\sum_{u,v=0}^{N-1} |R(u+\tau,v+\tau)R(u,v)| \leq \left[\sum_{u,v=0}^{N-1} R^2(u+\tau,v+\tau)\right]^{1/2}\left[\sum_{u,v=0}^{N-1} R^2(u,v)\right]^{1/2}.$$

But these sums of squared terms, such as $R^2(u+\tau, v+\tau)$, under the transformation $s = u - v$, $t = v$ become

$$\begin{aligned}\sum_{u,v=0}^{NT-1} R^2(u+\tau, v) &\le \sum_{s=-\infty}^{\infty} \sum_{t=-NT+1}^{NT-1} R^2(s+t+\tau, t+\tau) \\ &\le 2N \sum_{s=-\infty}^{\infty} \sum_{t=0}^{T-1} R^2(s+t+\tau, t+\tau).\end{aligned}$$

The other three sums are similarly bounded by this last quantity and so $E\{J^2_{k,NT}(\tau)\}$ is $O(1/N)$, hence establishing (d). ∎

Note all the conditions give consistency independently of $k = 0, 1, \ldots, T-1$ and conditions (c) and (d) are also independent of τ. All of these conditions are versions of the conditions of Proposition 9.11 but with uniformity.

Almost Sure Consistency. Conditions for almost sure consistency can also be obtained when $\{Z^\dagger_{t,\tau}\}$ is PC in t for fixed τ, for then $\{Z^\dagger_{t+jT,\tau}\}$ is stationary in j for fixed t, τ.

Proposition 9.16 *If X is PC-T with bounded fourth moments, then the following are sufficient for a.s. consistency of $\widehat{B}^\dagger_{k,N}(\tau)$:*

(a) *for some $\alpha > 0$*

$$\frac{1}{(NT)^2} \sum_{s=0}^{NT-1} \sum_{t=0}^{NT-1} R_{Z^\dagger}(s, t, \tau) \le \frac{K}{N^\alpha}, \tag{9.45}$$

where $R_{Z^\dagger}(s, t, \tau) = E\{Z^\dagger_{s,\tau} Z^\dagger_{t,\tau}\}$;

(b) *either $\sum_{u=-\infty}^{\infty} |R_{Z^\dagger}(t+u, t, \tau)| < \infty$ or $R_{Z^\dagger}(t+u, t, \tau) = O(u^{-\alpha})$ for $t = 0, 1, \ldots, T-1$ are sufficient for (a);*

(c) *if X_t is Gaussian, the conditions $\sum_{t=0}^{T-1} \sum_{\tau=-\infty}^{\infty} |R(t+\tau, t)|^2 < \infty$ or $\max_{t=0}^{T-1} |R(t+\tau, t)| = O(\tau^{-\alpha})$ are sufficient for those in (b).*

Proof. Item (a) is sufficient for mean-square consistency because the left quantity in (9.45) dominates $E\{J^2_{k,\tau,NT}\}$ (see (9.43)). Following the proof of Doob [49, Chapter X, Theorem 6.1], (9.45) implies $J_{k,\tau,N_m} \overset{m\to\infty}{\longrightarrow} 0$ a.s on a subsequence N_m and to make the general statement it is enough for $\sup_t R_{Z^\dagger}(t, t, \tau) < \infty$, a condition ensured by the bounded fourth moments. We note that (9.45) is equivalent to

$$\frac{1}{N} \sum_{j=-N+1}^{N-1} \left(1 - \frac{|j|}{N}\right) E\{Z^\dagger_{s+jT,\tau} Z^\dagger_{t,\tau}\} \le \frac{K}{N^\alpha}, \quad s, t = 0, 1, \ldots, T-1,$$

which is a little stronger than the uniformity of condition (a) of Proposition 9.12 with respect to t.

Both statements of item (b) are sufficient for item (a) and these are stronger than the uniformity of condition (b) in Proposition 9.12 with respect to t.

For item (c), application of the Isserlis formula (9.34) yields

$$\frac{1}{N}\sum_{j=-N+1}^{N-1}\left(1-\frac{|j|}{N}\right)E\{Z^{\dagger}_{t+jT,\tau}Z^{\dagger}_{t,\tau}\} \tag{9.46}$$
$$\leq \frac{1}{N}\sum_{j=-\infty}^{\infty}\Big[R(t+jT+\tau,t+\tau)R(t+jT,t)$$
$$+R(t+jT+\tau,t)R(t+jT,t+\tau)\Big]$$

and, for example, the first of these sums is bounded by

$$\left[\sum_{j=-\infty}^{\infty} R(t+jT+\tau,t+\tau)^2\right]^{1/2}\left[\sum_{j=-\infty}^{\infty} R^2(t+jT,t)\right]^{1/2}$$
$$\leq \left[\sum_{t=0}^{T-1}\sum_{j=-\infty}^{\infty} R^2(t+jT+\tau,t+\tau)\right]^{1/2}\left[\sum_{t=0}^{T-1}\sum_{j=-\infty}^{\infty} R^2(t+jT,t)\right]^{1/2}$$
$$\leq \sum_{t=0}^{T-1}\sum_{j=-\infty}^{\infty}|R(t+jT+\tau,t)|^2<\infty$$

and the other similarly. ∎

Note that whenever the conditions of Lemma 9.2 are also satisfied then there is also a.s. consistency for $\widehat{R}_N(t+\tau,t)$. Furthermore, $\max_t|R_Z(t+j,t,\tau)|=O(j^{-\alpha})$ and $\sum_{j=-\infty}^{\infty}\sum_{t=0}^{T-1}|R_Z(t+j,t,\tau)|<\infty$ suffice correspondingly for the conditions given in (b).

Asymptotic Normality. The first results on asymptotic normality for estimators of covariances of PC sequences were due to Pagano [175] for periodic autoregressions, and to Vecchia and Ballerini [219], who began with a one sided (causal) infinite moving average.

Here we will give only one result, which is a consequence of Proposition 9.13 and Corollary 9.13.1. Denote $\mathbf{e}_k^*$ as the conjugate transpose of the row vector $\mathbf{e}_k=(1,e^{i2\pi k/T},e^{i2\pi 2k/T},\ldots,e^{i2\pi(T-1)k/T})$.

Proposition 9.17 *Under the conditions of Proposition 9.13,*

$$\sqrt{N}(\widehat{B}^{\dagger}_{k,N}(\tau)-B_k(\tau))\Rightarrow \mathbf{N}\left(0,T^{-1}\mathbf{e}_k^*\boldsymbol{\Sigma}\mathbf{e}_k\right)$$

whenever $\mathbf{e}_k^* \mathbf{\Sigma} \mathbf{e}_k > 0$,

$$[\mathbf{\Sigma}]_{p,q} = \sum_{j=-\infty}^{\infty} E\{Z^{\dagger}_{t_p+jT,\tau} Z^{\dagger}_{t_q,\tau}\},\ t_p, t_q = 0, 1, \ldots, T-1,$$

and $V(\lambda)$ *is defined by (6.42).*

Proof. The result follows from the fact that

$$\widehat{B}^{\dagger}_{k,N}(\tau) = \frac{1}{T} \sum_{t=0}^{T-1} \widehat{R}^{\dagger}_N(t+\tau, t) e^{-i2\pi kt/T}$$

can be expressed as a linear combination of random variables $\widehat{R}^{\dagger}_N(t+\tau, t)$ that are each asymptotically normal under the assumptions, and so the limit will also be asymptotically normal provided the variance $\mathbf{e}_k^* \mathbf{\Sigma} \mathbf{e}_k > 0$. ∎

For real-valued discrete-time Gaussian PC sequences, Genossar, Lev-Ari, and Kailath [75] show that consistency for every k and τ occurs if and only if one of the following holds:

(a) the measure F_0 has no discrete component;

(b) $\lim_{A\to\infty} A^{-1} \sum_{\tau=0}^{A-1} |B_0(\tau)|^2 = 0$.

When X_t is Gaussian, the Isserlis formula allows us to estimate the rate of convergence of the covariance $\text{Cov}\,[\widehat{B}_{j,N}(\tau_1), \widehat{B}_{k,N}(\tau_2)]$ as $N \to \infty$. This result is needed to prove consistency of estimators for the spectral densities $f_k(\lambda)$ under the Gaussian assumption. A similar result will show consistency of these estimators under the assumption of ϕ-mixing.

Convergence of $\text{Cov}\,[\widehat{B}_{j,N}(\tau_1), \widehat{B}_{k,N}(\tau_2)]$ *for* X_t *Gaussian.* In the following lemma all the variables $a, b, \ldots, f, u, t, t_1, t_2$ are integers and to save space we employ subscripts, writing $B_{t,\tau} = R_{t+\tau,t}$.

Lemma 9.6 *If* X_t *is PC-T with* $\sum_{\tau=-\infty}^{\infty} \sum_{t=0}^{T-1} |R_{t+\tau,t}|^2 = K < \infty$, *then for* $t_2 > t_1$,

$$\begin{aligned} J_n(u, t_1, t_2) &= \sum_{t_1}^{t_2} R_{t+au+b,u+c} R_{t+du+e,u+f} e^{-i2\pi nt/T} \\ &= (t_2 - t_1) \sum_{p=0}^{T-1} \sigma_p + \epsilon_{u,t_1,t_2}, \end{aligned} \tag{9.47}$$

where

$$\sigma_p = B_{p,u+c} B_{n-p,u+f} e^{i2\pi p(au+b)/T} e^{i2\pi(n-p)(du+e)/T},$$

and the error sequence $\epsilon_{u,t_1,t_2}, u \in \mathbb{Z}$ *is bounded and summable uniformly in* t_1, t_2. *That is,*

$$|\epsilon_{u,t_1,t_2}| = K_1 < \infty$$

and

$$\sum_{u=-\infty}^{\infty} |\epsilon_{u,t_1,t_2}| \leq K_2 < \infty,$$

where K_1, K_2 *may be chosen to be independent of* t_1 *and* t_2.

Proof. Denote I_t as the summand appearing in the line above (9.47). Then by defining $t_2{}' = t_1 - mT$, where $m = [(t_2 - t_1)/T]$, the number of whole intervals of length T in the interval $[t_1, t_2]$, we obtain

$$\begin{aligned} J_n(u, t_1, t_2) &= \sum_{t_1}^{t_2{}'} I_t + \sum_{t_2{}'}^{t_2} I_t \\ &= mT \sum_{p=0}^{T-1} \sigma_p + \sum_{t_2{}'}^{t_2} I_t \\ &= (t_2 - t_1) \sum_{p=0}^{T-1} \sigma_p + \epsilon_{u,t_1,t_2}, \end{aligned} \tag{9.48}$$

where

$$\epsilon_{u,t_1,t_2} = \sum_{t_2'}^{t_2} I_t - (t_2 - t'_2) \sum_{p=0}^{T-1} \sigma_p$$

which is null if $(t_2 - t_1) \text{mod } (T) = 0$. To obtain the bound on $|\epsilon_{u,t_1,t_2}|$,

$$\begin{aligned} |\epsilon_{u,t_1,t_2}| &\leq \sum_{t=0}^{T-1} 2|I_t| \leq 2 \left[\sum_{t=0}^{T-1} R^2_{t+au+b,u+c} \right]^{1/2} \left[\sum_{t=0}^{T-1} R^2_{t+du+e,u+f} \right]^{1/2} \\ &\leq 2 \left[\sum_{t=0}^{T-1} M_4 \right]^{1/2} \left[\sum_{t=0}^{T-1} M_4 \right]^{1/2} = 2M_4 T. \end{aligned} \tag{9.49}$$

To show that $|\epsilon_{u,t_1,t_2}|$ is summable with respect to u, observe that

$$\alpha_u(a, b, c) = \left[\sum_{t=0}^{T-1} R^2_{t+au+b,u+c} \right]^{1/2},$$

is an ℓ_2 sequence in the variable u,

$$\sum_{u=-\infty}^{\infty} \alpha_u{}^2(a, b, c) = \sum_{u=-\infty}^{\infty} \sum_{t=0}^{T-1} R^2_{t+au+b,u+c} = K < \infty.$$

The final estimate $\sum_{u=-\infty}^{\infty} |\epsilon_{u,t_1,t_2}| \leq 2K$ is obtained by application of the Schwarz inequality for ℓ_2 to the top line of (9.49). ∎

This lemma is used in the next proposition to obtain estimates on the rate of convergence of the covariance $\text{Cov}[\widehat{B}_{j,\tau_1,N}, \widehat{B}_{k,\tau_2,N}]$ as $N \to \infty$ for Gaussian PC sequences. However, the square summability must be replaced with the stonger absolute summability.

Proposition 9.18 *If X_t is PC-T, Gaussian, and $\sum_{\tau=-\infty}^{\infty} \sum_{t=0}^{T-1} |R_{t+\tau,t}|^2 < \infty$, then for any $0 \leq \tau_1, \tau_2 \leq N$*

$$\begin{aligned} N\text{Cov}\,[\widehat{B}_{j,\tau_1,N}, \widehat{B}_{k,\tau_2,N}] &= \sum_{-N}^{N} U_N(u,\tau_1,\tau_2)[F_{j,k}(u,\tau_1,\tau_2) + G_{j,k}(u,\tau_1,\tau_2)] \\ &\quad + O(N^{-1}) \end{aligned} \tag{9.50}$$

where

$$\begin{aligned} F_{j,k}(u,\tau_1,\tau_2) &= \sum_{p=0}^{T-1} B_{p,u+\tau_1+\tau_2} B_{j-k-p,u} e^{-i2\pi(ju+pu)/T}, \\ G_{j,k}(u,\tau_1,\tau_2) &= \sum_{p=0}^{T-1} B_{p,u+\tau_1} B_{j-k-p,u-\tau_2} e^{-i2\pi(ju-j\tau_2+k\tau_2+p\tau_2)/T}, \end{aligned} \tag{9.51}$$

and the real function $U_N(u,\tau_1,\tau_2) \leq 1$ for $u \in [-N,N]$, and increases to 1 as $N \to \infty$.

Proof. The proof will be given for $\tau_2 \geq \tau_1 \geq 0$ in order to conserve space. From the definition of $\widehat{B}_{k,\tau,N}$ and the use of the Isserlis formula we find

$$\begin{aligned} &\text{Cov}\,[\widehat{B}_{j,\tau_1,N}, \widehat{B}_{k,\tau_2,N}] \\ &= \frac{1}{N^2} \sum_{s=0}^{N-\tau_1-1} \sum_{j=0}^{N-\tau_2-1} B_{t+\tau_2,s-t+\tau_1-\tau_2} B_{t,s-t} e^{-i2\pi(js-kt)/T} \\ &+ \frac{1}{N^2} \sum_{s=0}^{N-\tau_1-1} \sum_{j=0}^{N-\tau_2-1} B_{t,s-t+\tau_1-\tau_2} B_{t+\tau_2,s-t-\tau_2} e^{-i2\pi(js-kt)/T} \end{aligned} \tag{9.52}$$

and these two sums may be associated with the sums involving F and G in (9.50); we shall evaluate the second of these only. To perform the evaluation, partition the rectangle $E = [0, N-\tau_1-1] \times [0, N-\tau_2-1]$ into three sets,

$$\begin{aligned} A_1 &= \{(s,t) \in E : t > s\}, \\ A_2 &= \{(s,t) \in E : s \geq t \textit{ and } s \leq t + \tau_2 - \tau_1, \\ A_3 &= \{(s,t) \in E : s \geq t + \tau_2 - \tau_1. \end{aligned}$$

and denote the sums over these sets by S_1, S_2, and S_3.

To evaluate S_1, the transformation $u = s - t,\ v = s$ results in

$$S_1 = \sum_{u=-N+\tau_2}^{0} \sum_{v=0}^{N-\tau_2+u} B_{v-u,u+\tau_1} B_{v-u+\tau_2,u-\tau_2} e^{-i2\pi(jv-kv+ku)/T}$$

and the application of Lemma 9.6 to the inner sum yields

$$NS_1 = \sum_{u=-N+\tau_2}^{0} \left[1 - \frac{\tau_2 - u}{N}\right] G_{j,k}(u,\tau_1,\tau_2) + \frac{1}{N} \sum_{u=-N+\tau_2}^{0} \epsilon(u,0,N-\tau_2+u).$$

The expression $[1-(\tau_2-u)/N]$ in the preceding line may be identified with $U_N(u,\tau_1,\tau_2)$ in the interval $[-N+\tau_2, 0]$. The same transformation yields the other two parts,

$$NS_2 = \sum_{u=0}^{\tau_2-\tau_1} \left[1 - \frac{\tau_2}{N}\right] G_{j,k}(u,\tau_1,\tau_2) + \frac{1}{N} \sum_{u=0}^{\tau_2-\tau_1} \epsilon(u,u,N-\tau_2+u)$$

and

$$NS_3 = \sum_{u=\tau_2-\tau_1}^{N-\tau_1} \left[1 - \frac{\tau_1 + u}{N}\right] G_{j,k}(u,\tau_1,\tau_2) + \frac{1}{N} \sum_{u=\tau_2-\tau_1}^{N-\tau_1} \epsilon(u,u,N-\tau_1).$$

We then identify $[1-(\tau_2/N]$ and $[1-(\tau_1+u)/N]$ with $U_N(u,\tau_1,\tau_2)$ over the respective sets of indices appearing in these two lines and define $U_N(u,\tau_1,\tau_2) = 0$ for u not in $[-N+\tau_2, N+\tau_1]$. We may thus write

$$N(S_1+S_2+S_3) = \sum_{-N}^{N} U_N(u,\tau_1,\tau_2) G_{j,k}(u,\tau_1,\tau_2) + \frac{K_2}{N},$$

where

$$\begin{aligned} K_2 \quad = \quad & \sum_{u=-N+\tau_2}^{0} |\epsilon(u,0,N-\tau_2+u)| + \sum_{u=0}^{\tau_2-\tau_1} |\epsilon(u,u,N-\tau_2+u)| \\ & + \sum_{u=\tau_2-\tau_1}^{N-\tau_1} |\epsilon(u,u,N-\tau_1)|. \qquad \blacksquare \end{aligned}$$

Convergence of $\mathrm{Cov}\,[\widehat{B}_{j,N}(\tau_1), \widehat{B}_{k,N}(\tau_2)]$ *for* X_t ϕ*–Mixing.*

Lemma 9.7 *If X_t is periodically stationary with period T, $EX_t^4 < \infty$, $t \in \mathbb{Z}$ and uniformly ϕ-mixing with $\sum_{n=-\infty}^{\infty}(\bar{\phi}_n)^{1/2} < \infty$, then there are real numbers $C_1, C_2 \geq 0$ for which*

$$|\text{Cov}\,[\widehat{B}_{j,NT}(\tau_1), \widehat{B}_{k,NT}(\tau_2)]| \leq \frac{1}{NT}[C_1|\tau_1| + C_2]^{1/2}[C_1|\tau_2| + C_2]^{1/2}. \quad (9.53)$$

Proof. As in (9.43), set

$$J_{k,\tau,NT} = \widehat{B}_{k,NT}(\tau) - E\{\widehat{B}_{k,NT}(\tau)\} = \frac{1}{NT}\sum_{I(NT,\tau}Z_{t,\tau}^{\dagger}e^{-i2\pi kt/T}$$

with $I(NT,\tau) = \{0, 1, \ldots NT - \tau - 1\}$ and the result follows if we show $EJ_{k,\tau,NT}^2 \leq [C_1|\tau| + C_2]/NT$. To do this, note that $Z_{t,\tau} = X_{t+\tau}X_t$ implies for any s, t, τ that

$$\mathcal{B}(Z_u,\ u \leq t) \subset \mathcal{B}(X_u,\ u \leq t + \tau)$$

and

$$\mathcal{B}(Z_u,\ u \geq t + s) \subset \mathcal{B}(X_u,\ u \geq t + s).$$

Then we conclude that $Z_{t,\tau}$ is ϕ-mixing with mixing function $\phi_{t,s}^Z \leq \phi_{t,s-\tau}^X$ and so $\bar{\phi}_s^Z = \max_t \phi_{t,s}^Z \leq \max_t \phi_{t,s-\tau}^X = \bar{\phi}_{s-\tau}$ for $s > \tau$. Combining the like results for $\tau < 0$ gives $\bar{\phi}_s^Z \leq \bar{\phi}_{s-|\tau|}$ and for $s < 0$ gives $\bar{\phi}_s^Z \leq \bar{\phi}_{|s|-|\tau|}$ when $|s| > |\tau|$. Using the fact that (see (9.19)) $|R_{Z,\tau}(u,v)| \leq 2M_4\bar{\phi}^Z(u-v) \leq 2M_4\bar{\phi}_{|u-v|-\tau}$ when $|u - v| > |\tau|$ produces the estimate

$$\begin{aligned} EJ_{k,\tau,NT}^2 &\leq \frac{1}{(NT)^2}\sum_{u=0}^{NT-\tau}\sum_{v=0}^{NT-\tau}|R_{Z,\tau}(u,v)| \\ &\leq \frac{1}{(NT)^2}\sum_E + \frac{1}{(NT)^2}\sum_{F-E}2M_4\bar{\phi}^{1/2}, \end{aligned}$$

where $F = I(NT,\tau) \times I(NT,\tau)$, $E = \{(u,v) \in F : |u - v| < |\tau|\}$. The first sum is $O(C_1|\tau|/N)$ and the second is $O(C_2/N)$. ∎

9.4 ESTIMATION OF $R(t + \tau, t)$: PRACTICE

As in the case of m_t, we can describe the time dependent covariance of a PC sequence by $R(t + \tau, t)$ or by the Fourier coefficients, $B_k(\tau)$, and thus wish to estimate either. Here we describe implementations of the estimators

$\widehat{R}_N(t+\tau,t)$ and $\widehat{B}_{k,NT}(\tau)$, where the latter is estimated directly from the series by

$$\widehat{B}_{k,NT}(\tau) = \frac{1}{NT} \sum_{t \in I_{NT,\tau}} [X_{t+\tau} - \widehat{m}_{t+\tau,N}][X_t - \widehat{m}_{t,N}]e^{-i2\pi kt/T}, \tag{9.54}$$

as in the estimation of the mean m_t.

Again, as in the case of estimating m_t, the main reason $\widehat{B}_{k,NT}(\tau)$ is computed via (9.54) is that a test for significance based on neighboring frequencies becomes available as in Section 9.2.2. As mentioned earlier, $\widehat{B}_{k,NT}(\tau)$ can alternatively be computed from $\widehat{R}_{N_{t,\tau}}(t+\tau,t)$ by

$$\widehat{B}_{k,NT}(\tau) = \frac{1}{T} \sum_{t=0}^{T-1} \widehat{R}_{N_{t,\tau}}(t+\tau,t)e^{-i2\pi kt/T}, \tag{9.55}$$

but then tests for $B_k(\tau) = 0$ must be formulated differently, for example, by assuming that the series is stationary white noise.

9.4.1 Computation of $\widehat{R}_N(t+\tau,t)$

Given a sample of length NT, the general idea is to estimate $\widehat{R}_N(t+\tau,t)$ according to (9.25). One primary objective of this estimation is to determine if there is a τ for which $R(t+\tau,t)$ is properly periodic in t. As in the stationary case, the range of τ for which we can reliably estimate $R(t+\tau,t)$ is limited by the sample size. As $|\tau|$ increases, the sample size available for the correlation estimate diminishes toward zero. So our strategy must be, as it always is, to say as much as we can from the finite sample. In this case, we can only estimate the finite set of correlations $\{R(t+\tau,t),\ t = 0,1,\ldots,T-1,\ |\tau| \le \tau_{\max}\}$, and for each τ determine whether or not $R(t+\tau,t)$ has a periodic variation with respect to t. This effort can be lessened somewhat because for a real PC sequence, $B(t,-\tau) = B(t-\tau,\tau)$ and hence it is sufficient to test only $0 \le \tau \le \tau_{\max}$. If the tests for proper periodicity are negative for all $0 \le \tau \le \tau_{\max}$, then our conclusion is that the observed series is not consistent with PC-T structure. This statement is made only for the period T tested, so there is a natural question of whether T could be incorrect.

To further investigate the estimation of $R(t+\tau,t)$ we begin with $(\tau = 0)$ the estimation of $R(t,t) = \sigma^2(t)$. After testing for the presence of a periodic mean, this is the next natural step in determining whether a series has periodic structure with period T. Although there are some arguments to the contrary, we consider the rejection of $\sigma^2(t) \equiv \sigma^2$ (in favor of the proper T-periodicity of $\sigma^2(t)$) to signify presence of periodic correlation of period T. The first argument to the contrary is that tests for $\sigma^2(t) \equiv \sigma^2$ must rely on a single sample path, and so we cannot know if we are seeing a true PC sequence X_t or a randomly shifted $Y_t = X_{t+\theta}$ (with θ uniformly distributed on $\{0, 1, \ldots, T-1\}$) version of it. Using the principle of parsimony discussed in Section 6.8.3, the simplest approach is to always consider that the time we call $t = 0$ is some specific time that is not a random variable; that is, we do not allow random shifts to come into the probabilistic model we assume for our observational system. Another argument to the contrary is that we can construct random sequences with periodic variances (see the example at the end of Chapter 2) that are not PC. However, we do not believe they exist in any practical sense and thus we again come to deciding for the existence of PC-T structure whenever we can accept that $\sigma^2(t)$ is properly periodic with period T.

In order to help the perception of $\sigma^2(t) \equiv \sigma^2$, our realization of the estimator

$$\widehat{\sigma}_N(t)^2 = \frac{1}{N_t - 1} \sum_{p=0}^{N_t-1} [X_{t+pT} - \widehat{m}_{t,N_t}]^2, \tag{9.56}$$

introduced in (1.2), includes $1 - \alpha$ confidence intervals around each point estimate; these confidence intervals are based on the assumption that for the tth season, the random variables $\{X_{t+pT} - m_t, p = 0, 1, \ldots, N-1\}$ are Normal$(0, \sigma_t^2)$. Hence the confidence intervals are determined by a χ^2 distribution with $N - 1$ degrees of freedom. The existence of nonoverlapping confidence intervals, as in Figure 1.2, gives a preliminary clue that $\sigma^2(t) \equiv \sigma^2$. There are several possible tests for heterogeneous variance that can be employed, but we use Bartlett's test coded in MATLAB by Trujillo-Ortiz and Hernandez-Walls [216].

Program `persigest.m`. The program `persigest.m` implements the estimator $\widehat{\sigma}_N(t)$, but slightly more generally because the series may contain missing values and the length of the series may not be an integral number of periods. Given an input time series vector, and a specified period T, the program computes and returns $\widehat{\sigma}_N(t)$ based on all the values that are present (not missing) in the series; that is, for each t in the base period, $N_t = \text{card}\{p \in \{0, 1, \ldots, N-1\} : X_{t+pT} \text{ not missing }\}$. Using a specified α, the $1 - \alpha$ confidence intervals are computed at each $t = 1, 2, \ldots, T$. The p-value for Bartlett's test for homogeneity of variances is also computed and is present on the plots. The demeaned series $X_t - \widehat{m}_{t,N}$ is normalized by $\widehat{\sigma}_N(t)$ and returned to the calling program. When all the data for time t (modulo T) are missing, the missing points corresponding to these times are omitted from the time series plot, and $\widehat{\sigma}_N(t)$ is not plotted.

Figure 1.2 presented an application of `persigest` to a solar radiation series from station DELTA of the Taconite Inlet Project. The p-value from the Bartlett test for $\sigma(t) \equiv \sigma$ was computed by MATLAB to be zero. As in the case of the periodic mean, this is not too surprising considering the clarity of the periodicity and the size of the dataset (82 periods). After shortening the dataset to 4 periods we found that the p-value is ~ 0.6 and for 8 periods it is $\sim 3.7e-007$.

The estimation of $\sigma(t)$ and the result of the hypothesis test for $\sigma(t) \equiv \sigma$ represents only the simplest analysis of second order properties of X_t. Rejecting $\sigma(t) \equiv \sigma$ strongly suggests PC structure, but leaves open whether or not X_t is simply the result of a stationary process subjected to amplitude-scale modulation, as described in Section 2.1.3; to resolve this, we must estimate $R(t+\tau,t)$ for $\tau \neq 0$. And on the other hand, $\sigma(t) \equiv \sigma$ is possible for PC processes that are formed by various forms of time-scale modulation, as described in Section 2.1.4. So again we must estimate $R(t+\tau,t)$ for $\tau \neq 0$ in order to complete the analysis.

A test for $\sigma(t) \equiv \sigma$ based on the Fourier series representation of $R(t,t) = \sigma^2(t)$ will be given in Section 9.4.2.

Program `peracf.m`. The program `peracf.m` implements a slight modification of the $\widehat{\sigma}_{N,t}$ to accommodate missing values and the possibility that the length of the series may not be an integral number of periods. Given an input time series vector, and a specified period T, the program computes and returns

$$\widehat{R}_{N_{t,\tau}}(t+\tau,t) = \frac{1}{N_{t,\tau}} \sum_{k=0}^{N_{t,\tau}-1} [X_{t+kT+\tau} - \widehat{m}_{t+\tau,N_{t+\tau}}][X_{t+kT} - \widehat{m}_{t,N_t}], \quad (9.57)$$

where N_t is the cardinality of the indexes $I_t = \{k : X_{t+kT} \text{ not missing }\}$, and similarly for $I_{t+\tau}$. The quantity $N_{t,\tau}$ is the cardinality of $I_t \bigcap I_{t+\tau}$.

Denoting $\widehat{\sigma}^2_{N_t}(t) = \widehat{R}_{N_t}(t,t)$, the estimator of the autocorrelation (coefficient)

$$\widehat{\rho}_{N_{t,\tau}}(t+\tau,t) = \frac{\widehat{R}_{N_{t,\tau}}(t+\tau,t)}{\widehat{\sigma}_{N_{t+\tau}}(t+\tau)\widehat{\sigma}_{N_t}(t)} \quad (9.58)$$

is also provided. For each t,τ, confidence limits for $\widehat{\rho}_{N_{t,\tau}}(t+\tau,t)$ are computed by use of the Fisher transformation

$$Z = \frac{1}{2}\log\frac{1+\widehat{\rho}}{1-\widehat{\rho}}, \quad (9.59)$$

under which the Z are approximately $N(\mu_Z, \sigma_Z^2)$, where $\mu_Z \approx \zeta + (1/2N_{t,\tau})\rho$, with ζ the Fisher transformation of the true ρ and $\sigma_Z^2 = 1/(N_{t,\tau}-3)$ (see Cramér [34, page 399]). Assuming that the term $(1/2N_{t,\tau})\rho$ can be ignored, the confidence limits for ρ are determined simply from those of Z by the inverse of the Fisher transformation. A test for equality of correlations $\rho(t+\tau,t) \equiv \rho(\tau)$, where $\rho(\tau)$ is some unknown constant, may be made from the variable

$$S_\rho^2(\tau) = \sum_{t=1}^{T} \frac{(Z_t - \widehat{Z_t})^2}{1/(N_{t,\tau}-3)},$$

which, under the null hypothesis, is (approximately) $\chi^2(T-1)$.

Sometimes it is also of interest to test for $\rho(t+\tau,t) \equiv 0$ for some specific τ. Then we take $\mu_Z = 0$ so the test for $\rho(t+\tau,t) \equiv 0$ becomes that of testing the Zs for $\mu_Z = 0$.

Figure 9.1 presents the estimates of $\widehat{\rho}_{N_{t,1}}(t+1,t)$ for the solar radiation data of Figures 1.1 and 1.2. The sample sizes are $N_{t,1} = 82$ for $t = 1, ..., 23$ and $N_{t,1} = 81$ for $t = 24$. The 0.99 confidence intervals for true $\rho(t+1,t)$ are shown and the test for $\rho(t+1,t) \equiv \rho(1)$ yields a p-value of $5.3e-005$. The evidence for periodicity in $\rho(t+1,t)$ is strong although not as strong as for σ_t. For lags $\tau = 2, 3, 4$, the p-values for the test of $\rho(t+\tau,t) \equiv \rho(\tau)$ were $0.16, 0.58, 0.91$, but the specific test $\rho(t+\tau,t) \equiv 0$ gave computed p-values of zero; for these lags, significant nonzero correlation is present, but its time variation is insignificant.

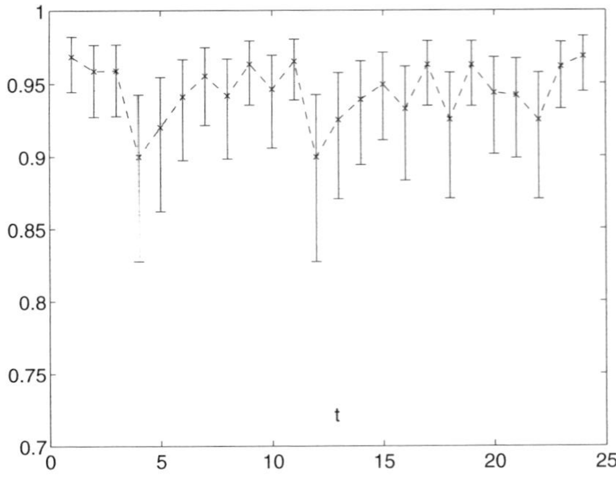

Figure 9.1 Values of $\widehat{\rho}_{t+1,t,N_{t,1}}$ versus t for solar radiation from station DELTA (see Figures 1.1 and 1.2) using $T = 24$. The sample sizes are $N_{t,1} = 82$ for $t = 1, ..., 23$ and $N_{t,1} = 81$ for $t = 24$. The error bars are the 99% confidence limits determined by the Fisher transformation and its inverse. The test for $\rho(t+1,t) \equiv \rho_1$ yields a p-value of $5.3e-005$, but for $\tau = 2, 3, 4$, the smallest was 0.16.

Recall from Sections 2.2.6 and 2.2.7 that we can construct simple PAR and PMA models whose variances are constant with respect to time so that a periodogram of squares does not detect the presence of PC structure. But we can detect it using $\widehat{\rho}_{N_{t,1}}(t+1,t)$ as demonstrated by the following examples.

Figure 9.2 presents the estimates of $\widehat{\rho}_{N_{t,1}}(t+1,t)$ for the solar radiation data of Figure 2.16, where $\sigma(t) \equiv \sigma$ could not be rejected by the periodogram of the squares. Here, using $N = 512$, the Bartlett test for $\sigma(t) \equiv \sigma$ gives a p-value of 1 (in Figure 2.16, $N = 5120$ was used). The test for $\rho(t+1,t) \equiv \rho(1)$, also based on $N = 512$, yields a p-value of 0. However, the more specific test

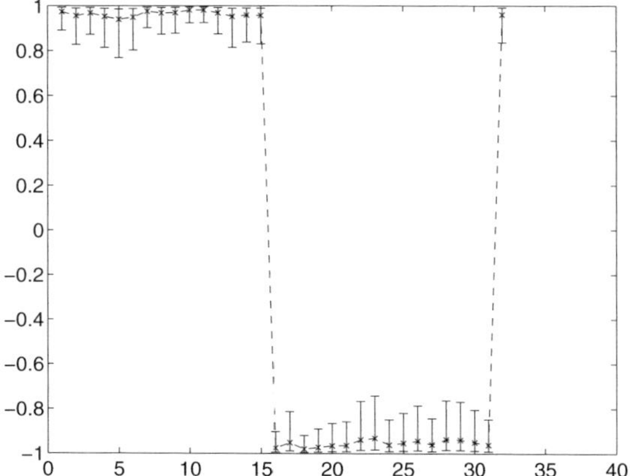

Figure 9.2 Values of $\widehat{\rho}_{t+1,t,N_{t,1}}$ versus t for the simulated switching $AR(1)$ of Figure 2.16, where $\phi(t) = 0.95$ for $0 \le t < T/2$ and $\phi(t) = -0.95$ for $T/2 \le t < T$, with $T = 32$; ξ_t is white noise. The sample sizes are $N_{t,1} = 16$ for $t = 1, ..., 31$ and $N_{t,1} = 15$ for $t = 32$. The error bars are the 99% confidence limits determined by the Fisher transformation and its inverse. The test for $\rho(t+1,t) \equiv \rho_1$ yields a p-value of 0, but for $\rho(t+1,t) \equiv 0$, the p-value was 0.7. See the text for an explanation.

$\rho(t+1,t) \equiv 0$ produced a p-value of 0.7, although the sample correlations are clearly not zero. This is caused by half the sample correlations being roughly 0.95 and the other half being roughly -0.95, giving an average that is consistent with 0. On the other hand, the $\rho(t+1,t) \equiv \rho(1)$ test uses a χ^2 and is very powerful in this situation.

This example illustrates that the perception of PC structure can be just one lag away.

In another example of a constant variance PC sequence, here we reexamine the case of the PMA(2) sequence $X_t = \xi_t + \cos(2\pi t/T)\xi_{t-1} + \sin(2\pi t/T)\xi_{t-2}$ presented in Figure 2.18, where $\sigma(t) \equiv \sigma$ could not be rejected by by the periodogram of the squares. Nor can it be rejected by the Bartlett test which gives a p-value of 0.6. However, the hypothesis $\rho(t+1,t) \equiv \rho(1)$ is strongly rejected (even visually, as seen in Figure 9.3) by a p-value of 0 based on $N = 600$; the more specific test $\rho(t+1,t) \equiv 0$ produced a p-value of $3.9e-07$. So the PC structure is clearly detected by the tests provided in program `peracf.m`.

The reader may wish to change parameters to make the contrast less extreme, or experiment with the other constant variance PC sequences discussed in Chapter 2.

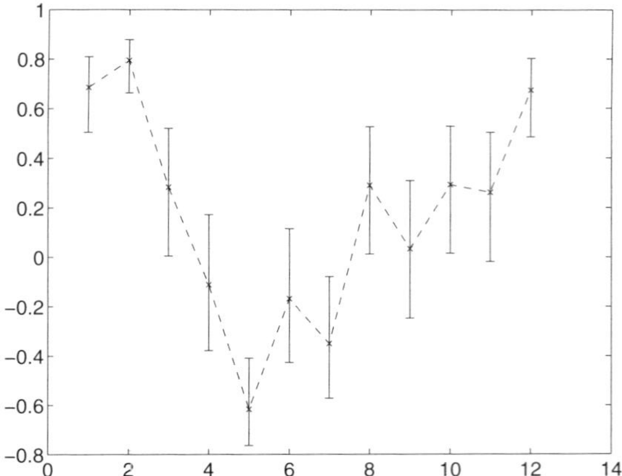

Figure 9.3 Values of $\widehat{\rho}_{t+1,t,N_{t,1}}$ versus t for the PMA(2) sequence $X_t = \xi_t + \cos(2\pi t/T)\xi_{t-1} + \sin(2\pi t/T)\xi_{t-2}$ presented in Figure 2.18. ξ_t is white noise and the resulting variance is constant $R_X(t,t) \equiv 2$. The sample sizes are $N_{t,1} = 50$ for $t = 1, ..., 11$ and $N_{t,1} = 49$ for $t = 12$. The error bars are the 99% confidence limits determined by the Fisher transformation and its inverse. The test for $\rho(t+1,t) \equiv \rho_1$ yields a p-value of 0, and for $\rho(t+1,t) \equiv 0$, the p-value was $3.9e-07$.

9.4.2 Computation of $\widehat{B}_{k,NT}(\tau)$

As in the case of computing $\widehat{\widehat{m}}_{k,N}$, our approach is to apply the Fourier transform, as in (9.22), to the product series

$$Y_{t,\tau} = [X_{t+\tau} - \widehat{m}_{t+\tau,N}][X_t - \widehat{m}_{t,N}]$$

in order to compute

$$\widehat{B}_{k,NT}(\tau) = \frac{1}{NT} \sum_{t \in I_{NT,\tau}} [X_{t+\tau} - \widehat{m}_{t+\tau,N}][X_t - \widehat{m}_{t,N}] e^{-i2\pi kt/T}.$$

Continuing with the Fourier transform method, we make the computation for each fixed τ of interest and note that the set $I_{NT,\tau}$, defined by (9.36), is not necessarily an integral number of periods in length. Thus for any τ we shall always cut the set $I_{NT,\tau}$ to be of length $N'T$, where $N' = \lfloor \text{card}\{I_{NT,\tau}\}/T \rfloor$, so the frequency $\lambda = 2\pi k/T$ occurring in the estimation of $B_k(\tau)$ is actually a Fourier frequency for an FFT of length $N'T$. The same algorithm discussed in Section 9.2.2 is then applied. In addition, we can see that, for $\tau = 0$,

$$\widehat{B}_{k,NT}(0) = \frac{1}{NT} \sum_{t=0}^{NT-1} [X_t - \widehat{m}_{t,N}]^2 e^{-i2\pi kt/T}$$

are given by the Fourier transform of the squares evaluated at certain Fourier frequencies. Hence the hypothesis test $\sigma_t^2 \equiv \sigma^2$ is the same as $B_k(0) = 0$ for all $k = 1, 2, ..., T-1$.

Program `Bcoeff.m`. For a real series and specified period T, the program `Bcoeff.m` implements the estimator $\widehat{B}_{k,NT}(\tau)$, but slightly more generally because the series may contain missing values and the series may not have a length of an integral number of periods. Missing values are set to the sample mean of the nonmissing values and the series is cut to the largest number N of whole periods in the series.

For each specified value of τ, the values of $\widehat{B}_{k,NT}(\tau)$ are computed only for $k = 0, 1, \ldots, \lfloor (T-1)/2 \rfloor$ because for real series, $\widehat{B}_{k,NT}(\tau) = \overline{\widehat{B}_{T-k,NT}(\tau)}$. In addition, the p-values for the test $B_k(\tau) = 0$, based on the variance contrast method, are returned. These p-values should be treated with caution because the requisite assumptions may not be met. Here are the considerations.

For large NT, the sample Fourier transform of $Y_{t,\tau} = [X_{t+\tau} - \widehat{m}_{t+\tau,N}][X_t - \widehat{m}_{t,N}]$,

$$\widetilde{Y_\tau}(\lambda_j) = \frac{1}{NT} \sum_{t=0}^{NT-1} Y_{t,\tau} e^{-\lambda_j t},$$

can be considered a long average, and so the Fourier coefficients $\widetilde{Y}(\lambda_j)$ would be expected to tend to normality. However, even if the series $\{X_t, t = 0, 1, \ldots NT-1\}$ is i.i.d. Normal, the series $Y_{t,\tau}$ may not even be i.i.d., except when $\tau = 0$, and so simple central limit results cannot be used; see the supplements at the end of the chapter for a problem. However, if we assume that the spectral density of $Y_{t,\tau}$ is smooth in the neighborhood of the frequencies $2\pi k/T$, then the p-values based on local estimates of the background variance are thought to be reasonable.

The results of `Bcoeff.m` when applied to the solar radiation data of Figures 1.1 and 1.2 are presented in Table 9.2.

The $\tau = 0$ results presented in Table 9.2(a) show that $B_0(0) = 0$ is strongly rejected, an expected result because $B_0(0)$ is the average variance of the sequence and must be nonzero for nontrivial sequences. The strong rejection of $B_1(0) = 0$ indicates the periodicity in the variance, a result consistent with the result of program `persigest`. The seemingly stronger rejection of $B_1(0) = 0$ relative to $B_0(0) = 0$ is attributed to the differences in the degrees of freedom in the variance samples. The $\tau = 1$ results presented in Table 9.2(b) show that $B_0(1) = 0$ and $B_1(1) = 0$ are strongly rejected, showing very significant $\tau = 1$ average correlation and very significant $\tau = 1$ periodic variation in the correlation. The $k = 1$ coefficient is also extremely significant at larger lags; analysis of larger k is not warranted until we remove the effect of the periodic variance.

Table 9.2 Result of program `Bcoeff.m` applied to solar radiation data of Figures 1.1 and 1.2. $NT = 1968$. Local variance estimates based on $n_A = 16$ neighboring Fourier coefficients.

(a) $\tau = 0$

k	$\lvert\widehat{B}_{k,NT}(\tau)\rvert$	$n1$	$n2$	Variance ratio	p-value
0	1.11e+004	1	16	1.63e+002	8.16e-010
1	4.42e+003	2	32	2.95e+002	0.00e+000
2	5.41e+002	2	32	5.08e+000	1.21e-002
3	2.89e+002	2	32	2.86e+000	7.18e-002
4	1.57e+002	2	32	1.43e+000	2.53e-001
5	9.08e+001	2	32	8.32e-001	4.44e-001
6	1.75e+001	2	32	3.73e-002	9.63e-001
7	7.25e+001	2	32	7.89e-001	4.63e-001
8	5.43e+001	2	32	8.38e-001	4.42e-001
9	5.99e+001	2	32	8.89e-001	4.21e-001
10	6.63e+001	2	32	8.54e-001	4.35e-001
11	6.86e+001	2	32	1.21e+000	3.12e-001

(b) $\tau = 1$

k	$\lvert\widehat{B}_{k,NT}(\tau)\rvert$	$n1$	$n2$	Variance ratio	p-value
0	1.01e+004	1	16	1.57e+002	1.11e-009
1	4.00e+003	2	32	2.23e+002	0.00e+000
2	4.72e+002	2	32	3.95e+000	2.93e-002
3	2.49e+002	2	32	2.78e+000	7.74e-002
4	1.14e+002	2	32	9.69e-001	3.90e-001
5	4.19e+001	2	32	3.77e-001	6.89e-001
6	2.49e+001	2	32	1.46e-001	8.65e-001
7	7.86e+001	2	32	1.36e+000	2.70e-001
8	5.89e+001	2	32	1.27e+000	2.96e-001
9	6.16e+001	2	32	1.54e+000	2.29e-001
10	7.71e+000	2	32	2.99e-002	9.71e-001
11	6.79e+001	2	32	3.31e+000	4.94e-002

To see if the Fourier coefficient method indicates $\rho_{t+\tau,t} \not\equiv \rho_\tau$, the program `Bcoeff.m` was applied to the series $X_t - \widehat{m}_{t,N}$ scaled by $\widehat{\sigma}_N(t)$, its sample periodic standard deviation. If the series is the result of an amplitude-scale modulation of a stationary series, then we expect that $\rho_k(\tau) = 0$ will be rejected for $k = 0$ and $\tau = 0$ and possibly some other τ; and it will never be rejected for any other k (i.e., for all $k > 0$) and τ.

Tables 9.3(a) and 9.3(b) first indicate that $\rho_k(\tau)=0$ is strongly rejected for $k=0$ and $\tau=1,2$, meaning that there are large average correlation coefficients at lags $\tau=1,2$. But also the coefficients $\rho_k(\tau)$ are never rejected

Table 9.3 Result for $\tau=1$ of program `Bcoeff.m` applied to $[X_t-\widehat{m}_{t,N}]/\widehat{\sigma}(t)$ for solar radiation data of Figures 1.1 and 1.2. $N'T=1944$ because one full period must be cut, giving $N'=81$. Local variance estimates based on $n_A=16$ neighboring Fourier coefficients.

(a) $\tau=1$					
k	$\lvert\widehat{\rho}_{k,NT}(\tau)\rvert$	$n1$	$n2$	Variance ratio	p-value
0	9.06e-001	1	16	1.25e+002	5.56e-009
1	2.90e-003	2	32	3.02e-002	9.70e-001
2	4.21e-003	2	32	4.48e-002	9.56e-001
3	7.93e-003	2	32	6.67e-001	5.20e-001
4	4.32e-003	2	32	2.35e-001	7.92e-001
5	2.97e-003	2	32	4.24e-001	6.58e-001
6	3.12e-003	2	32	3.18e-001	7.30e-001
7	4.58e-003	2	32	7.15e-001	4.97e-001
8	2.28e-003	2	32	4.10e-001	6.67e-001
9	3.70e-003	2	32	1.07e+000	3.55e-001
10	1.22e-003	2	32	1.29e-001	8.80e-001
11	2.27e-003	2	32	6.94e-001	5.07e-001
(b) $\tau=2$					
0	8.57e-001	1	16	1.11e+002	1.32e-008
1	5.74e-003	2	32	1.34e-001	8.75e-001
2	7.96e-003	2	32	2.38e-001	7.90e-001
3	9.47e-003	2	32	1.59e+000	2.20e-001
4	6.61e-003	2	32	1.68e+000	2.03e-001
5	2.71e-003	2	32	2.87e-001	7.52e-001
6	5.08e-003	2	32	1.26e+000	2.96e-001
7	3.86e-003	2	32	9.29e-001	4.05e-001
8	3.75e-003	2	32	5.59e-001	5.77e-001
9	7.83e-004	2	32	2.37e-002	9.77e-001
10	4.50e-003	2	32	7.50e-001	4.80e-001
11	2.93e-003	2	32	4.02e-001	6.72e-001

for $k>0$ and any $\tau=1,2$; that is, we cannot reject $\rho_{t+\tau,t}\equiv\rho_\tau$ for $\tau=1,2$ using the Fourier coefficient method using $n_A=16$ neighboring Fourier coefficients for the variance contrast. The most significant coefficient for the additional (not shown) analysis of $\tau = 3, 4, \ldots, 20$ yielded the p-value $6e-003$; but when this is corrected for multiple hypotheses (20 values of τ, 11 values of k) it becomes insignificant. So program `Bcoeff.m` does not reject the model of an amplitude modulated stationary sequence whereas the direct method does reject it with a p-value of $5.3e-005$ obtained from the test for $\rho(t+1,t) \equiv \rho_1$ based on $\widehat{\rho}_{t+1,t,N_{t,1}}$ (see Figure 9.1). It is not unexpected because the direct method examines the sample time-dependent correlations $\widehat{\rho}_{t+\tau,t,N_{t,\tau}}$ (or $\widehat{R}_{N_{t,\tau}}(t+\tau,t)$) for t in the base period, whereas $\rho_k(\tau)$ or $\widehat{B}_{k,NT}(\tau)$ are estimators for Fourier coefficients. If time sequences as in the plot of $\widehat{\rho}_{N_{t,1}}(t+1,t)$ in Figure 9.1 do not have a strong projection onto any particular one of the Fourier basis vectors, then deviation from constancy will be more easily observed in the time sequence. Thus we can reject $\rho(t+1,t) \equiv \rho_1$ better in the time domain than in the frequency domain. Of course, this is not always the case and we advocate use of both methods for determining if $\rho(t+\tau,t) \equiv \rho_\tau$.

PROBLEMS AND SUPPLEMENTS

9.1 Show if $\sum_p |R_{j+pT,k}| < \infty$ that

$$
\begin{aligned}
NE\{[\widehat{\widetilde{m}}_{j,N} - \widetilde{m}_j][\widehat{\widetilde{m}}_{k,N} - \widetilde{m}_k]\} &= f_Z^{jk}(0) \\
&= \frac{1}{T} f_{j-k}(2\pi j/T).
\end{aligned}
\tag{9.60}
$$

9.2 Show that if $NE[\widehat{m}_{s,N} - m_s][\widehat{m}_{t,N} - m_s] \to f_{st}(0)$, a sufficient condition for which is $\sum_{k=-\infty}^{\infty} |R_{s+kT,t}| < \infty$ for all $(s,t) \in \{0, 1, \ldots, T-1\}^2$, then using

$$
[\widehat{\widetilde{m}}_{k,N} - \widetilde{m}_k]^2 = \frac{1}{T^2} \sum_{s=0}^{T-1} \sum_{t=0}^{T-1} [\widehat{m}_{s,N} - m_s][\widehat{m}_{t,N} - m_t] e^{-i2\pi k(s-t)/T}
$$

we obtain

$$
\begin{aligned}
\lim_{N\to\infty} NE\{[\widehat{\widetilde{m}}_{k,N} - \widetilde{m}_k]^2 &= \frac{1}{T^2} \sum_{s=0}^{T-1} \sum_{t=0}^{T-1} f_{st}(0) e^{-i2\pi k(s-t)/T} \\
&= \frac{1}{T^2} [V^{-1}(0)]_k \mathcal{F}(0) [V(0)]_k \\
&= \frac{1}{T} f_Z^{kk}(0)
\end{aligned}
$$

by taking the kkth entry of (6.46). (Note that $[V^*(0)]_k$ is the kth row of $V^*(0)$ (or kth column of $V(0)$) defined in (6.42).) Finally, we recall that $f_Z^{kk}(\lambda) = f_0(2k\pi/T + \lambda/T)$, $\lambda \in [0, 2\pi)$. So this agrees with (9.10) of Proposition 9.3.

9.3 Show directly the following L_2 version of Proposition 9.4. Specifically, show that if X_t is PC-T with zero mean and Θ is independent of X_t, and uniformly distributed on $0, 1, \ldots, T-1$, then $\lim_{N\to\infty} E\{|e^{-i\lambda\Theta} J_{Y,N}(\lambda) - J_{X,N}(\lambda)|^2\} = 0$.

9.4 Here we construct a PC sequence with $\sigma^2_{t,\tau} = \sum_{j=-\infty}^{\infty} E\{Z_{t+jT,\tau} Z_{t,\tau}\} = 0$, where $Z_{t,\tau} = [X_{t+\tau} - m_{t+\tau}][X_t - m_t] - R_{t+\tau,t}$. Let $X_t = C(\omega)$ for t odd and $\xi_{t/2}$ for t even, where $C(\omega)$ is a random variable with $E\{C^4\} = [E\{C^2\}]^2$ and ξ_j is white noise. Then $t = 1, \tau = 2$ gives $Z_{t,\tau} = C^2(\omega) - E\{C^2\}$ and thus $E\{Z_{t+jT,\tau} Z_{t,\tau}\} = E\{[C^2(\omega) - E\{C^2\}]^2\} = 0$ for all j.

9.5 The following is a periodic extension of the Brockwell–Davis [28] Theorem 7.3.1. Suppose X_t is a linear PC-T sequence

$$X_t = \sum_{j=-\infty}^{\infty} \psi_j(t) \xi_{t-j}, \tag{9.61}$$

where $\psi_j(t) = \psi_j(t+T)$, $j \in \mathbb{Z}$, $\sum_j |\psi_j(t)| < \infty$, $t = 0, 1, \ldots, T-1$ and ξ_t is an i.i.d. mean zero, variance σ^2 sequence with $E\{\xi_t^4\} = \eta\sigma^4$. Then for $p \geq 0$, $q \geq 0$,

$$\begin{aligned}
&\lim_{N\to\infty} N \operatorname{Cov}[\widehat{R}_{t+p,t,N}, \widehat{R}_{t+q,t,N}] = \\
&\quad (\eta - 3)\sigma^4 \sum_{i=-\infty}^{\infty} \psi_{i+p}(t+p)\psi_i(t) S_{i,q,t} + \sum_{\nu=-\infty}^{\infty} R_{t+\nu T+p,t+q} R_{s+\nu T,t} \\
&\quad + \sum_{\nu=-\infty}^{\infty} R_{t+\nu T+p,t} R_{s+\nu T,t+q},
\end{aligned}$$

where $S_{i,q,t} = \sum_{\nu=-\infty}^{\infty} \psi_{i+q+\nu T}(t+q)\psi_{i+\nu T}(t)$.
Proof. First, the relation

$$E\{\xi_s \xi_t \xi_u \xi_v\} = \begin{cases} \eta\sigma^4 & s = t = u = v \\ \sigma^2 & s = t \neq u = v (\text{ plus two other ways}) \\ 0 & \text{none of } s, t, u, v \text{ are equal} \end{cases}$$

leads to

$$\begin{aligned}
E\{\widehat{R}_{s+p,s,N} \widehat{R}_{t+q,t,N}\} &= \frac{1}{N^2} \sum_{\mu=1}^{N-1} \sum_{\nu=1}^{N-1} E\{X_{s+\mu T+p} X_{s+\mu T} X_{t+\mu T+q} X_{t+\mu T}\} \\
&= \frac{1}{N^2} \sum_{\mu=1}^{N-1} \sum_{\nu=1}^{N-1} \Big[(\eta - 3)\sigma^4 A_{\mu,\nu}(s,t,p,q) + \\
&\quad R_{s+\mu T+p,s+\mu T} R_{t+\nu T+q,t+\nu T} + R_{s+\mu T+p,t+\nu T+q} R_{s+\mu T,t+\nu T} + \\
&\quad R_{s+\mu T+p,t+\nu T} R_{s+\mu T,t+\nu T+q}\Big]
\end{aligned}$$

where

$$A_{\mu,\nu}(s,t,p,q)$$
$$= \sum_{i=-\infty}^{\infty} \Big[\psi_{i+p}(s+\mu T+p)\psi_i(t+\mu T)\psi_{i+t-s+\nu T-\mu T+q}(t+\nu T+q)$$
$$\times \psi_{i+t-s+\nu T-\mu T}(t+\nu T)\Big].$$

When computing $N\text{Cov}\,[\widehat{R}_{t+p,t,N}, \widehat{R}_{t+q,t,N}]$ (where $s=t$) the term

$$R_{s+\mu T+p,s+\mu T}R_{t+\nu T+q,t+\nu T}$$

is cancelled by the subtraction of the mean. The remaining two terms having products of R collapse to a single sum due to the stationarity in the variables (μ,ν). As an example, one of the remaining terms is

$$\frac{1}{N}\sum_{\mu=1}^{N-1}\sum_{\nu=1}^{N-1} R_{s+\mu T+p,t+\nu T+q}R_{s+\mu T,t+\nu T} = \sum_{\mu=-N+1}^{N-1} (1-\frac{|\mu|}{N})R_{s+\mu T+p,t+q}R_{s+\mu T,t}.$$

The term involving $S_{i,q,t}$ is established from (again $s=t$)

$$\frac{1}{N}\sum_{\mu=1}^{N-1}\sum_{\nu=1}^{n-1} A_{\mu,\nu}(s,t,p,q)$$
$$= \sum_i \psi_{i+p}(t+p)\psi_i(t) \sum_{\eta=-N+1}^{N-1} \left(1-\frac{|\eta|}{N}\right)\psi_{i+q+\eta T}(t+q)\psi_{i+\eta T}(t)$$

and the Lebesgue convergence theorem. ∎

9.6 Under the hypotheses of the proposition above, the estimators $\widehat{R}_{t+\tau,t,N}$ and $\widehat{\rho}_{t+\tau,t,N}$ are asymptotically normal.

9.7 Here is a sketch of the proof of Proposition 9.14. The existence of $\sigma^2_{t,\tau} = \sum_{j=-\infty}^{\infty} R_{Z^\dagger,\tau}(t+jT,t)$ may be verified for every t,τ using Isserlis' formula, (9.47), and $\sum_{t=0}^{T-1}\sum_{\tau=-\infty}^{\infty}|R_{t+\tau,t}|^2 < \infty$. But since we must also have $lim_{\tau\to\infty}|R_{t+\tau,t}| = 0$, results of Maruyama [151] and Kolmogorov and Rozanov [133] show that X_t is α mixing (no rate specified). We get existence of all moments due to the fact that X_t is Gaussian. Finally, apply Theorem 1.4 of Ibragimov [120] to the stationary sequence $\zeta_j = Z_{t+jT,\tau}$.

9.8 Give an example to show that if X_1, X_2, X_3 are i.i.d., then the random variables X_1X_2 and X_2X_3 are not necessarily independent. This applies to the discussion of program `Bcoeff.m`.

9.9 Almost PC processes. For results on consistency and asymptotic normality for covariance and spectral estimators for almost PC processes (continuous and discrete time) see references [30, 39–41, 43, 44, 46, 67, 108, 109, 135].

9.10 Almost sure consistency of $\widehat{B}_{k,N}(\tau)$ via the random shift. Assume $\{Z_{t,\tau}\}$ is PC in t for fixed τ and denote

$$B_{Z,k}(\tau,\tau') = \frac{1}{T}\sum_{t=0}^{T-1} R_Z(t,\tau,\tau')e^{-i2\pi kt/T},$$

where $R_Z(t,\tau,\tau') = E\{X_{t+\tau+\tau'}X_{t+\tau'}X_{t+\tau}X_t\} - R(t+\tau+\tau', t+\tau')R(t+\tau, t)$. Show the following are sufficient for a.s. consistency of $\widehat{B}_{k,N}(\tau)$:

(a) For some $\alpha > 0$

$$\sigma_N^2(2\pi k/T) = \frac{1}{N}\sum_{j=-N+1}^{N-1}\left(1-\frac{|j|}{N}\right)B_{Z,0}(\tau,j)e^{-i2\pi jk/T} \le \frac{K}{N^\alpha};$$

(b) $\sum_{j=-\infty}^{\infty}|B_{Z,0}(\tau,j)| < \infty$ or $B_{Z,0}(\tau,j) = O(j^{-\alpha})$ are sufficient for (a) with $k = 0,1,\ldots,T-1$.

CHAPTER 10

SPECTRAL ESTIMATION

Suppose X_t is a zero mean stationary sequence and we wish to estimate, from a single sample path, the spectral density function f, where $R(\tau) = \int_0^{2\pi} e^{i\lambda\tau} f(\lambda)d\lambda$. The typical approach to the nonparametric estimation of a continuous f is through use of the sample Fourier transform

$$\widetilde{X}_N(\lambda)) = \sum_{t=0}^{N-1} X_t e^{-i\lambda t} \tag{10.1}$$

from which the periodogram

$$I_N(\lambda) = \frac{1}{2\pi N} |\widetilde{X}_N(\lambda)|^2 \tag{10.2}$$

is formed. Practically, if X_t is not known to be of zero mean, then we replace X_t with $X_t - \widehat{m}$. The periodogram is an asymptotically unbiased estimator for f although it is not consistent. Consistent estimators for f are obtained by tapering (or weighting) the covariance estimator by a multiplicative kernel

Periodically Correlated Random Sequences:Spectral Theory and Practice. By H.L. Hurd and A.G. Miamee

$k(\cdot) \in l_1(-\infty, \infty)$ that is bounded and even with $k(0) = 1$. This estimator can equivalently be expressed as a smoothing of the periodogram by $W(\cdot)$, the Fourier transform of $w(\cdot)$; thus

$$\begin{aligned} \widehat{f}_N(\lambda) &= \frac{1}{2\pi} \sum_{-(N-1)}^{N-1} w(\mu_N v) \widehat{R}_N(v) e^{-i\lambda v} \\ &= \frac{1}{\mu_N} \int_0^{2\pi} W\left(\frac{\gamma - \lambda}{\mu_N}\right) I_N(\gamma) d\gamma, \end{aligned} \tag{10.3}$$

where $\mu_N \to 0$ and $N\mu_N \to \infty$ as $N \to \infty$. This procedure gives smoothing kernels that are dependent on N in a manner that produces the consistency of $\widehat{f}_N(\lambda)$. For example, if a stationary process X is Gaussian and $R(\cdot) \in l_1(-\infty, \infty)$, then $\widehat{f}_N(\lambda)$ is a consistent estimator at every λ. Details may be found in many places [8, 27, 28, 88, 177, 198].

The consistency can be explained from the fact that neighboring periodogram values, say, $I_N(\lambda_1)$ and $I_N(\lambda_2)$ with $\lambda_1 \neq \lambda_2$, become asymptotically independent as $N \to \infty$. Hence in a fixed neighborhood $(\lambda_0 - \epsilon, \lambda_0 + \epsilon)$ of some λ_0, the number of independent periodogram values increases proportionally to N, suggesting that on the interval $(\lambda_0 - \epsilon, \lambda_0 + \epsilon)$ we can consistently estimate the average spectral density $(2\epsilon)^{-1} \int_{\lambda_0-\epsilon}^{\lambda_0+\epsilon} f(\lambda) d\lambda$. Generally this represents a bias in the estimation of $f(\lambda_0)$. However, if the smoothing is controlled as $N \to \infty$ in the manner prescribed above, the number of independent samples still increases fast enough to reduce the variation but yet the width of the smoothing kernel, and hence the bias, shrinks to zero, producing the consistency.

Here we are concerned with the estimation of the possibly complex density functions $f_k(\lambda)$ for real PC sequences when the $F_k(\cdot)$ in (6.15) are absolutely continuous with respect to Lebesgue measure so that

$$B_k(\tau) = \int_0^{2\pi} e^{i\lambda\tau} f_k(\lambda) d\lambda. \tag{10.4}$$

Since the measures $F_k(\cdot)$ can be identified (see Section 6.2 and Figures 6.1–6.3) with the concentration on $S_k = \{(\lambda_1, \lambda_2) : \lambda_1 \in [0, 2\pi), \lambda_2 = \text{mod } (\lambda_1 - k2\pi/T, 2\pi)\}$ of the spectral measure F in the harmonizable representation (1.15), then an estimation of $f_k(\lambda)$ for $k = 0, 1, 2, \ldots, T-1$ comprises a spectral analysis of the PC sequence.

But F can also be composed by splicing together the measures $\{\mathcal{F}_{pq}\}$ from the matrix valued spectral measure $\mathcal{F}$ associated with the multivariate sequence $Z_t^k, k = 0, 1, \ldots, T-1$ (see Proposition 6.8 and Figure 6.5). Or to obtain F_k, splice together the $\mathcal{F}_{pq}$ for which $(p-q) \text{mod } T = k$. Using the relationship

$$\mathbf{F}(d\lambda) = T\mathbf{V}(\lambda)\mathcal{F}(d\lambda/T)\mathbf{V}^{-1}(\lambda) \tag{10.5}$$

(see Proposition 6.9 for the definition of the unitary and continuous $\mathbf{V}(\lambda)$) between $\mathcal{F}$ and the matrix valued spectral measure $\mathbf{F}$ of the blocked multivariate sequence $\mathbf{X}_n$, we arrive at the relation between the densities, namely,

$$f_{\mathbf{X}}(\lambda) = T\mathbf{V}(\lambda) f_Z\Big(\frac{\lambda}{T}\Big)\mathbf{V}^{-1}(\lambda) \tag{10.6}$$

when $\mathbf{F}$ and $\mathcal{F}$ are absolutely continuous with respect to Lebesgue measure. Recall from Corollary 6.9.1 that $|\mathbf{F}| \ll |\mathcal{F}|(\cdot/T)| \ll |\mathbf{F}|$. Then any condition on $\mathbf{X}_n$ that permits the estimation of $f_{\mathbf{X}}(\lambda)$ gives a way to estimate $f_k(\lambda)$ via inverting (10.6) and splicing. But here we will mainly take the approach of estimating $f_k(\lambda)$ directly from the sample Fourier transform of an observed series, noting that these methods extend to continuous time and APC sequences.

Spectral analysis for PC processes seems to have begun with Gudzenko [84] and was treated in the continuous time case, using the same methods employed here, by Hurd [101, 105]. This treatment is essentially an adaptation of Parzen's work (e.g., see [177]) and we note that Parzen's work on spectral density estimation for stationary processes having periodically missing observations [180] essentially treats the estimation of $f_0(\lambda)$ for a particular type of PC process. This problem has also been treated in a nonprobabilistic context by Gardner [66, 67, 69] who discusses the connection between the probabilistic and non-probabilistic treatments.

An estimator for $f_k(\lambda)$ is constructed quite naturally from an extension of the periodogram.

10.1 THE SHIFTED PERIODOGRAM

The principal idea for the PC case is based on the formation of a two-dimensional periodogram

$$f_N(\lambda_1, \lambda_2) = \frac{1}{2\pi N}\widetilde{X}_N(\lambda_1)\overline{\widetilde{X}_N(\lambda_2)} \tag{10.7}$$

from the tensor product of the sample Fourier transform (10.1) with itself. Since $I_N(\lambda) = f_N(\lambda, \lambda)$, the usual estimators for the stationary spectral density are formed by smoothing $f_N(\lambda_1, \lambda_2)$ along the main diagonal $\lambda_1 = \lambda_2$. Now taking the clue from the support of the spectral covariance measure of a PC-T sequence (Figure 1.3) we can guess that estimators of $f_k(\lambda)$ can be formed by smoothing $f_N(\lambda_1, \lambda_2)$ along the line of support of F in $[0, 2\pi) \times [0, 2\pi)$ that corresponds to F_k, that is, along $\lambda_2 = \lambda_1 - k2\pi/T, 2\pi$ (modulo 2π) for $\lambda_1 \in [0, 2\pi)$.

Now to complete the details, we first show the *shifted periodogram*,

$$f_{k,N}(\lambda) = f_N(\lambda, \lambda - 2\pi k/T), \tag{10.8}$$

is the Fourier transform of $\widehat{B}_{k,N}(\tau)$ and is an asymptotically unbiased but inconsistent estimator for $f_k(\lambda)$. Then we examine the smoothing of the shifted periodogram $f_{k,N}(\lambda)$ by a family of kernels, depending on N, to yield consistent estimators for $f_k(\lambda)$.

Lemma 10.1 *If $\{X_n, n = 0, ..., N-1\}$ is any real sequence, $f_{k,N}(\lambda)$ is the Fourier transform of $\widehat{B}_{k,N}(\tau)$.*

Proof. First write

$$\begin{aligned} f_{k,N}(\lambda) &= \frac{1}{2\pi N}\left[\sum_{s=0}^{N-1} X_s e^{-i\lambda s}\right]\left[\sum_{t=0}^{N-1} X_t e^{it(\lambda-2\pi k/T)}\right] \\ &= \frac{1}{2\pi N}\sum_{s=0}^{N-1}\sum_{t=0}^{N-1} X_s X_t e^{-i\lambda(s-t)} e^{-i2\pi kt/T} \end{aligned} \tag{10.9}$$

We now perform the indicated sum over the region $[0, N-1] \times [0, N-1]$ in two parts denoted by S_1 for $u = m - n < 0$ and S_2 for $u \geq 0$. This produces

$$\begin{aligned} S_1 &= \frac{1}{2\pi}\sum_{u=-N+1}^{-1} \widehat{B}_{k,N}(u)e^{-i\lambda u}, \\ S_2 &= \frac{1}{2\pi}\sum_{u=0}^{N-1} \widehat{B}_{k,N}(u)e^{-i\lambda u}, \end{aligned}$$

which shows that

$$f_{k,N}(\lambda) = \frac{1}{2\pi}\sum_{u=-N+1}^{N-1} \widehat{B}_{k,N}(u)e^{-i\lambda u}, \tag{10.10}$$

where we take $\widehat{B}_{k,N}(u) = 0$ for u not in the interval $[-N+1, N-1]$. ∎

As in the stationary case, the shifted periodogram is asymptotically unbiased.

Proposition 10.1 *If X_t is PC-T and $\sum_{\tau=-\infty}^{\infty}\sum_{t=0}^{T-1} |R(t+\tau, t)| < \infty$, then $f_{k,N}(\lambda)$ is an asymptotically unbiased estimator for $f_k(\lambda)$.*

Proof. Use (9.39) of Lemma 9.5 to write

$$\begin{aligned} E\{f_{k,N}(\lambda)\} &= \frac{1}{2\pi}\sum_{u=-N+1}^{N-1} E\{\widehat{B}_{k,N}(u)\}e^{-i\lambda u} \\ &= \frac{1}{2\pi}\sum_{u=-N+1}^{N-1} B_k(u)\left[1-\frac{|u|}{N}\right]e^{-i\lambda u} + \frac{1}{2\pi N}\sum_{u=-N+1}^{N-1} \epsilon(N,u). \end{aligned} \tag{10.11}$$

The condition $\sum_{\tau=-\infty}^{\infty}\sum_{t=0}^{T-1}|R(t+\tau,t)|<\infty$ ensures that $\sum_{\tau=-\infty}^{\infty}|B_k(\tau)|<\infty$ so that

$$f_k(\lambda)=\frac{1}{2\pi}\sum_{u=-\infty}^{\infty}B_k(u)e^{-i\lambda u}$$

is continuous in λ. As $N\to\infty$, the first term in (10.11) converges to $f_k(\lambda)$ and the second term converges to 0 by Lemma 9.5. ∎

Since the periodogram is an inconsistent estimator for $f(\lambda)$ in the stationary case, we certainly expect inconsistency of $f_{k,N}(\lambda)$ as an estimator for $f_k(\lambda)$. The following proposition leads to a similar conclusion for the PC case, assuming that X_t is Gaussian.

Proposition 10.2 *If X_t is a Gaussian PC-T sequence for which*

$$\sum_{\tau=-\infty}^{\infty}\sum_{t=0}^{T-1}|R(t+\tau,t)|<\infty,$$

then

$$\lim_{N\to\infty}\operatorname{Var}[f_{k,N}(\lambda)]=\begin{cases} f_0(\lambda)f_0(\lambda-2\pi k/T), & \lambda\neq\pi n/T\\ f_0(\lambda)f_0(\lambda-2\pi k/T)+|f_{n-k}(\pi n/T)|^2, & \lambda=\pi n/T\end{cases}. \tag{10.12}$$

Proof. Using the definition (10.8) and Isserlis' formula again, we obtain

$$\begin{aligned}\operatorname{Var}[f_{k,N}(\lambda)] &= I_1+I_2 \qquad (10.13)\\ &= \frac{1}{(2\pi N)^2}\sum_{s=0}^{N-1}\sum_{u=0}^{N-1}\sum_{t=0}^{N-1}\sum_{v=0}^{N-1}R_{s,u}R_{t,v}e^{-i\lambda(s-u+v-t)-i2\pi k(t-v)/T}\\ &+ \frac{1}{(2\pi N)^2}\sum_{s=0}^{N-1}\sum_{u=0}^{N-1}\sum_{t=0}^{N-1}\sum_{u=0}^{N-1}R_{s,v}R_{t,u}e^{-i\lambda(s+v-u-t)-i2\pi k(t-v)/T}.\end{aligned}$$

The computations used in obtaining (10.11) from (10.10) and (9.39) may be used to evaluate the two terms appearing in (10.13). Setting $\tau_1=s-u$ and $\tau_2=t-v$, the first of the two terms may be written

$$\begin{aligned}I_1 &= \frac{1}{(2\pi)^2}\sum_{-N+1}^{N-1}\left[B_0(\tau_1)\left(1-\frac{|\tau_1|}{N}\right)e^{-i\lambda\tau_1}+\frac{\epsilon_1(N,\tau_1)}{N}\right]\\ &\times\sum_{-N+1}^{N-1}\left[B_0(\tau_2)\left(1-\frac{|\tau_2|}{N}\right)e^{i\lambda\tau_2-i2\pi k\tau_2/T}+\frac{\epsilon_2(N,\tau_2)}{N}\right]\end{aligned} \tag{10.14}$$

where $\epsilon_1(N,\tau_1)$ and $\epsilon_2(N,\tau_2)$ are summable by virtue of Lemma 9.5. From the summability of $B_0(\tau)$ we may write

$$f_0(\lambda) = \frac{1}{2\pi}\sum_{-\infty}^{\infty} B_0(\tau)e^{-i\lambda\tau}$$

and so conclude

$$\lim_{N\to\infty} I_1 = f_0(\lambda)f_0(\lambda - 2\pi k/T). \tag{10.15}$$

Similarly, for $\lambda \neq \pi n/T$, an application of the Riemann–Lebesgue lemma yields $\lim_{N\to\infty} I_2 = 0$. For $\lambda = \pi n/T$

$$\begin{aligned} \lim_{N\to\infty} I_2 &= f_{n-k}(\pi n/T)f_{k-n}(-\pi n/T) \\ &= |f_{n-k}(\pi n/T)|^2. \quad \blacksquare \end{aligned} \tag{10.16}$$

Regarding the issue of consistency, at any λ for which $f_k(\lambda) \neq 0$, it is necessary from the domination by the diagonal (see Section 6.3)

$$|f_k(\lambda)|^2 \leq f_0(\lambda)f_0(\lambda - 2\pi k/T)$$

that both $f_0(\lambda) \neq 0$ and $f_0(\lambda - 2\pi k/T) \neq 0$ and thus from (10.12) the asymptotic variance is positive.

It may also be seen that the limits in (10.12) reduce to the familiar result (see Parzen [177]) for the stationary case by setting $k = n = 0$,

$$\lim_{N\to\infty} \operatorname{Var}[f_0(N,\lambda)] = \begin{cases} f_0^2(\lambda), & \lambda \neq 0 \\ 2f_0^2(\lambda), & \lambda = 0 \end{cases}.$$

10.2 CONSISTENT ESTIMATORS

We now show that consistent estimators for the $f_k(\lambda)$ may be obtained by smoothing $f(N,\lambda_1,\lambda_2)$ along the support line in $[0,2\pi)\times[0,2\pi)$ that corresponds to F_k, that is, along $\lambda_2 = \lambda_1 - 2\pi k/T$. In view of (10.8) and Lemma 10.1, these estimators may also be expressed as a weighting of $\widehat{B}_{k,N}(\tau)$ by kernel $w(\cdot)$; that is,

$$\begin{aligned} \widehat{f}_{k,N}(\lambda) &= \frac{1}{2\pi}\sum_{-N+1}^{N-1} w(\mu_N v)\widehat{B}_{k,N}(\tau)e^{-i\lambda v} \\ &= \frac{1}{\mu_N}\int_0^{2\pi} W((\sigma-\lambda)/\mu_N)f_{k,N}(\sigma)d\sigma, \end{aligned} \tag{10.17}$$

where μ_N is a positive sequence with $\mu_N \to 0$ and $N\mu_N \to \infty$ as $N \to \infty$. The sequence $w(j)$ is taken here to be even, bounded, summable and $w(0) = 1$;

thus also $w(j)$ is square summable and the spectral smoothing kernel $W(\lambda)$ is the $L_2[0, 2\pi)$ function determined by

$$W(\lambda) = \sum_{j=-\infty}^{\infty} w(j)e^{ij\lambda}.$$

We now give two types of consistency results. The first is for Gaussian processes, where the Isserlis formula is the crucial ingredient, and the second is for ϕ-mixing processes.

Gaussian Sequences. We now show that $\text{Cov}[\widehat{f}_{j,N}(\lambda_1), \widehat{f}_{k,N}(\lambda_2)]$ is $O(1/N\mu_N)$ for arbitrary j, k and λ_1, λ_2 and hence the consistency is established.

Proposition 10.3 *If X_t is a Gaussian PC-T sequence for which*

$$\sum_{\tau=-\infty}^{\infty} \left[\sum_{t=0}^{T-1} |R(t+\tau, t)|^2 \right]^{1/2} < \infty,$$

then there exists a $K > 0$ such that for any $j, k \in \{0, 1, ..., T-1\}$ and $\lambda_1, \lambda_2 \in [0, 2\pi)$

$$N\mu_N \text{Cov}\,[\widehat{f}_{j,N}(\lambda_1), \widehat{f}_{k,N}(\lambda_2)] \le K$$

for N sufficiently large.

Proof. From the definition of $\widehat{f}_{k,N}(\lambda_2)$ we write

$$\begin{aligned} &N\mu_N \text{Cov}\,[\widehat{f}_{j,N}(\lambda_1), \widehat{f}_{k,N}(\lambda_2)] \\ &= \frac{N\mu_N}{(2\pi)^2} \sum_{-N+1}^{N-1} \sum_{-N+1}^{N-1} w(\mu_N\tau_1)w(\mu_N\tau_2)\text{Cov}\,[\widehat{B}_{j,N}(\tau_1)\widehat{B}_{k,N}(\tau_2)]e^{-i\lambda_1\tau_1+i\lambda_2\tau_2} \\ &= S_F + S_G + S_O, \end{aligned} \tag{10.18}$$

where the three terms S_F, S_G, and S_O result from the expression (9.50) for $N\text{Cov}\,[\widehat{B}_{j,N}(\tau_1)\widehat{B}_{k,N}(\tau_2)]$. Recall the real function $U_N(u, \tau_1, \tau_2) \le 1$ for $u \in [-N, N]$ increases to 1 as $N \to \infty$.

First, the sum S_O involving the $O(1/N)$ term in (9.50) when transformed by $\nu_j = \mu_N\tau_j, j = 1, 2$ is bounded by

$$|S_O| \le \frac{O(N^{-1})}{\mu_N(2\pi)^2} \sum_{-N\mu_N}^{N\mu_N} \sum_{-N\mu_N}^{N\mu_N} |w(\nu_1)|\,|w(\nu_2)|,$$

which converges to 0 as $N \to \infty$ because $w(j)$ is summable and $N\mu_N \to \infty$ as $N \to \infty$.

Now for the term S_F containing $F(u, \tau_1, \tau_2)$, first set

$$S^2(\tau) = \sum_{j=0}^{T-1} B^2(j, \tau)$$

and note by hypothesis that $\sum_{-\infty}^{\infty} S(\tau) < \infty$. Then

$$|S_F| \le \frac{\mu_N}{(2\pi)^2} \sum_{-N+1}^{N-1} \sum_{-N+1}^{N-1} \sum_{-N+1}^{N-1} |w(\mu_N \tau_1)||w(\mu_N \tau_2)|S(u + \tau_1 - \tau_2)S(u)$$

and then the transformation $\nu_1 = \tau_1 - \tau_2$, $\nu_2 = \mu_N \tau_2$ yields

$$\begin{aligned} |S_F| &\le \frac{1}{(2\pi)^2} \sum_{-\mu_N N}^{\mu_N N} |w(\nu_2)| \sum_{-N-\nu_2/\mu_N}^{N-\nu_2/\mu_N} |w(\mu_N \nu_1 + \nu_2)| \\ &\quad \times \sum_{\nu=-N+1}^{N-1} S(\nu + \nu_1)S(\nu) \\ &\le \frac{1}{(2\pi)^2} \sum_{-\infty}^{\infty} \sum_{-\infty}^{\infty} \sum_{-\infty}^{\infty} |w(\nu_2)||w(\mu_N \nu_1 + \nu_2)|S(\nu + \nu_1)S(\nu). \quad (10.19) \end{aligned}$$

An application of the Schwarz inequality for square summable sequences gives, for any N,

$$\sum_{-\infty}^{\infty} |w(\nu_2)||w(\mu_N \nu_1 + \nu_2)| \le \sum_{-\infty}^{\infty} |w(j)|^2 < \infty.$$

The summability of $S(\tau)$ produces

$$\sum_{-\infty}^{\infty} \sum_{-\infty}^{\infty} S(\nu + \nu_1)S(\nu) < \infty$$

and the result for S_F follows from these facts. A similar analysis produces the result for S_G. ∎

We leave it as a problem to show condition $\sum_{\tau=-\infty}^{\infty} \sum_{t=0}^{T-1} |R(t+\tau, t)| < \infty$ is sufficient for $\sum_{\tau=-\infty}^{\infty} \left[\sum_{t=0}^{T-1} |R(t + \tau, t)|^2\right]^{1/2} < \infty$.

ϕ-Mixing Sequences. Consistency of $\widehat{f}_{k,N}\lambda)$ under ϕ-mixing assumptions is interesting because it can be established with considerably less effort than even for the case of Gaussian processes. The result here relies on Lemma 9.7, which

gave a simple bound on the rate of convergence for $|\text{Cov}[\widehat{B}_{j,NT}(\tau_1), \widehat{B}_{k,NT}(\tau_2)]|$.

Proposition 10.4 *If X_t is periodically stationary with period T, $EX_t^4 < \infty$, $t \in \mathbb{Z}$ and uniformly ϕ-mixing with $\sum_{n=-\infty}^{\infty}(\bar{\phi}_n)^{1/2} < \infty$, $k(j)$ is any sequence with $\sum_{j=-\infty}^{\infty} k(j)|j|^{1/2} < \infty$, then*

$$\lim_{N\to\infty} \text{Cov}\,[\widehat{f}_{j,N}(\lambda_1), \widehat{f}_{k,N}(\lambda_2)] = 0$$

if $\mu_N \to 0$, $N\mu_N^3 \to \infty$ as $N \to \infty$.

Proof. The proof is a result of the estimates

$$\begin{aligned}
&|\text{Cov}\,[\widehat{f}_j(\lambda_1), \widehat{f}_k(\lambda_2)]| \\
&\quad\leq \frac{1}{(2\pi)^2} \sum_{\tau_1=-N}^{N} \sum_{\tau_2=-N}^{N} |k(\mu_N\tau_1)||k(\mu_N\tau_2)||\text{Cov}\,[\widehat{B}_{j,NT}(\tau_1), \widehat{B}_{k,NT}(\tau_2)]| \\
&\quad\leq \frac{1}{NT(2\pi)^2} \sum_{-N}^{N}\sum_{-N}^{N} |k(\mu_N\tau_1)||k(\mu_N\tau_2)|[C_1|\tau_1| + C_2]^{1/2}[C_1|\tau_2| + C_2]^{1/2} \\
&\quad= \frac{1}{NT(2\pi)^2}\left[\sum_{\tau=-N}^{N} |k(\mu_N\tau)|[C_1|\tau| + C_2]^{1/2}\right]^2 \\
&\quad= \frac{1}{NT\mu_N^2(2\pi)^2}\left[\sum_{u=-N\mu_N}^{N\mu_N} |k(u))|[C_1|u/\mu_N| + C_2]^{1/2}\right]^2 \\
&\quad\leq \frac{1}{NT\mu_N^2(2\pi)^2}\left[\sum_{-\infty}^{\infty} |k(u))|[C_1'|u/\mu_N|^{1/2} + C_2']\right]^2 \\
&\quad\leq \frac{1}{NT\mu_N^2(2\pi)^2}\left[\frac{K_1}{\mu_N^{1/2}} + K_2\right]^2.
\end{aligned}$$

The next to last line follows from considering N fixed, and then there must be a u_0 (which may depend on N) and a C_1' for which $[C_1|u/\mu_N|+C_2] \leq C_1'|u/\mu_N|$ for $|u| > |u_0|$. And then for $|u| \leq |u_0|$, since $[C_1|u/\mu_N|+C_2]$ is bounded, there is a C_2' with $[C_1|u/\mu_N| + C_2] \leq C_2'$. Combining these observations with the hypotheses gives the result. ∎

10.3 ASYMPTOTIC NORMALITY

Here we sketch some results on asymptotic normality of $\widehat{f}_{k,N}(\lambda)$ for linear PC sequences. We omit detailed proofs but indicate the path to the result via stationary results presented in [28].

Proposition 10.5 *Suppose $X_t = \sigma(t)\epsilon_t$ for ϵ_t real, i.i.d., mean zero with variance σ_ϵ, and $\sigma(t) = \sigma(t+T) > 0$ for all t. Then for any $0 \le \lambda < \pi$, and $0 \le k < T$, we have that* Re $\widehat{f}_{k,N}(\lambda)$ *and* Im $\widehat{f}_{k,N}(\lambda)$ *are asymptotically normal.*

This follows easily from Brockwell and Davis [28, Proposition 10.3.2], where we only need to see that the periodic variances $\sigma^2(t)$ have a minimum, and this controls the Lindeberg condition.

We leave it as a problem to compute $\text{Var}[\widehat{f}_{k,N}(\lambda)]$ and $\text{Cov}(\widehat{f}_{k,N}(\lambda_1), \widehat{f}_{k,N}(\lambda_2))$ for $\lambda, \lambda_1, \lambda_2$ in the Fourier frequencies for sample size N. Recall for PC white noise that $f_k(\lambda) = B_k(0)/2\pi$, where $B_k(0) = \frac{1}{T}\sum_{t=0}^{T-1}\sigma^2(t)e^{-i2\pi kt/T}$, $0 \le \lambda < 2\pi$.

In the following, we assume X_t is Gaussian in order to use Propositions 10.1, 10.2, and 10.3.

Proposition 10.6 *Suppose X_t is a Gaussian, zero mean real linear PC-T sequence*

$$X_t = \sum_{j=-\infty}^{\infty} \psi_j(t)\xi_{t-j}, \tag{10.20}$$

where $\psi_j(t) = \psi_j(t+T)$, $j \in \mathbb{Z}$ are real, $\sum_j |\psi_j(t)| < \infty$, $t = 0, 1, \ldots, T-1$, and $\{\xi_j\}$ is a real zero mean Gaussian i.i.d. sequence. Then with $\widehat{f}_{k,N}(\lambda)\}$ defined as in (10.17), where μ_N is a positive sequence with $\mu_N \to 0$ and $N\mu_N \to \infty$ as $N \to \infty$, we have

(a) $\lim_{N\to\infty} E\{\widehat{f}_{k,N}(\lambda)\} = f_k(\lambda)$;

(b)

$$\lim_{N\to\infty} \text{Var}\left[f_{k,N}(\lambda)\right] \tag{10.21}$$

$$= \begin{cases} f_0(\lambda)f_0(\lambda - 2\pi k/T), & \lambda \ne \pi n/T \\ f_0(\lambda)f_0(\lambda - 2\pi k/T) + |f_{n-k}(\pi n/T)|^2, & \lambda = \pi n/T \end{cases};$$

(c) *for any $j, k \in \{0, 1, ..., T-1\}$ and $\lambda_1, \lambda_2 \in [0, 2\pi)$, there exists a $K > 0$ for which*

$$N\mu_N \text{Cov}\left[\widehat{f}_{j,N}(\lambda_1), \widehat{f}_{k,N}(\lambda_2)\right] \le K$$

for N sufficiently large;

(d) $\widehat{f}_{k,N}(\lambda)$ *is asymptotically normal.*

Proof. For (a)–(c), we first note that $\sum_{j=-\infty}^{\infty} |\psi(t)|^2 < \infty$ because ℓ^1 sequences are also ℓ^2. Then

$$R(t+\tau, t) = \sum_{j=-\infty}^{\infty} \psi_j(t+\tau)\psi_{j-\tau}(t)$$

and

$$\begin{aligned}
\sum_{\tau=-\infty}^{\infty} |R(t+\tau,t)| &\leq \sum_{\tau=-\infty}^{\infty} |\sum_{j=-\infty}^{\infty} \psi_j(t+\tau)\psi_{j-\tau}(t)| \\
&\leq \sum_{j=-\infty}^{\infty} \sum_{\tau=-\infty}^{\infty} |\psi_j(t+\tau)||\psi_{j-\tau}(t)| \\
&\leq |\sum_{j=-\infty}^{\infty} \psi_j^* \sum_{\tau=-\infty}^{\infty} |\psi_{j-\tau}(t)| < \infty,
\end{aligned}$$

where $\psi_j^* = \max_{t=0,1,\ldots,T-1} |\psi_j(t+\tau)|$ and $\sum_{j=-\infty}^{\infty} \psi_j^* < \infty$. Propositions 10.1, 10.2, and 10.3 give the results. For (d) the same argument used in [28, Section 11.7] for asymptotic normality of smoothed estimators based on cross spectral densities is applicable here. In summary, a discrete frequency version of the smoothed periodogram estimator (10.17) is constructed from a shifted periodogram based on a sample Fourier transform at the Fourier frequencies $j2\pi/N$, $j = 0, 1, \ldots, N-1$. As N increases, item (c) says that the number of effectively independent samples accumulated in the smoothing increases, giving the asymptotic normality. ∎

Another route to asymptotic results can be based on the fact that any linear PC sequence, when blocked, gives a linear T-variate stationary sequence (see (9.16). Thus the asymptotic normality of spectral estimators for $f_{\mathbf{X}}(\lambda)$, the spectral density of the blocked sequence $\mathbf{X}_n$, can be transformed to estimators for $f_{\mathbf{Z}}(\lambda/T)$ via the inversion of (10.6), and estimates for $f_k(\lambda)$ are produced by splicing the latter. Figures 6.1–6.5 may be helpful. We do not give the details here, but see [28, Section 11.7] and [88, page 289] for a detailed proof where fourth moments are required. Also see Nematollahi and Rao [167] for a treatment of spectral analysis for PC sequences based on $\mathbf{X}_n$.

Confidence limits for $\widehat{f}_{k,N}(\lambda)$ will be discussed in a later section.

10.4 SPECTRAL COHERENCE

Since we can get consistent estimators of $f_k(\lambda)$, a natural question is whether we can use the estimator $\widehat{f}_k(\lambda)$ given by (10.17) to produce a test for the presence of PC-T structure. That is, if an observed value of $\widehat{f}_k(\lambda)$ is significantly non-zero, we would declare that PC-T structure is present. Since $f_k(\lambda)$ can be identified with cross spectral densities, the notion of coherence (or coherency) provides a natural framework for making such tests. Recall coherency [27, Chapters 7 and 8] indicates the linear relation between the random spectral measures ξ_1 and ξ_2 of two stationary series. For PC sequences, we wish to measure the linear relation between the random amplitudes $\xi(d\lambda)$ and $\xi(d\lambda - \lambda_0)$, where $\lambda_0 = 2\pi k/T$. Thus *spectral coherence* refers to coherence statistics applied to the random spectral measure of a possibly nonstationary sequence; in order to obtain empirical measurements, we apply it to the sample spectra, namely, the sample Fourier transform. Many of the properties of complex random variables that are pertinent to spectral analysis of time series were initiated by N. R. Goodman in his thesis and subsequently (see [55, 80–82]), including the idea of testing for various nonstationary structure based on correlations among FFT ordinates.

10.4.1 Spectral Coherence for Known T

For a PC-T sequence whose spectral measures F_k, $k = 0, 1, \dots, T-1$ are absolutely continuous with respect to Lebesgue measure, the domination of the diagonal of F, namely, $|f_k(\lambda)|^2 \le f_0(\lambda) f_0(\lambda - 2\pi k/T)$, gives a natural way to measure if $f_k(\lambda)$ is large. That is, for $f_0(\lambda) \neq 0$ and $f_0(\lambda - 2\pi k/T) \neq 0$, we define the theoretical complex coherence (coherency) between the random amplitudes at frequencies λ and $\lambda - 2\pi k/T$ by

$$\gamma(\lambda, \lambda - 2\pi k/T) = \frac{f_k(\lambda)}{f_0^{1/2}(\lambda) f_0^{1/2}(\lambda - 2\pi k/T)}. \tag{10.22}$$

Assuming the conditions of Proposition 10.1, since

$$E\{f_{k,N}(\lambda)\} = (2\pi N)^{-1} \mathrm{Cov}\,[\widetilde{X}_N(\lambda), \widetilde{X}_N(\lambda - 2\pi k/T)],$$

we can also write

$$\begin{aligned} \gamma(\lambda, \lambda - 2\pi k/T) &= \lim_{N\to\infty} \frac{\mathrm{Cov}\,[\widetilde{X}_N(\lambda), \widetilde{X}_N(\lambda - 2\pi k/T)]}{\mathrm{Var}^{\,1/2}[\widetilde{X}_N(\lambda)]\mathrm{Var}^{\,1/2}[\widetilde{X}_N(\lambda - 2\pi k/T)]} \\ &= \lim_{N\to\infty} \mathrm{Corr}\,[\widetilde{X}_N(\lambda), \widetilde{X}_N(\lambda - 2\pi k/T)]. \end{aligned}$$

Adapting this to sample quantities, we replace the quantities in the first line of (10.22) with their estimates, $\widehat{f}_{k,N}(\lambda)$, $\widehat{f}_{0,N}(\lambda)$, and $\widehat{f}_{0,N}(\lambda - 2\pi k/T)$, and then express these estimates in terms of the values of a sample Fourier transform evaluated at the usual Fourier frequencies. From (10.17), setting $\lambda_j = j2\pi/N$, this leads to

$$\begin{aligned}\widehat{f}_k(\lambda_j) &= \frac{1}{\mu_N}\int_0^{2\pi} W((\sigma-\lambda_j)/\mu_N) f_{k,N}(\sigma)d\sigma \\ &\approx \sum_{m=1}^{M} W_m(N)\widetilde{X}_{j-M/2+m}\overline{\widetilde{X}}_{j-M/2-kN'+m}, \qquad (10.23)\end{aligned}$$

where in the discrete frequency approximation, the weights $W_m(N)$ incorporate the denominator term $1/\mu_N$ from the first line. Thus we obtain an approximate squared coherence

$$\begin{aligned}&|\widehat{\gamma}(\lambda_j, \lambda_j - 2\pi k/T, W)|^2 \\ &= \frac{|\sum_m W'_m \widetilde{X}_{j-M/2+m}\overline{\widetilde{X}}_{j-M/2-kN'+m}|^2}{\sum_m W'_m |\widetilde{X}_{j-M/2+m}|^2 \sum_m W'_m |\widetilde{X}_{j-M/2-kN'+m}|^2}. \qquad (10.24)\end{aligned}$$

Since the smoothing sequence W_m is typically concentrated over a small interval of frequencies, we simplify to the case where W_m is constant, and thus

$$|\widehat{\gamma}(\lambda_j, \lambda_j - 2\pi k/T)|^2 = \frac{|\sum_{m=1}^{M} \widetilde{X}_{j-M/2+m}\overline{\widetilde{X}}_{j-M/2-kN'+m}|^2}{\sum_{m=1}^{M} |\widetilde{X}_{j-M/2+m}|^2 \sum_{m=1}^{M} |\widetilde{X}_{j-M/2-kN'+m}|^2}. \qquad (10.25)$$

The quantity $|\widehat{\gamma}(\lambda_j, \lambda_j - 2\pi k/T)|^2$ is called the sample magnitude squared coherence (or *spectral coherence*), where the dependence on M is suppressed here. A slightly more general version (see (10.30)) of sample magnitude squared coherence was one of the tests proposed by Goodman for testing of various for nonstationary structures [82] based on correlations among FFT ordinates. It turns out to be perfectly matched to perceiving the spectral correlations of PC sequences.

Under the null case, $\widetilde{X}_j$ are complex Gaussian with uncorrelated real and imaginary parts for each j and $E\{\widetilde{X}_j\overline{\widetilde{X}_{j'}}\} = 0$, $j \neq j'$, the sample squared coherence $|\widehat{\gamma}|^2$ has probability density

$$p(|\gamma|^2) = (M-1)(1-|\gamma|^2)^{M-2}, \quad 0 \leq |\gamma|^2 \leq 1. \qquad (10.26)$$

Setting $X = |\gamma|^2$, it is easily determined that $P[X \leq x] = 1-(1-x)^{M-1}$, which for a Type I error of α leads to the solution for the α-threshold of

$$x_\alpha = |\gamma|^2_\alpha = 1 - e^{\log(\alpha)/(M-1)}. \qquad (10.27)$$

Note that (10.26) depends only on the length M of the smoothing window. Since M is the number of complex products, it should be identified with $2M$ real products or degrees of freedom in the usual sense.

If, in (10.24) or (10.25), the $\widetilde{X}_j$ do not have mean zero, but share a common mean $E\{\widetilde{X}_j\} = \mu$, then (10.25) may be replaced by

$$|\widehat{\gamma}(\lambda_j, \lambda_j - 2\pi k/T)|^2 = \frac{|\sum_{m=1}^{M}[\widetilde{X}_{j-M/2+m} - \widehat{\mu}][\overline{\widetilde{X}}_{j-M/2-kN'+m} - \widehat{\mu}]|^2}{\sum_{m=1}^{M}|\widetilde{X}_{j-M/2+m} - \widehat{\mu}|^2 \sum_{m=1}^{M}|\widetilde{X}_{j-M/2-kN'+m} - \widehat{\mu}|^2} \tag{10.28}$$

and then $p(|\gamma|^2) = (M-2)(1-|\gamma|^2)^{M-3}$, leading to the solution for the αthreshold,

$$x_\alpha = |\gamma|^2_\alpha = 1 - e^{\log(\alpha)/(M-2)}. \tag{10.29}$$

In summary, if $|\widehat{\gamma}(\lambda_j, \lambda_j - 2\pi k/T)|^2$ exceeds the threshold, then we can declare that evidence of PC structure exists for these specific T, λ and k.

We repeat for emphasis that the preceding test is for a specific T, λ and k. Frequently, T can be considered known, and then a test for presence of PC-T structure can be constructed from a family of hypothesis tests defined by a set $\mathbf{H}_T$ of pairs (λ_j, k). If there is no prior knowledge, $\mathbf{H}_T$ should cover the set $[0, 2\pi) \times \{1, 2, \ldots, T-1\}$ by finite collection of points that accounts for X_t being real. In this case we suggest $\mathbf{H}_T = \Lambda \times [1, 2, \ldots, \lfloor (T-1)/2 \rfloor]$, where $\Lambda = \{\lambda_j = jM\pi/N, j = 1, 2, \ldots, \lfloor N/M \rfloor\}$. If none of the null hypotheses in $\mathbf{H}_T$ is rejected, then there is no evidence within the family $\mathbf{H}_T$ that X_t is PC-T. However, the thresholds for the individual tests require adjustment for multiple hypotheses. This problem has not been systematically studied for the hypothesis testing problems connected with determining the presence of PC structure. Our elementary approach has been to use the Bonferroni correction together with a simply reasoned estimate of the number of independent tests in the family.

10.4.2 Spectral Coherence for Unknown T

In our previous discussion of spectral coherence, T was assumed known. Here we show how spectral coherence can be used as a basis for testing for the presence of PC structure when T is unknown. Rather than restricting the computation of the coherence statistic to a specific support line determined by T, we compute

$$|\widehat{\gamma}(\lambda_p, \lambda_q, M)|^2 = \frac{|\sum_{m=1}^{M} \widetilde{X}_{p-M/2+m}\overline{\widetilde{X}}_{q-M/2+m}|^2}{\sum_{m=1}^{M}|\widetilde{X}_{p-M/2+m}|^2 \sum_{m=1}^{M}|\widetilde{X}_{q-M/2+m}|^2} \tag{10.30}$$

for (p, q) in a square array and determine the (p, q) for which $|\widehat{\gamma}(\lambda_p, \lambda_q, M)|^2$ is significant. The perception of this empirical spectral coherence is aided by plotting the coherence values only at points where a threshold is exceeded, where the threshold is determined by the null distribution of $|\widehat{\gamma}(\lambda_p, \lambda_q, M)|^2$

under the assumption that the $\widetilde{X}_j$ are i.i.d. complex Gaussian variates, so that (10.26) may be used for setting thresholds.

The spectral coherence statistic of (10.30) is sometimes called *diagonal spectral coherence* because it may be seen as a smoothing of the two-dimensional periodogram

$$f(N, j, k) = \frac{1}{2\pi N} \widetilde{X}_j \overline{\widetilde{X}_k}$$

along a diagonal line (having unity slope) starting at the coordinate (p, q), and then normalizing by the product of the smoothed diagonal terms. Since the support of the spectral measure F for PC sequences consists of straight lines of unity slope, the diagonal spectral coherence computation gives a test for the presence of PC structure [107].

Effects of Parameter M. From the viewpoint of sensitivity, (10.27) shows that the threshold $|\gamma|^2_\alpha$ decreases as M increases, meaning smaller values of true coherence will be called significant. So sensitivity argues for larger M. But since the parameter M controls the length of a smoothing window applied to some diagonal line, we see that choosing M too small relative to the smoothness of the underlying coherence will diminish our ability to detect the presence of significant coherence. On the other hand, if the underlying coherence varies rapidly along some diagonal line, then choosing M too large will cause the coherence statistic to be diluted with terms $\widetilde{X}_j \overline{\widetilde{X}_k}$ having low values of true coherence. This causes the effective M to be smaller, and hence the threshold for significant coherence is set too low, producing too many erroneous rejections of the null. As in most nonparametric smoothing procedures, it is thus useful to observe the results of a collection of smoothing parameters. We typically use $M = 8, 16$, and 32 to begin.

If the underlying coherence along some line were very smooth (in the limit, uniform as in PC white noise) then we are motivated to consider making M as large as permitted (i.e., $M = N$) so that only one value of spectral coherence is determined for each separation d from the main diagonal. It is not difficult to show that for $M = N$ the numerator of (10.30) is given by

$$\sum_{m=0}^{M-1} \widetilde{X}_{p-M/2+m} \overline{\widetilde{X}}_{q-M/2+m} = N \sum_{n=0}^{N-1} |X_n|^2 e^{i2\pi(q-p)n/N}; \tag{10.31}$$

thus it may be seen that $|\gamma(\lambda_p, \lambda_q, M)|^2$, as a function of $d = p - q$, is proportional to the magnitude of the normalized Fourier transform of $|X_n|^2$. That is, it becomes the usual periodogram of the squares, whose utility (and limitations) for recognizing the presence of PC structure we have already seen (see Sections 2.2.6 and 2.2.7). Specifically, the periodogram of squares is useful only when $\sigma^2(t)$ is properly periodic, which corresponds to $B_k(0) \neq 0$ for some $k > 0$. If $B_k(0) = 0$ for some $k > 0$, since $B_k(0) = \int_0^{2\pi} f_k(\lambda) d\lambda$, we conclude that the density $f_k(\lambda)$ along the kth support line integrates to 0. From (10.22) we can see that the theoretical coherence $\gamma(\lambda, \lambda - 2\pi k/T)$ can thus change sign along the support line.

10.5 SPECTRAL ESTIMATION: PRACTICE

The practical computations of $\widehat{f}_{k,N}(\lambda)$ and $|\widehat{\gamma}(\lambda_j, \lambda_j - 2\pi k/T)|^2$ follow exactly as described above. The sample Fourier transform $X_N(\lambda)$ is computed for a finite collection of λ and for given T, the shifted periodograms are computed and smoothed to produce estimates $\widehat{f}_{k,N}(\lambda)$ of $f_k(\lambda)$. Using the distribution of $|\widehat{\gamma}(\lambda_j, \lambda_j - 2\pi k/T)|$ we make a test to see if $f_k(\lambda)$ is significant in comparison to $f_0^{1/2}(\lambda) f_0^{1/2}(\lambda - 2\pi k/T)$. In contrast, confidence intervals for $f_k(\lambda)$ or $\gamma(\lambda, \lambda - 2\pi k/T)$ tell us something about the estimation errors.

10.5.1 Confidence Intervals

First note that $f_k(\lambda)$ and $\widehat{f}_{k,N}(\lambda)$ are typically not real when $k > 0$. As in Brillinger [26] we treat the real and imaginary parts separately, using a Student's t to describe the distribution of Re $[\widehat{f}_k(\lambda_j) - f_k(\lambda_j)]$ and Im $[\widehat{f}_k(\lambda_j) - f_k(\lambda_j)]$ relative to their sample variances $\widehat{\sigma}^2_{\mathrm{re}}$ and $\widehat{\sigma}^2_{\mathrm{im}}$. Then, setting

$$\Delta_{\mathrm{re}} = \frac{\widehat{\sigma}_{\mathrm{re}}}{\sqrt{M}} t_{M-1}(1 - \alpha/2), \tag{10.32}$$

the confidence interval for Re $f_k(\lambda)$ is [Re $\widehat{f}_k(\lambda_j) - \Delta_{re}$, Re $\widehat{f}_k(\lambda_j) + \Delta_{re}$]. The confidence interval for Im $f_k(\lambda)$ is [Im $\widehat{f}_k(\lambda_j) - \Delta_{im}$, Im $\widehat{f}_k(\lambda_j) + \Delta_{im}$].

Confidence intervals for $|\gamma(\lambda, \lambda - 2\pi k/T)|$ can also be based on the Fisher transformation. Enochson and Goodman [55] show that

$$z = \frac{1}{2} \log \frac{1 + |\hat{\gamma}|}{1 - |\hat{\gamma}|} = \tanh^{-1} |\hat{\gamma}| \tag{10.33}$$

and show that for large values $(0.4 < |\gamma|^2 < 0.95)$ z is close to $N(\mu_z, \sigma_z^2)$, where

$$\mu_z = \zeta + \frac{1}{2(M-1)}, \quad \sigma_z^2 = \frac{1}{2M},$$

and

$$\zeta = \frac{1}{2} \log \frac{1 + |\gamma|}{1 - |\gamma} = \tanh^{1-} \gamma.$$

Nutall and Carter [173] show graphically the errors in the transformation for a few values of M. But with modern computing, the confidence intervals can be computed exactly, as in Wang and Tang [220].

The estimators for $f_k(\lambda)$ and $|\gamma(\lambda, \lambda - 2\pi k/T)|^2$ have been implemented in programs `fkest.m` and `scoh.m`. Although the programs were constructed for real series, only slight modifications should be required to make them applicable to complex series.

Program `fkest.m`. For specified period T and k, the program `fkest.m` implements the estimator $\widehat{f}_{k,N}(\lambda)$. The Fourier transform of the sample is computed by the FFT program at the $\Lambda_F = \{\lambda_j = j2\pi/N,\ j = 0, 1, \ldots, N-1\}$. The estimator $\widehat{f}_{k,N}(\lambda)$ is computed for a subset of Λ_F using a specified window W_m. The quantities Re $\widehat{f}_{k,N}(\lambda)$ and Im $\widehat{f}_{k,N}(\lambda)$ are plotted along with their confidence intervals determined by Student's t as discussed earlier. In addition, the values of $|\gamma(\lambda, \lambda - 2\pi k/T)|^2$ are plotted along with the threshold (10.27) for significant coherence.

Program `scoh.m`. First, an FFT of length N is computed at the frequencies $\Lambda_F = \{\lambda_j = j2\pi/N,\ j = 0, 1, \ldots, N-1\}$. Next, the magnitude squared coherence $|\gamma(\lambda_p, \lambda_q, M)|^2$ is computed in a specified square set of (p, q) and using a specified smoothing window W_m. Only values of $|\gamma|^2$ exceeding the threshold (10.27) are plotted. For multiple hypothesis corrections, we replace the specified α with α/N_s, where N_s is the number of points on one side of the plotting square. A provision is also made to smooth the resulting image with a two-dimensional smoothing window.

10.5.2 Examples

10.5.2.1 White Noise Starting with a very simple case, we examine $\widehat{f}_{k,N}(\lambda_j)$ assuming the sequence is PC with $T = 16$, when it is just white noise. Therefore we know the support of resulting spectral measure F is just the main diagonal of $[0, 2\pi) \times [0, 2\pi)$ and hence $f_k(\lambda) \equiv 0$ for all λ when $k > 0$ and $B_k(\tau) \equiv 0$ for all τ when $k > 0$.

Figure 10.1 presents $\widehat{f}_{1,N}(\lambda_j)$, $N = 1024$ for $\lambda_j = 2j\pi/N$, $j = 0, 1, \ldots, 255$, in the case that X_t is just stationary white noise and hence $E\{\widetilde{X}_j \overline{\widetilde{X}_{j'}}\} = 0$ for $j \neq j'$. The smoothing weights are uniform $W_m' = 2\pi/NM$, with $M = 16$. It is clear that $B_k(\tau) \equiv 0$ and hence $f_k(\lambda) \equiv 0$ for $k \neq 0$; in addition, for $k = 0$ we have $B_0(0) = \sigma_0^2$ and $f_0(\lambda) \equiv \sigma_0^2/2\pi$. The confidence intervals for Re $\widehat{f}_{1,N}(\lambda)$ and Im $\widehat{f}_{k,N}(\lambda)$ are determined by the t statistic method (10.32) with $\nu = M - 1$ degrees of freedom.

The spectral coherence images in Figure 10.2 show, for white Gaussian noise, the speckled character of the image for a small smoothing window ($M = 4$) and the effect of increasing the window to $M = 16$.

Since a spectral coherence image is a presentation of many values of $|\gamma|^2$, the issue of multiple hypothesis correction naturally arises. By counting only

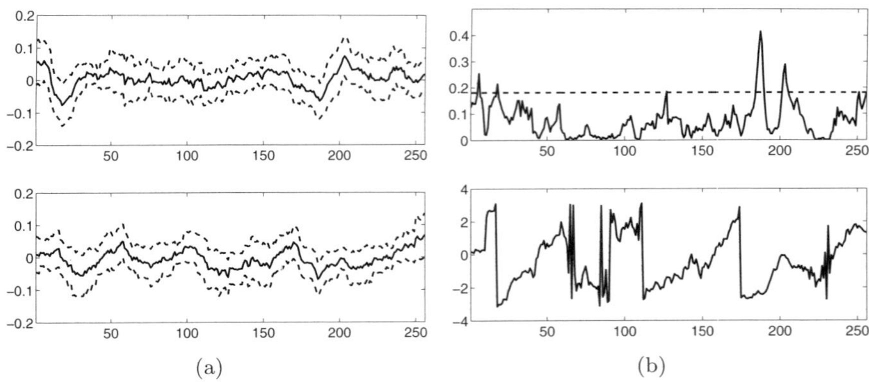

Figure 10.1 Estimate $\widehat{f}_{1,N}(\lambda)$ and $|\gamma(\lambda, \lambda - 2\pi/T)|^2$ for white noise, using FFT of length 1024, $T = 16$ and a uniform smoothing window of length 16. (a) Top is Re $\widehat{f}_{1,N}(\lambda)$; bottom is Im $\widehat{f}_{k,N}(\lambda)$. Confidence intervals based on the t statistic (95%). (b) Top is $|\gamma(\lambda, \lambda - 2\pi/T)|^2$; bottom is arg $\gamma(\lambda, \lambda - 2\pi/T)$. Coherence threshold based on (10.27) with $\alpha = 0.05$.

one side of the diagonal (say, $j > i$) for an image of side n, we obtain $n(n-1)/2$ distinct values of $|\gamma|^2$. But these are clearly not independent random variables so using the Bonferroni correction $\alpha/(n(n-1)/2)$ in place of α is too harsh. Using α/n, although still conservative, is considered more reasonable. The bottom two spectral coherence images of Figure 10.2 show the same images as the top row, but the plot threshold is determined by (10.27) with α/n in place of α.

10.5.2.2 PC White Noise If X_t is PC white noise (see Sections 2.2.2 and 6.4.1), $X_t = \sigma(t)\epsilon_t$, where ϵ_t is stationary white noise and $\sigma(t) = \sigma(t+T)$, then from the remarks following (6.59), we have

$$f_k(\lambda) = \frac{B_k(0)}{2\pi} \text{ for } 0 \leq \lambda < 2\pi,$$

where $B_k(0) = T^{-1}\sum_{t=0}^{T-1} \sigma^2(t) e^{-i2\pi kt/T}$. Taking the simple $\sigma(t) = \sigma_0[1 + \alpha\cos(2\pi t/T)]$ we obtain $B_0(0) = \sigma_0^2(1+\alpha^2/2)$, $B_1(0) = \alpha\sigma_0^2$, $B_2(0) = \alpha^2\sigma_0^2/4$, and since the sequence is real, $B_{T-1}(0) = B_1(0)$, $B_{T-2}(0) = B_2(0)$. All other values of k produce $B_k(0) = 0$. It follows that $f_0(\lambda) = (1+\alpha^2/2)/2\pi$, $f_1(\lambda) = \alpha/2\pi$, and $f_2(\lambda) = \alpha^2/2\pi$. All other values of k produce $f_k(\lambda) = 0$. The support of resulting spectral measure F is shown in Figure 10.3 for $T = 16$.

Figure 10.4(a) presents $\widehat{f}_{1,N}(\lambda_j)$, $N = 1024$ for $\lambda_j = 2j\pi/N$, $j = 0, 1, \ldots, 255$, for PC white noise with $\alpha = 1$. The smoothing weights are uniform $W_m' =$

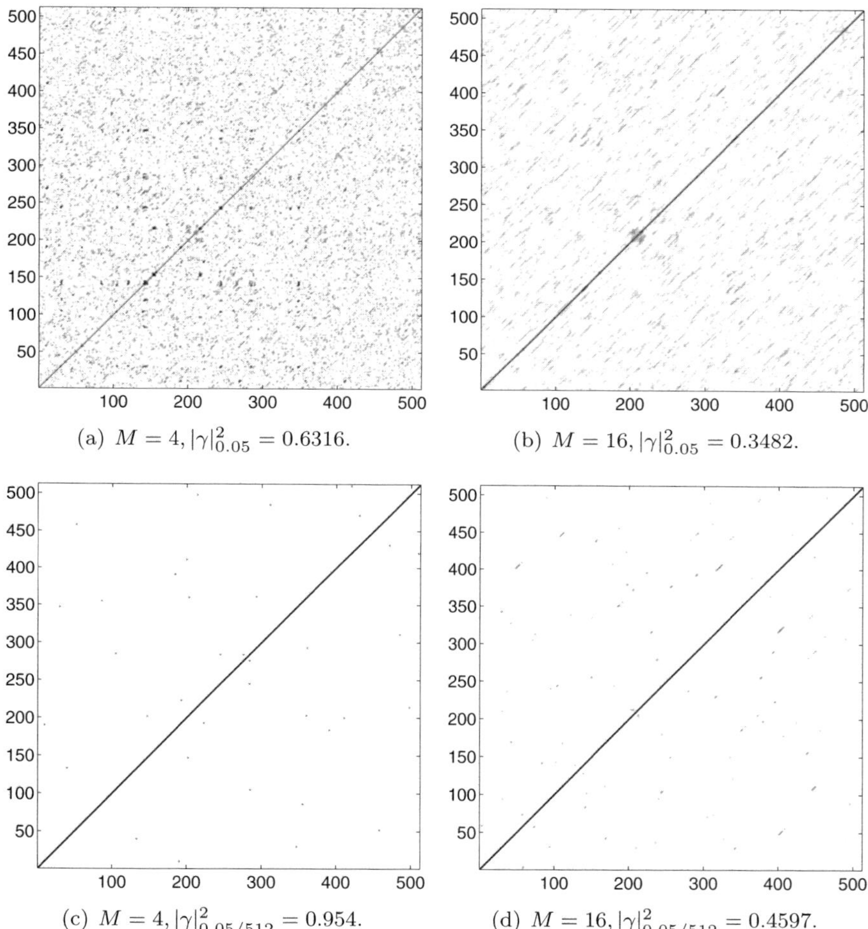

(a) $M = 4, |\gamma|^2_{0.05} = 0.6316$.

(b) $M = 16, |\gamma|^2_{0.05} = 0.3482$.

(c) $M = 4, |\gamma|^2_{0.05/512} = 0.954$.

(d) $M = 16, |\gamma|^2_{0.05/512} = 0.4597$.

Figure 10.2 Spectral coherence image of white noise based on an FFT of length 1024. Only values exceeding the α thresholds $|\gamma|^2_\alpha$ are plotted. For (a) and (b), $\alpha = 0.05$. For (c) and (d), the Bonferroni correction is used, $\alpha = 0.05/512$.

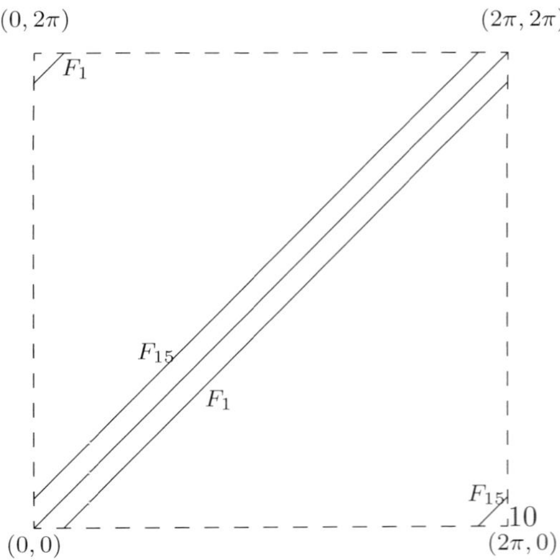

Figure 10.3 Support set S_1 and S_{15} for real $X_t = \sigma_0[1 + \alpha \cos 2\pi t/16]\epsilon_t$. Since X_t is real, $B_1(\tau) \equiv B_{15}(\tau)$ and $f_1(\lambda) \equiv f_{15}(\lambda)$.

$2\pi/NM$, with $M = 16$. It is clear from the top of Figure 10.4(b) that almost all of the $|\gamma|^2$ values are significant relative to the 0.05 threshold. The estimates and confidence intervals for Re $\widehat{f}_{1,N}(\lambda)$ and Im $\widehat{f}_{k,N}(\lambda)$ show that the values of Im $f_k(\lambda)$ evidently are not contributing to the large values of $|\gamma(\lambda, \lambda - 2\pi/T)|^2$. This is correct as the true values of Im $f_1(\lambda)$ are zero.

Figures 10.4(c) and 10.4(d) show, for the PC white noise simulation with $\alpha = 1$, spectral coherence images based on an FFT of length 1024. Note the indices run from 1 to 512, corresponding to $\lambda_j = 2j\pi/N$, $j = 0, 1, \ldots, 511$. This corresponds to the lower left quarter square $[0, \pi) \times [0, \pi)$ of $[0, 2\pi) \times [0, 2\pi)$, and so the support sets of F_1 and F_{15} outside this quarter square are not seen. See Figure 10.3. But the value of the period T may easily be inferred from these images by determining the least value of $p - q$ for which there is significant coherence. The plot threshold for Figure 10.4(c) is determined by p-value= 0.05 whereas for Figure 10.4(d) the plot threshold is determined by p-value= 0.05/512.

10.5.2.3 PAR(1) We demonstrate the use of spectral coherence on some simulated PAR(1) series, where we can compute the true spectral density.

Figure 10.5 shows 1200 points of a simulated PAR(1) series $X_t = \phi(t)X_{t-1} + \xi_1$ for which $\phi(t) = 0.6 + 0.3 \cos 2\pi t/12$.

Figure 10.6(a) show the estimates Re $\widehat{f}_{0,N}(\lambda)$ and Im $\widehat{f}_{0,N}(\lambda)$ due to smoothing by $M = 32$ points along with the true densities computed according to (8.39). Note the non-zero values of Im $\widehat{f}_{0,N}(\lambda)$ are essentially computational noise and may be considered zero. The two panels of Figure 10.6(b) show the

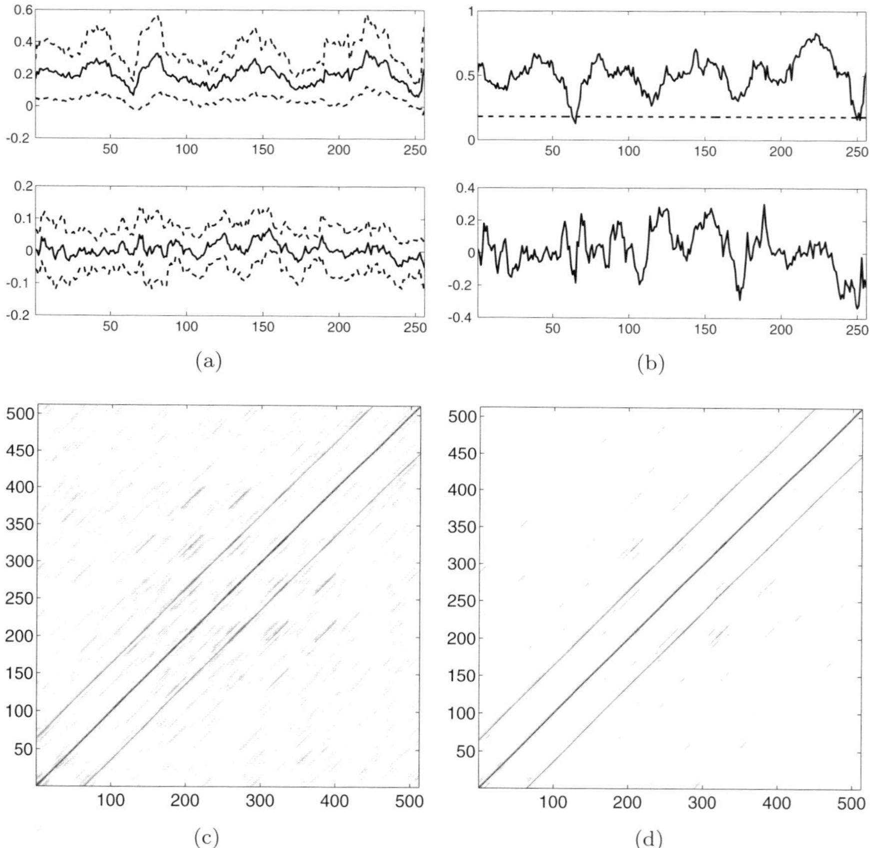

Figure 10.4 Simulated periodic white noise, $X_t = [1+\cos(2\pi t/T)]\epsilon_t$ with $T = 16$. Spectral estimates based on an FFT of length 1024 and $M = 16$. (a) Top is $\text{Re}\,\widehat{f}_{1,N}(\lambda)$ and bottom is $\text{Im}\,\widehat{f}_{1,N}(\lambda)$, both with 95% confidence intervals. (b) Top is $|\gamma(\lambda, \lambda - 2\pi/T)|^2$ with $\alpha = 0.05$ threshold. Bottom is $\arg[\text{Re}\,\widehat{f}_{1,N}(\lambda)+i\text{Im}\,\widehat{f}_{1,N}(\lambda)]$ (radians). In (c) and (d), only values exceeding the $|\gamma|^2$ thresholds are plotted. (c) $M = 32, |\gamma|^2_{.05} = 0.0921$. (d) $M = 32, |\gamma|^2_{.05/512} = 0.2576$

estimates $\text{Re}\,\widehat{f}_{1,N}(\lambda)$ and $\text{Im}\,\widehat{f}_{1,N}(\lambda)$. The confidence intervals contain the theoretically computed densities.

Figure 10.7 presents a spectral coherence image for this series based on one FFT of length 1200. Since $T = 12$, there are 100 periods in the sample and so the F_1 line should occur at a shift of 1200/12 from the diagonal. The line is present but it does not extend very far because $|\gamma(\lambda, \lambda - 2\pi/T)|^2$ quickly diminishes as λ increases.

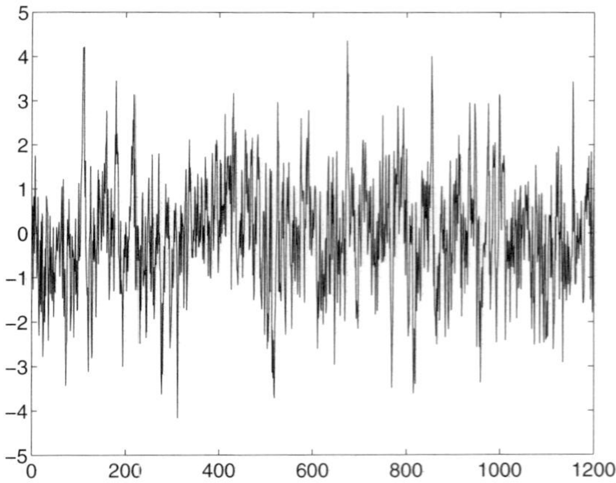

Figure 10.5 Simulated PAR(1) series $X_t = \phi(t)X_{t-1}+\xi_1$ for which $\phi(t) = 0.6 + 0.3\cos(2\pi t/12)$.

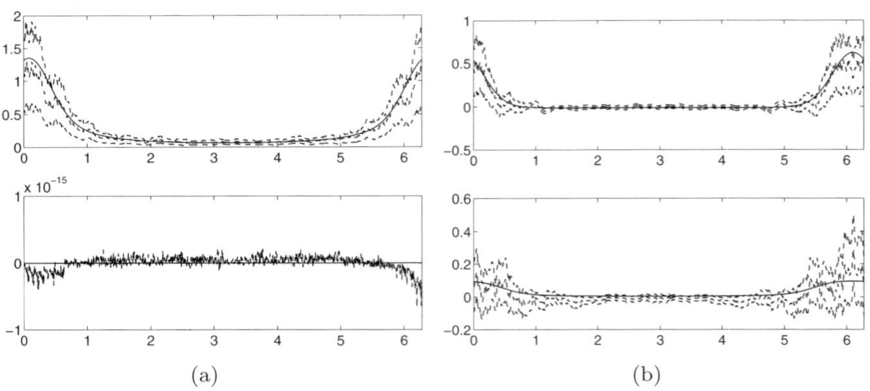

Figure 10.6 Estimates and true values of $f_0(\lambda)$ and $f_1(\lambda)$ for simulated PAR(1) with $\phi(t) = 0.6 + 0.3\cos(2\pi t/12)$. The FFT length was 1200 and smoothing was uniform with $M = 32$. Solid lines are true values; dashed lines are estimates and 95% confidence intervals. (a) Top is Re $\widehat{f}_{0,N}(\lambda)$ and bottom is Im $\widehat{f}_{0,N}(\lambda)$. (b) Top is Re $\widehat{f}_{1,N}(\lambda)$ and bottom is Im $\widehat{f}_{1,N}(\lambda)$.

10.5.2.4 Polarity Modulated AR(1) We begin with the zero mean AR(1) sequence $Y_t = \phi Y_{t-1} + \xi_t$, where ξ_t is an orthonormal sequence and $\phi < 1$. Then

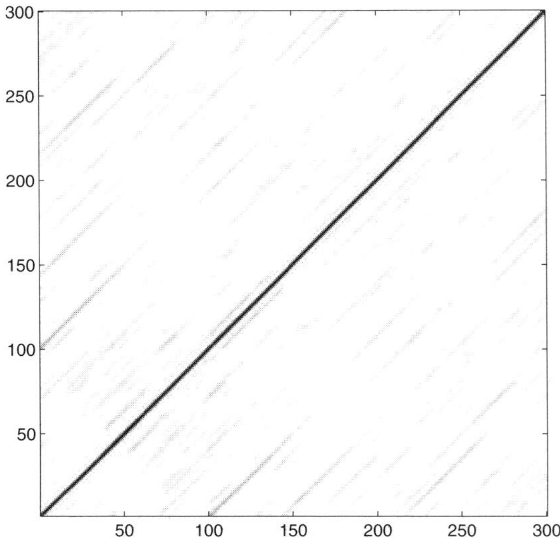

Figure 10.7 Spectral coherence image using one FFT of length 1200 and $M = 64$ for simulated PAR(1) for which $\phi(t) = 0.6 + 0.3\cos(2\pi t/12)$.

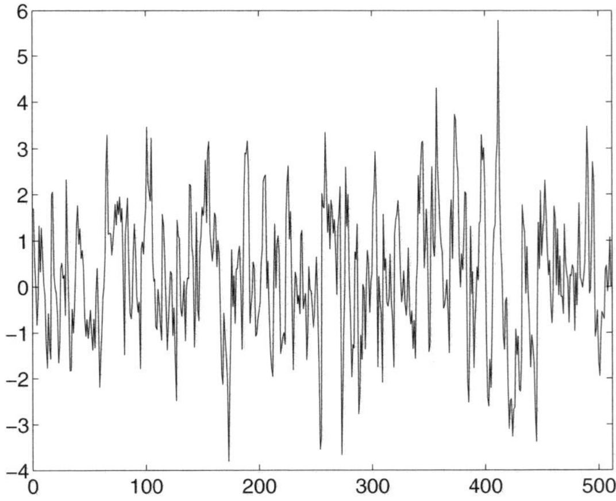

Figure 10.8 Sample of $X_t = P_t Y_t$ where $Y_t = 0.95Y_{t-1} + \xi_t$ and $P_t = 1$ when $t \bmod T < \lfloor T/2 \rfloor$ and $P_t = -1$ otherwise.

we form

$$X_t = \begin{cases} Y_t, & 0 \le t \bmod T < [T/2] \\ -Y_t, & [T/2] \le t \bmod T < T \end{cases} . \tag{10.34}$$

Since this is an amplitude scale modulation of a stationary sequence by a periodic sequence, then X_t is PC-T. But since $E\{X_t^2\} = E\{Y_t^2\}$, it is another case where X_t is PC-T but the variance is not properly periodic. Figure 10.8 shows a 512 point sample of the simulated X_t for $\phi = 0.95$ and $T = 32$. Although we do not show the plot of $\widehat{\sigma}_N^2(t)$, the p-value for the Bartlett test for $\sigma(t) \equiv \sigma$ produces a p-value of 0.99. The main point of interest here is the spectral coherence image shown in Figure 10.9. Although the occurrence of significant spectral coherence on the support lines S_T is remarkable, the same phenomenon occurs for other models having constant variance.

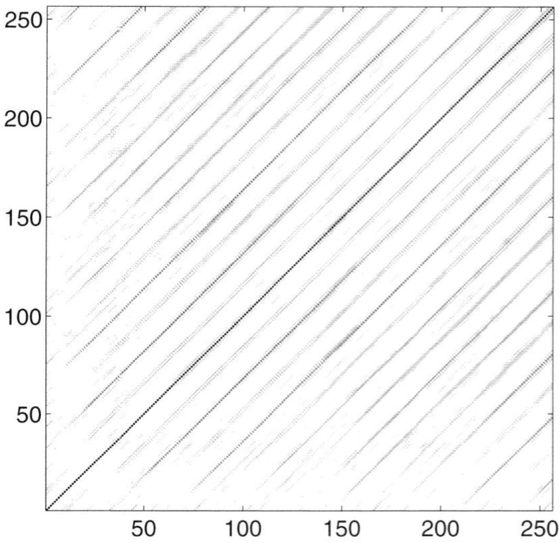

Figure 10.9 Spectral coherence image for switching AR(1) described in Figure 10.8 based on one FFT of length 512. $M = 32$.

Bird Call. Figure 10.10 presents a series of 4096 points from an audio recording of the Eastern screech owl [85]. During this segment of 0.93 second, the modulation was very regular, appearing periodic, but with unknown period. The periodograms of X_t and X_t^2, shown in Figure 10.11 reveal no significant periodicity in X_t (top panel) but very significant periodicity in the squares X_t^2 (bottom panel). Although the threshold line is drawn using $\alpha = 0.01$, the strongest harmonics are also significant at $\alpha = 0.01/128$, the threshold for Bonferroni correction based on the 128 displayed values. The periodicity is also very clear in the spectral coherence image of Figure 10.12 because the apparent amplitude modulation is very severe. Since the strongest harmonic is at frequency index of 16, corresponding to $16 - 1$ periods in the record, the period is estimated to be $4096/(16 - 1) = 273.07$. Although an isolated bird call is clearly a signal of finite duration, (and hence theoretically unable to be a PC sequence), the sample observed is consistent with the PC structure. We consider it not to be a paradox, but a fact related to the inferences possible from a finite sample.

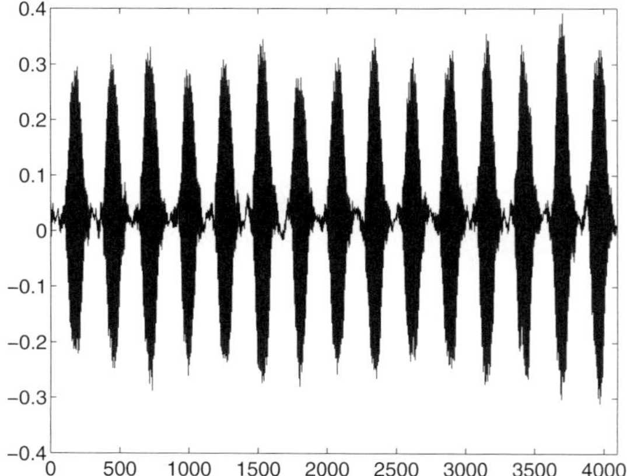

Figure 10.10 Sample (0.93 second) of audio recording of an Eastern screech owl . See [85].

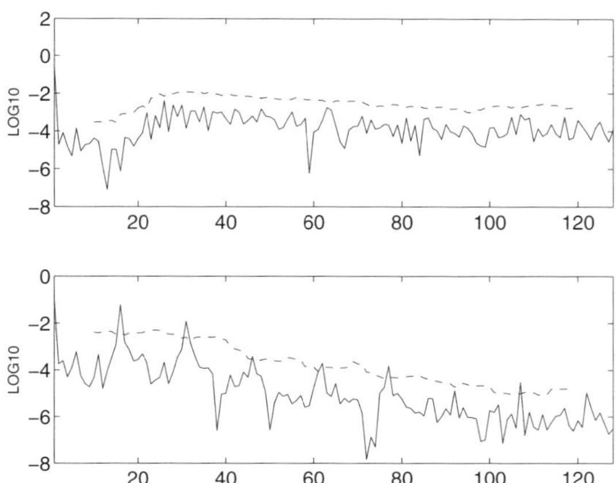

Figure 10.11 Periodograms of Eastern screech owl audio signal based on one FFT of length 4096. Dashed lines are the $\alpha = 0.01/128$ (Bonferroni correction for 128 points plotted) threshold for test of variance contrast based on a half neighborhood of size $m = 8$. Top : X_t. Bottom : X_t^2.

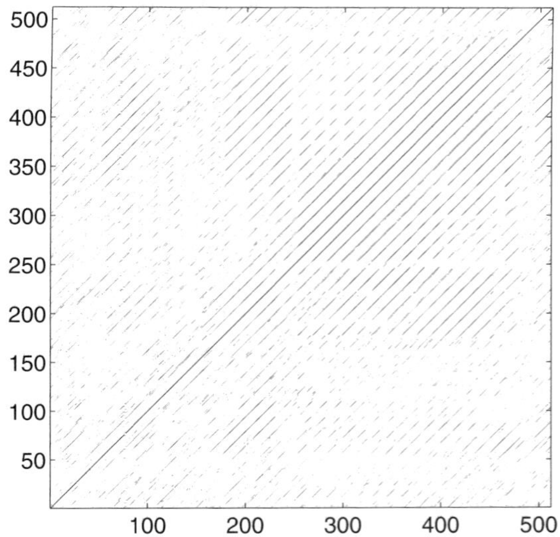

Figure 10.12 Spectral coherence image of Eastern screech owl audio signal based on one FFT of length 4096 and $M = 16$.

10.6 EFFECTS OF DISCRETE SPECTRAL COMPONENTS

Let us observe that the presence of any discrete spectral components in X_t will spoil every one of the consistency results for $\widehat{R}_N^*(t+\tau, t), \widehat{B}_{k,NT}^*(\tau), \widehat{f}_{k,N}(\lambda)$ given in Chapters 9 and 10. This applies to random discrete spectral components as well as nonrandom ones, such as the ones associated with the periodic mean. In addition, discrete spectral components can seriously distort and bias the estimation of spectral coherence. To see this, suppose for a harmonizable X_t we have $\xi(\{\lambda_a\}) = A(\omega)e^{i\lambda_a}$ and $\xi(\{\lambda_b\}) = B(\omega)e^{i\lambda_b}$. Now from a realization of X_t consider the computation of $|\gamma(\lambda_p, \lambda_q, M)|^2$, where we assume for simplicity that $\lambda_p = \lambda_a$ and $\lambda_q = \lambda_b$. Then

$$\begin{aligned} |\gamma(\lambda_p, \lambda_q, M)|^2 &= \frac{|\sum_{m=1}^{M} \widetilde{X}_{p-M/2+m}\overline{\widetilde{X}}_{q-M/2+m}|^2}{\sum_{m=1}^{M} |\widetilde{X}_{p-M/2+m}|^2 \sum_{m=1}^{M} |\widetilde{X}_{q-M/2+m}|^2} \\ &= \frac{|A(\omega)B(\omega) + \zeta_1|^2}{|A^2(\omega) + \zeta_2||B^2(\omega) + \zeta_3|}, \end{aligned}$$

where ζ_1, ζ_2, and ζ_3 are the remaining random parts in the respective sums. In the realization, if $A(\omega)$ and $B(\omega)$ are large compared to the remaining parts, then $|\gamma(\lambda_p, \lambda_q, M)|^2$ will be near 1. So to perceive the presence of PC-T structure with absolutely continuous spectra we must first detect and remove any discrete spectral components, both the random and the nonrandom.

Our approach for removing discrete spectral components is to first remove the periodic mean and then the remaining components that are detectable.

10.6.1 Removal of the Periodic Mean

The removal of the sample periodic mean by $Y_t = X_t - \widehat{m}_{t,N}$ has already been described and utilized in the results of Chapter 9. We note that for a sequence to be PC-T, the *only* nonrandom discrete spectral components are those of the periodic mean, for any others would produce a mean function that did not satisfy $m(t) = m(t + T)$.

Recall from Section 6.7 that the periodic mean produces atoms in the measure F at $(\lambda_1, \lambda_2) = (2\pi j/T, 2\pi k/T)$ of weight $\widetilde{m}_j\overline{\widetilde{m}_k}$ for $j, k = 0, 1, \ldots, T-1$. Since from (9.7)

$$\lim_{N\to\infty} \widehat{m}_{t,N} = \sum_{k=0}^{T-1} \xi(\{2\pi k/T\})e^{i2\pi kt/T} = M(t) \tag{10.35}$$

(in the mean square sense), removal of the sample mean eliminates more than just the $\widetilde{m}_k$, it removes the effects of random discrete spectral components having the same frequencies as the sample mean. That is, we may have $\xi(\{2\pi k/T\}) = \widetilde{m}_k + A_k$, where $\widetilde{m}_k$ is the Fourier coefficient of the (nonrandom) mean m_t and A_k is the random but zero mean remainder. Note that $E\{|A_k|^2\} = 0$, $k = 0, 1, \ldots, T-1$ is the condition for the lifted sequence $\mathbf{X}_n$ to be L^2 mean ergodic (see Propositions 9.2 and 9.3). Finally, since $X'_t = X_t - M(t)$ will clearly have $F_{X'}(j2\pi/T, j2\pi/T) = 0$, $j = 0, 1, \ldots, T-1$ and $\| X_t - \widehat{m}_{t,N} \| \to \| X_t - M(t) \|$, then the spectral measure of $X_t - \widehat{m}_{t,N}$ must converge to zero at frequencies $(j2\pi/T, j2\pi/T)$ for $j = 0, 1, \ldots, T-1$. We leave it as a problem to show this using the spectral representation of X_t.

10.6.2 Testing for Additive Discrete Spectral Components

From the previous paragraphs, the other random periodic components necessarily must have random amplitudes with zero means, since the mean m_t is explained completely by the $\widetilde{m_k}$. Assuming X_t is harmonizable (as it would be if X_t were PC-T or stationary), the other periodic components can be written

$$X''_t = \sum_{j=1}^{\infty} \xi(\lambda_j)e^{i\lambda_j t}.$$

In order for X''_t to be PC-T it is necessary[1] for $(\lambda_j, \lambda_k) \in S_T$ whenever $E\{\xi(\lambda_1)\overline{\xi(\lambda_2)}\} \neq 0$. Although detecting and removing all the remaining dis-

[1]This requirement does not exist if X_t is almost PC. See the supplements in Chapter 1 for the definition.

crete spectral components is not feasible with a finite sample, we can still detect and remove or suppress the larger ones to reduce their effect on the tests for PC structure.

The problem of detecting the presence of a periodic sequence added to a random sequence is a very old one and has a large literature; see, for example, Schuster [208, 209] and Fisher [57]. Thorough treatments can be found in books by Hannan [88, page 463], Anderson [8], Brockwell and Davis [28, page 334] and Quinn [186].

Here we use a method based on the periodogram, whose original purpose was finding the frequencies and amplitudes of additive sine and cosine components in a time series (see Schuster [209]). The periodogram can also be viewed as the least-squares estimation of the amplitudes of the sine and cosine components when the frequencies are fixed, as in the FFT algorithm. Its use for estimating spectral densities was described earlier in this chapter.

The method we use may be motivated by the simple case of detecting the discrete frequency terms comprising a trigonometric polynomial in the presence of additive uncorrelated (white) noise. That is,

$$X_t = f_t + \xi_t, \qquad \text{where} \qquad f_t = \sum_{p=1}^{P} a_p e^{i\lambda_p} \tag{10.36}$$

and ξ_t is real and i.i.d. $N(0, \sigma_\xi^2)$ and $\lambda_p > 0,\ p = 1, 2, \ldots, P$. The usual periodogram for a sample of length N is defined simply as

$$I_N(\lambda) = \frac{1}{2\pi N} |\widetilde{X}_N(\lambda)|^2, \tag{10.37}$$

where

$$\widetilde{X}_N(\lambda) = \sum_{t=0}^{N-1} X_t e^{-i\lambda t}. \tag{10.38}$$

Since we can only compute the transform (10.38) for a finite set of λ, we will typically use the set of *Fourier* frequencies $\{0, 2\pi/N, 4\pi/N, \ldots, (N-1)2\pi/N\}$, which are precisely the frequencies computed by the FFT. Under the null hypothesis that $f_t \equiv 0$, X_t is white noise with zero mean and variance σ_ξ^2, so for $\lambda_j = j2\pi/N$ (a Fourier frequency) with $j \geq 1$, the random variables

$$\operatorname{Re}\{\widetilde{X}_N(\lambda_j)\} \quad \text{and} \quad \operatorname{Im}\{\widetilde{X}_N(\lambda_j)\}$$

are of zero mean, of variance $N\sigma_\xi^2/2$, and orthogonal to $\operatorname{Re}\{\widetilde{X}_N(\lambda_{j'})\}$ and $\operatorname{Im}\{\widetilde{X}_N(\lambda_{j'})\}$ for $j \neq j'$.

Hence $\widetilde{X}_N(2\pi j/NT)$, $j = 1, ..., N-1$ will be complex Gaussian with zero mean and variance $N\sigma_\xi^2$. The random variable $\widetilde{X}_N(0)$ is always real for a real series. Hence for $j > 0$ the random variables

$$Z_j = \frac{1}{N}\frac{|\widetilde{X}_N(2\pi j/N)|^2}{\sigma_\xi^2}, \quad j = 1, ..., N-1 \tag{10.39}$$

are independent and distributed $\chi^2(2)$. For the case $j = 0$, $(N)^{-1}\widetilde{X}_N(0)$ is the sample mean and $Z_0 = |(N)^{-1}\widetilde{X}_N(0)|^2$ is $\chi^2(1)$ since $E\{X_t\} = 0$ under the null.

Finally, as a test for the presence of a harmonic component at some $\lambda_{k_0} = 2\pi k_0/N \neq 0$, we use the *variance contrast* ratio, defined as the ratio of Z_{k_0} to the average of Z_j over a deleted neighborhood $A(k_0)$ of k_0. Sometimes we specify the deleted neighborhood in terms of its half-width $m = n_A/2$, where $n_A = \text{card}\,(A)$. Under the null hypothesis stated above,

$$\begin{aligned} Z_{k_0} &= \frac{|\widetilde{X}_N(\lambda_{k_0})|^2}{N\sigma_\xi^2} \sim \chi^2(2), \\ \sum_{k\in A} Z_k &= \frac{\sum_{k\in A}|\widetilde{X}_N(\lambda_k)|^2}{N\sigma_\xi^2} \sim \chi^2(2n_A). \end{aligned}$$

Thus the distribution of the *variance contrast* ratio

$$\eta(k_0, A, T) = \frac{Z_{k_0}}{\sum_{k\in A} Z_k} \times \frac{2n_A}{n_1} \tag{10.40}$$

is $F(n_1, 2n_A)$, where $n_1 = 2$ except when $k_0 = 0$, and then Z_0 has an imaginary part of zero, making $n_1 = 1$. This is a slight variant of the test described by Anderson [8, Section 4.3.3]. Assume under the alternative hypothesis that the frequencies of f_t in (10.36) are among the Fourier frequencies so at such a frequency, say, $\lambda_p = 2\pi k_p/N$, the random variables Z_{k_p} will each be noncentral $\chi^2(2)$ with noncentrality parameter $N|a_p|^2/\sigma_\xi^2$ and the variance contrast ratio is a noncentral F. For power calculations of related tests in a signal detection application, see Whalen [221] and Robertson [194].

The efficacy of the *variance contrast* can be argued even when the model (10.36) of the null can be relaxed to permit Y_t to be a stationary sequence with a sufficiently smooth spectral density, for then the denominator of the contrast ratio remains a good estimate of the variance in a neighborhood of λ_{k_p}.

This method is used as a basis for testing $\widetilde{m}_k = 0$ and $B_k(\tau) = 0$ in the programs `permcoeff.m` and `Bcoeff.m` described in Sections 9.2.2 and 9.4.2. In this case the frequencies to be tested are $2\pi j/T$, $j = 0, 1, \ldots, T-1$. In the more general case, when no specific frequency is known, we test all the Fourier frequencies λ_{k_0} possible by using a neighborhood of typically 8 or 16 points centered at λ_{k_0} with some points deleted near the center. This is done for all k_0 for which the neighborhoods can be obtained from the periodogram. Again, the threshold is chosen (using the F) and discrete components declared at frequencies where the threshold is exceeded.

The preceding defines the basic idea, but there is another issue to discuss in getting to a practical algorithm. Suppose there is a large discrete component at λ_{k_0} and none within the deleted neighborhood $A(k_0)$, then the algorithm behaves as expected when the neighborhood is centered at k_0. But when the neighborhood is centered at k_1 with $k_0 \in A(k_1)$, then $\sum_{k \in A(k_1)} I_N(\lambda_k)$ will be too large due to the large discrete component at λ_{k_0}. To alleviate this problem we eliminate from $\sum_{k \in A(k_1)} I_N(\lambda_k)$ values of $I_N(\lambda_k)$ that are too far from their median, where decision thresholds are determined by the probability of incorrectly eliminating large values. When large values are "trimmed" out of the background, the value $2n_A$ is adjusted and the threshold for a specified α is recomputed.

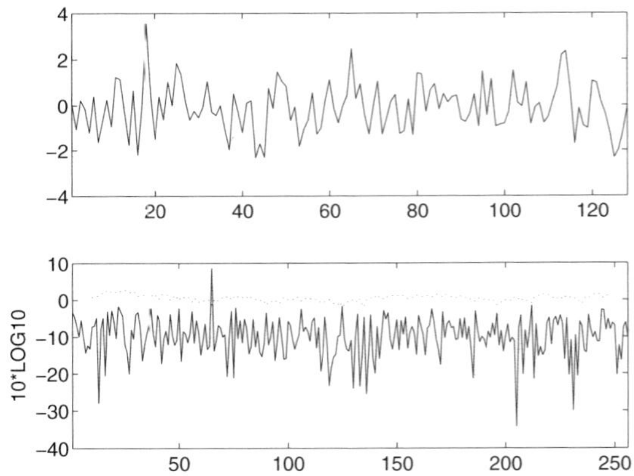

Figure 10.13 (Top) Simulated series Y_t. (Bottom) Periodogram based on 512 point sample. Dotted line is $\alpha = 0.001$ threshold for test of variance contrast using half-width $m = 8$.

The top trace of Figure 10.13 presents 128 consecutive values of the sum

$$Y_t = A\cos(\lambda_1 t) + \xi_t, \tag{10.41}$$

where $E\{\xi_t\} = 0$, $\text{Var}\{\xi_t\} = 1$, $A = 0.5$, and $\lambda_1 = 2\pi/8$.

The bottom trace of Figure 10.13 shows the periodogram values along with a $\alpha = 0.001$ threshold for the test of variance contrast when $n_A = 2m = 16$. Only at (or very near) frequency $\lambda = 2\pi/8$ (for which there are 64 periods in 512 samples) is the observed periodogram value significantly above the average of the 16 neighboring values. The computed p-value of the variance contrast at $\lambda = 2\pi/8$ is 2.5×10^{-7}. Note this p-value is for one test at a specific fixed and known frequency. The simple but conservative Bonferroni adjustment for multiple hypotheses (see Westfall and Young [222, Section 2.3.1]) yields $250 \times 2.5 \times 10^{-7} = 6.25 \times 10^{-4}$, still extremely significant.

10.6.3 Removal of Detected Components

The use of the periodogram to detect the presence of harmonic components of the form $A \cos \lambda_{k_0} t + B \sin \lambda_{k_0} t$ in a real series suggests how to remove such terms from a series. For if a discrete component is detected at some λ_{k_0}, then the Fourier components $\text{Re}\,\{2\widetilde{X}_N(\lambda_{k_0})/n\}$ and $\text{Im}\,\{2\widetilde{X}_N(\lambda_{k_0})/n\}$ are the ordinary least-squares coefficients of regression of the observed series on the functions $\cos \lambda_k t$ and $-\sin \lambda_k t$. Hence if a discrete frequency component is detected at λ_{k_0}, the residual from the regression is

$$Y_t - \text{Re}\,\{2\widetilde{X}_N(\lambda_{k_0})/n\} \cos \lambda_{k_0} t + \text{Im}\,\{2\widetilde{X}_N(\lambda_{k_0})/n\} \sin \lambda_{k_0} t$$

Figure 10.14 shows that subtracting the detected discrete component in this manner effectively suppresses completely the component at $\lambda = 2\pi/8$, not

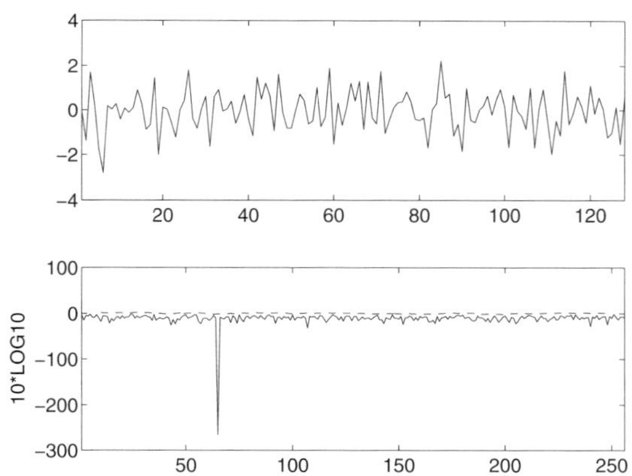

Figure 10.14 (Top) $Y_t - \text{Re}\,\{2\widetilde{X}_N(2\pi/8)/n\} \cos(2\pi t/8) + \text{Im}\,\{2\widetilde{X}_N(2\pi/8)/n\} \sin(2\pi t/8)$. (Bottom) Periodogram based on 512 point sample. Dotted line is $\alpha = 0.001$ threshold for test of variance contrast using $m = 8$.

just the discrete part. This effect can be mitigated by adding back a random component $a \cos \lambda_k t + b \sin \lambda_k t$, where a and b are distributed $N(0, \widehat{\sigma}^2)$ and $\widehat{\sigma}^2$ is essentially the denominator

$$\widehat{\sigma}^2 = \frac{1}{N \text{card}\,(A)} \sum_{k \in A} |\widetilde{X}_N(\lambda_k)|^2$$

in the variance contrast ratio (10.40).

PROBLEMS AND SUPPLEMENTS

10.1 The two spectra. Here we would like to point out the relationship between an empirical spectral analysis (what we can measure) and the spectral representation of the operator that propogates the process. First let us suppose we have a Hilbert space $\mathcal{H}$ on which there is defined some unitary operator U (more than one unitary operator can be defined on $\mathcal{H}$). Let $X \in \mathcal{H}$ and consider the stationary sequence $X_n = U^n X$ (we are taking $X_0 = X$) whose correlation sequence is given by

$$R(\tau) = \langle X_{n+\tau}, X_n \rangle = \int_0^{2\pi} e^{i\lambda\tau} F_{X_0}(d\lambda),$$

where the existence of F and the truth of this representation is guaranteed by the Herglotz theorem or via the spectral theorem for unitary operators (Chapter 4). Clearly it is possible to define many stationary sequences on a Hilbert space $\mathcal{H}$ on which we have a unitary operator.

The quantities we can observe and measure are determined by the action of U on specific vectors. Indeed, for two points $X_1, X_2 \in \mathcal{H}$ it is possible for F_{X_1} to be absolutely continuous with respect to Lebesgue measure while F_{X_2} is discrete.

This same observation carries over to PC-T sequences. Given $\mathcal{H}$ and unitary operator U on $\mathcal{H}$, a PC-T sequence may be formed by starting first with a set of T vectors $\mathbf{X} = (X_0, X_1, ..., X_{T-1})'$ and then for any $n = j + kT$ with $0 \le j \le T-1$ we define $X_{kT+j} = U^k[x_j]$. This PC-T sequence will have a correlation $R_{\mathbf{X}}(m,n)$ and spectral distribution functions $f_{\mathbf{X},k}$, $k = \{0, 1, ..., T-1\}$. These quantities are specific to the starting collection $\mathbf{X}$ just as the correlation and spectrum are specific to X in the stationary case. And so all we can do empirically is attempt to estimate these quantities for the specific $\mathbf{X}$ we happen to receive in our experiment.

10.2 Show the condition $\sum_{\tau=-\infty}^{\infty} \sum_{t=0}^{T-1} |R(t+\tau, t)| < \infty$ is sufficient for $\sum_{\tau=-\infty}^{\infty} \left[\sum_{t=0}^{T-1} |R(t+\tau,t)|^2\right]^{1/2} < \infty$. As a hint, note that for fixed τ,

$$[\sum_{t=0}^{T-1} |R(t+\tau,t)|^2]^{1/2} \le T^{1/2} \max_{t=0,1,\dots,T-1} |R(t+\tau,t)| = T^{1/2} |R(t_\tau + \tau, t_\tau)|,$$

where $t_\tau \in \{0, 1, \dots, T-1\}$. Then show $\sum_{\tau=-\infty}^{\infty} |R(t_\tau + \tau, t_\tau)| < \infty$.

10.3 Compute Var $[\widehat{f}_{k,N}(\lambda)]$ and Cov $(\widehat{f}_{k,N}(\lambda_1), \widehat{f}_{k,N}(\lambda_2))$, beginning with Proposition 10.5, for $\lambda, \lambda_1, \lambda_2$ in the Fourier frequencies. Recall for PC white noise that $f_k(\lambda) = B_k(0)/2\pi$, where $B_k(0) = T^{-1}\sum_{t=0}^{T-1} \sigma^2(t) e^{-i2\pi kt/T}$, $0 \le \lambda < 2\pi$.

10.4 Estimation of spectral densities for two-dimensional PC fields has been studied by Alekseev [6] and Dehay and Hurd [47].

10.5 Observing random periodic components. Consider the application of the periodogram test for discrete spectral components to the three following random sequences, where we have included the dependence on $\omega \in \Omega$:

1. $Y(t, \omega) = 0.5 \cos(2\pi t/16) + \xi(t, \omega)$; the periodic component of $Y(t, \omega)$ is a nonrandom function as constructed in the simulation.

2. $Y(t, \omega) = X(\omega) \cos(2\pi t/16) + \xi(t, \omega)$ with $\text{Var}[X] < \infty$; the periodic component of $Y(t, \omega)$ is then a nonstationary random function.

3. $Y(t, \omega) = 0.5 \cos(2\pi(t + \Theta(\omega))/16) + \xi(t, \omega)$, where $\Theta(\omega)$ is a random variable independent of $\xi(t, \omega)$ and uniformly distributed on $[0, 1, ..., 15]$; the periodic component of $Y(t, \omega)$ is a stationary random function.

Our observations of time series generated by these three cases will be very much the same. In particular, if we happen to experience $X(\omega) = 0.5$, then the observed $Y(t, \omega)$ will appear essentially the same in all three cases except for the shift of phase due to Θ. And the periodogram test for periodicity will produce the same result because, for all three cases, the forming of $I_N(\lambda)$ by (10.2) destroys the dependence on $\Theta(\omega)$. Our point is that from a single sample we can perceive whether a periodic component is present, but we cannot perceive whether it arises from a nonrandom function (such as a periodic mean), from a nonstationary random periodic function, or from stationary random periodic function. The manner in which we view it thus becomes an assumption about the underlying model. The most general (inclusive) view, without additional side information, is that it is a nonstationary random periodic function.

10.6 As pointed out in Section 10.4.2, choosing $M = N$ can sometimes be a detriment because it is possible for the phase of the spectral correlation measure F_k to change along its support line. Thus it is possible that the quantity (10.30) may average to zero. This argues that small values of M also have their purpose. To obtain a more powerful test for small values of M that still requires persistence along diagonal lines, we can *incoherently* average [107] all the values of $|\gamma(p, q, M)|^2$ from a coherence plot for a fixed value of $d = p - q$. To be precise, the *incoherent* statistic is the average

$$\delta(d, M) = \frac{1}{L+1} \sum_{p=0}^{L} |\gamma(pM, pM + d, M)|^2, \tag{10.42}$$

where we typically use $L = \lfloor (N - 1 - d)/M \rfloor$. The quantity $\delta(d, M)$ is plotted as a function of the difference frequency d. In essence, this statistic was utilized by Bloomfield, Hurd and Lund [19] for determining whether residuals of time series after model fitting are cyclostationary.

10.7 For a harmonizable X_t, show that

$$X_t - \widehat{m}_{t,N} = \int_0^{2\pi} e^{i\lambda t} W_N(\lambda T)\xi(d\lambda),$$

where $W_N(\lambda T)$ is continuous in λ, $W_N(\lambda T) = 0$ for $\lambda = j2\pi/T$, $j = 0, 1, \ldots, T-1$, and all $N \geq 1$. Also $\lim_{N\to\infty} W_N(\lambda T) = 1$ for $\lambda \neq j2\pi/T$, $j = 0, 1, \ldots, T-1$.

CHAPTER 11

A PARADIGM FOR NONPARAMETRIC ANALYSIS OF PC TIME SERIES

Suppose one is given a sample of a time series and asked the question: Does this series exhibit the PC property? If so, what can we say about it? This chapter summarizes and organizes the methods discussed in previous chapters into a procedural outline, or paradigm, for answering these questions. This is done only within the scope of mean, correlation, and spectral measurements. Obviously, our ability to answer these questions, especially regarding characterization, will substantially improve by the inclusion of PARMA time series analysis, a topic to be addressed in future writings.

These questions are to be answered by the observation of a single realization, and there are two distinct subcases to be considered- when the period T is *known* or when it is *unknown*.

Period is known. This is the case when the periodic effects are suspected or known to be coupled to a physical system that determines the period. The most obvious examples are the diurnal period in hours ($T = 24$) and the annual and quarterly periods, say in months so the periods

Periodically Correlated Random Sequences:Spectral Theory and Practice. By H.L. Hurd and A.G. Miamee

are $T = 12$ and $T = 4$, respectively. Known periods may also arise in mechanical or electrical systems.

When T is known, a strategy for determining the presence of PC structure can be based on the consistency results discussed in the preceding chapters. The tools we have at our disposal are estimators for m_t or its Fourier coefficients $\widetilde{m}_k$, for $R(t+\tau, t)$ or its Fourier coefficients $B_k(\tau)$, and for the densities $f_k(\lambda)$. When the sequence is found to be PC-T, there remains the question of whether or not it is just a periodically scaled (amplitude modulated) stationary sequence.

Period is unknown. This situation arises when one has no prior information about the presence of PC structure in the data or about its period. The general strategy is to use the multiple hypothesis test:

$\mathbf{H}_0$: observed series is not PC;
$\mathbf{H}_j, j = 1, ..., N$: observed series is PC with period T_j.

Although multiple alternatives can be tested using the same tools, we prefer spectral methods for detection and removal of harmonic components and for determining the presence of the PC property.

11.1 THE PERIOD T IS KNOWN

We build our understanding of the series of steps, roughly in the order given below. In all these tests, some adjustment for multiple hypotheses is likely required. Our programs implement simple multiple hypothesis corrections as appropriate.

The mean m_t. Use `permest.m` to form and plot $\widehat{m}_{t,N}$, to test for $m_t \equiv m$ and to return $X_t - \widehat{m}_{t,N}$ for later analysis. Rejection of $m_t \equiv m$ indicates a properly periodic mean. In physical systems it is a clue that periodic non-stationarity in the covariance structure (more generally, in the probability law) may be present. Although it is theoretically possible for the sample mean to converge to a random limit, $\lim_{N\to\infty} \widehat{m}_{t,N} = \sum_{k=0}^{T-1} \xi(\{2\pi k/T\})$, we usually assume the limit is nonrandom and so $\xi(\{2\pi k/T\}) = \widetilde{m}_k$.

Removal of mean and discrete spectral components. Using `pgram.m`, identify frequencies of discrete spectral components and remove them by the method indicated in Section 10.6.3.

Fourier coefficients of m_t. Use `permcoeff.m` to form $\widehat{\widetilde{m}}_{k,N}$ and test for $\widetilde{m}_k = 0$, for $k = 0, 1, \ldots, T-1$. $\widetilde{m}_0 = 0$ corresponds to the average

$T^{-1}\sum_{t=0}^{T-1} m_t = 0$. $\widetilde{m}_k \neq 0$ for $k \geq 1$ corresponds to $m_t \not\equiv m$, a properly periodic mean m_t. The interpretation regarding nonstationarity is the same as the preceding.

The variance $\sigma^2(t)$. Use `persigest.m` to form and plot $\widehat{\sigma}_N(t)$, to test $\sigma(t) \equiv \sigma$, and to return the demeaned and scaled series $[X_t - \widehat{m}_{t,N}]/\widehat{\sigma}_N(t)$ for later analysis. Rejection of $\sigma(t) \equiv \sigma$ indicates a properly periodic variance but leaves open whether or not X_t is simply the result of a stationary process subjected to amplitude-scale modulation. Another test, based on the Fourier coefficients $B_k(0)$ of $R(t,t) = \sigma^2(t)$, is given below.

The covariance $R(t+\tau, t)$. Use `peracf.m` to form and plot $\widehat{R}_{N_{t,\tau}}(t+\tau, t)$, to test for (a) $R(t+\tau,t) \equiv R(\tau)$ (for given lag τ, the covariance is constant with respect to t) and (b) $R(t+\tau,t) \equiv 0$ (for given lag τ, the covariance is zero for all t). Rejection of $R(t+\tau,t) \equiv R(\tau)$ indicates the sequence is properly PC (there exist lags for which $R(t+\tau,t)$ is properly periodic). Rejection of $R(t+\tau,t) \equiv 0$ for some $\tau \neq 0$ indicates the sequence is not PC white noise.

The correlation $\rho(t+\tau, t)$. The application of `peracf.m` to the demeaned and normalized $[X_t - \widehat{m}_{t,N}]/\widehat{\sigma}_N(t)$ produces $\widehat{\rho}_{N_{t,\tau}}(t+\tau,t)$. Using test (a) above, rejection of $\rho(t+\tau,t) \equiv \rho(\tau)$ indicates that X_t is properly PC and is not just an amplitude modulated stationary sequence. That is, there exist lags for which $\rho(t+\tau,t)$ is properly periodic. Test (b) gives the same information as it does for the covariance: rejection of $\rho(t+\tau,t) \equiv 0$ for some $\tau \neq 0$ indicates the sequence is not PC white noise.

The coefficients $B_k(\tau)$. Use `Bcoeff.m` to form $\widehat{B}_{k,NT}(\tau)$ for a collection of τ and for $k = 0, 1, \ldots, \lfloor (T-1)/2 \rfloor$. For these k and τ, test $B_k(\tau) = 0$ (we use the variance contrast method). Recall $B_0(0) > 0$ is expected. Rejection of $B_k(0) = 0$ for some $k > 0$ indicates a properly periodic $\sigma(t)$ (recall $B_k(0) = T^{-1} = \sum_{t=0}^{T-1} \sigma^2(t) e^{i2\pi kt}$). Rejection of $B_k(\tau) = 0$ for some $k > 0$, $\tau \neq 0$ indicates $R(t+\tau,t)$ is properly periodic at lag τ with frequency $2\pi k/T$.

Note that testing $B_k(\tau) = 0$ for some specific τ can be a more sensitive (more power) detector of the presence of PC structure if only a few $B_k(\tau)$ dominate the rest. We leave the quantification of this statement as a problem.

The coefficients $\rho_k(\tau)$. The application of `Bcoeff.m` to the demeaned and normalized $[X_t - \widehat{m}_{t,N}]/\widehat{\sigma}_N(t)$ produces $\widehat{\rho}_{t+\tau,t,N_{t,\tau}}$ for a collection of τ and for $k = 0, 1, \ldots, \lfloor (T-1)/2 \rfloor$. For these k and τ, test $\rho_k(\tau) = 0$ (we use the variance contrast method). Since the normalized series has constant unit variance, we expect $\rho_0(0) > 0$ and $\rho_k(0) = 0$ for $k > 0$. Rejection of $\rho_k(\tau) = 0$ for some $k > 0$, $\tau \neq 0$ indicates $\rho(t + \tau, t)$ is properly periodic at lag τ with frequency $2\pi k/T$, and that X_t is properly PC-T and is not just an amplitude modulated stationary sequence.

The densities $f_k(\lambda)$ for $B_k(\tau)$. Use `fkest.m` to form $\widehat{f}_k(\lambda)$ based on $X_t - \widehat{m}_t$ for $k = 0, 1, \ldots, \lfloor (T-1)/2 \rfloor$ and a subset of the Fourier frequencies $\Lambda_F = \{\lambda_j = j2\pi/N,\ j = 0, 1, \ldots, N-1\}$, where N is the sample size. For these k and λ, test $f_k(\lambda) = 0$ via sample magnitude squared coherence $|\hat{\gamma}(\lambda_j, \lambda_j - 2\pi k/T)|^2$. We expect $f_0(\lambda) \geq 0$ for all λ. Rejection of $f_k(\lambda) = 0$ for some k and λ indicates the spectral measure F is not zero on the line $\lambda_2 = \lambda_1 - 2\pi k/T$, and hence X_t is PC-T.

The densities $\check{f}_k(\lambda)$ for $\rho_k(\tau)$. Use `fkest.m` to form $\widehat{\check{f}}_k(\lambda)$ based on $[X_t - \widehat{m}_t]/\widehat{\sigma}_N(t)$ for $k = 0, 1, \ldots, \lfloor (T-1)/2 \rfloor$ and a subset of the Fourier frequencies $\Lambda_F = \{\lambda_j = j2\pi/N,\ j = 0, 1, \ldots, N-1\}$, where N is the sample size. For these k and λ, test $\check{f}_k(\lambda) = 0$ via sample magnitude squared coherence $|\hat{\gamma}(\lambda_j, \lambda_j - 2\pi k/T)|^2$. We expect $\check{f}_0(\lambda) \geq 0$ for all λ. Rejection of $\check{f}_k(\lambda) = 0$ for some k and λ indicates the spectral measure $\check{F}$ is not zero on the line $\lambda_2 = \lambda_1 - 2\pi k/T$, meaning $[X_t - m_t]/\sigma(t)$ is PC-T, and so X_t is properly PC-T and is not just an amplitude modulated stationary sequence.

The spectral measures F and $\check{F}$. Use `scoh.m` to make quick check that the support of F and $\check{F}$ is in the expected location, namely in the set S_T defined by (1.18) and shown in Figure 1.3.

11.2 THE PERIOD T IS UNKNOWN

When T is unknown we are faced with either conducting the fixed T tests for a range of T or conducting some tests organized for the more general case of determining an unknown T.

The steps below, although few, are useful for finding PC structure when the value of T is unknown. The idea is to get some candidate values of T, or in other words, to identify some diagonals in $[0, 2\pi) \times [0, 2\pi)$ where there is significant spectral coherence. We can then estimate the densities on these lines. Some of these can also be used for inference on almost PC sequences, a topic beyond our current scope. The tests we employ are somewhat more "spectral" in nature, as in nonparametric spectral estimation. As before our programs implement simple multiple hypothesis corrections as appropriate.

The mean m_t. Use `pgram.m` to plot the periodogram and identify significantly large harmonic terms. A program `spermean.m` (not demonstrated here) forms form $X_t - \widehat{m}_t$, the residual from the regression on Fourier frequency components. Without the knowledge of T, this is the only method we currently have to remove large components, whether they arise from a time varying mean or from random-amplitude components.

The variance $\sigma^2(t)$. Use `pgram.m` and `spermean.m` on the squares $[X_t - \widehat{m}_t]^2$ to plot the periodogram, to identify significantly large harmonic terms. Use the significantly large terms to form $\widehat{\sigma}(t)$ and $[X_t - \widehat{m}_t]/\widehat{\sigma}(t)$ for later analysis.

The spectral measure F. Use `scoh.m` applied to $X_t - \widehat{m}_t$ to determine if there are support lines on which there is significant coherence. Once some candidate support lines $\lambda_2 = \lambda_1 - \delta_k,\ k = 1, 2, \ldots, n$ are identified, the density f_{δ_k} on those lines can be estimated using `fkest.m`. Repeat the process for $[X_t - \widehat{m}_t]/\widehat{\sigma}_N(t)$ to find lines $\check{\delta}_k$ on which to estimate $\check{f}$.

The densities f on the support lines defined by $\delta_1, \delta_1, \ldots, \delta_n$. Use `fkest.m` to form $\widehat{f}$ based on $X_t - \widehat{m}_t$ on the given lines and for a subset of the Fourier frequencies $\Lambda_F = \{\lambda_j = j2\pi/N,\ j = 0, 1, \ldots, N-1\}$, where N is the sample size. Note we have already rejected that the density is zero on the line $\lambda_2 = \lambda_1 - \delta_k$.

The densities $\check{f}$ on the support lines defined by $\check{\delta}_1, \check{\delta}_1, \ldots, \check{\delta}_n$. Use `fkest.m` to form $\widehat{\check{f}}$ based on $[X_t - \widehat{m}_t]/\widehat{\sigma}_N(t)$ on the given lines and for a subset of the Fourier frequencies $\Lambda_F = \{\lambda_j = j2\pi/N,\ j = 0, 1, \ldots, N-1\}$, where N is the sample size. Note we have already rejected that the density is zero on the line $\lambda_2 = \lambda_1 - \check{\delta}_k$.

For unknown T the forming of $\widehat{m}_t$ and $\widehat{\sigma}(t)$ by identification of large components may require experimentation with thresholds. It is part of the price of dropping the constraint of a fixed T.

REFERENCES

1. I. L. Abreu, "A note on harmonizable and stationary sequences," *Bol. Soc. Mat. Mexicana,* **15**, pp. 48–51, 1970.
2. N. I. Akheizer and I. M. Glazman, *Theory of Linear Operators in Hilbert Space.* Fredrick Unger, 1961 and Dover, New York, 1993.
3. L.V. Ahlfors, *Complex Analysis*, 2nd ed. McGraw–Hill, New York, 1966.
4. E. J. Akutowicz, "On an explicit formula in least square prediction," *Math. Scan.*, **5**, pp. 261–266, 1957.
5. V. G. Alekseev, "Estimating the spectral densities of a Gaussian periodically correlated stochastic process," *Prob.Inf.Transm.*, **24**, pp. 109–115, 1988.
6. V. G. Alekseev, "On spectral density estimates of a Gaussian periodically correlated random fields," *Prob. Math. Stat.*, **11**, pp. 157–167, 1991.
7. J. Allen and S. Hobbs, "Detecting target motion by frequency-plane smoothing," in *Proceedings of the Twenty-Sixth Asilomar Conference on Systems and Computers*, Pacific Grove, CA, pp. 1042–1047, 1992.
8. T. W. Anderson, *The Statistical Analysis of Time Series*, Wiley, Hoboken, NJ, 1971.
9. T. W. Anderson, *An Introduction to Multivariate Analysis*, 2nd ed. ,Wiley, Hoboken, NJ. 1984.

10. P. L. Anderson, M. M. Meerschaert, and A. Vecchia, "Innovations algorithm for periodically stationary time series," *Stoch. Proc. Appl.*, **83**, pp.149–169, 1999.

11. C. F. Ansley, "An algorithm for the exact likelihood of a mixed autoregressive moving average process, *Biometrika*, **66**, pp. 59–65, 1979.

12. W. R. Bennett, "Statistics of regenerative digital transmission, " *Bell Syst Tech. J.*, **37**, pp. 1501–1542, 1958.

13. P. Billingsley, *Convergence of Probability Measures*, Wiley, Hoboken, NJ, 1968.

14. P. Billingsley, *Probability and measure*, 2nd ed., Wiley-Interscience, Hoboken, NJ, 1986.

15. S. Bittanti and G. De Nicolao, "Markovian representations of cyclostationary processes," in *Lecture Notes in Control and Information Sciences No. 161*, L. Gerencsér and P. E. Caines Eds., Springer, New York, 1991.

16. S. Bittanti, P. Bolzern, L. Piroddi, and G. De Nicolao, "Representation, prediction and identification of cyclostationary processes—a state-space approach," in *Cyclostationarity in Communications and Signal Processing*, W. A. Gardner, Ed., IEEE Press, New York, 1993.

17. S. Bittanti and P. Colaneri, "Invariant representations of discrete-time periodic sysytems," *Automatica*, **36**, pp. 1777–1793, 2000.

18. P. Bloomfield, *Fourier Analysis of Time Series: An Introduction*, Wiley, Hoboken, NJ, 1976.

19. P. Bloomfield, H.L. Hurd and R. Lund, "Periodic correlation in stratospheric ozone time series," *J. Time Series Anal.*, **15**, pp. 127–150, 1994.

20. S. Bochner, "A theorem on Fourier–Stieltjes integrals," *Bull. AMS*, **40**, pp. 272–276, 1934.

21. G. E. P. Box and G. M. Jenkins, *Time Series Analysis: Forecasting and Control*, Holden Day, San Francisco, 1970.

22. G. E. P. Box, G. M. Jenkins, and G. Reinsel, *Time Series Analysis*, 3rd ed., Prentice-Hall, Englewood Cliffs, NJ, 1994.

23. R. A. Boyles and W. A. Gardner, "Cycloergodic properties of discrete-parameter nonstationary stochastic processes," *IEEE Trans. Inf. Theory*, **IT–29**, pp. 105–114, 1983.

24. F. J. Beutler, "On stationary conditions for certain periodic random processes," *J. Math. Anal. Appl.*, **3**, pp. 25–36, 1961.

25. W. M. Brelsford, "Probability predictions and time series with periodic structure," PhD Dissertation, Johns Hopkins University, Baltimore, MD, 1967.

26. D. R. Brillinger, *Time Series: Data Analysis and Theory*, Holt, Rinehart and Winston, New York, 1965.

27. D. R. Brillinger, *Time Series: Data Analysis and Theory*, Holt, Rinehart and Winston, New York, 1975.

28. P. J. Brockwell and R. A. Davis, *Time Series: Theory and Methods*, 2nd ed., Springer, New York, 1991.

29. S. Cambanis and C. H. Houdré, "On the continuous wavelet transform of second order random processes," *IEEE Trans. Inf. Theory*, **IT-41**, pp. 628-642, 1995.

30. S. Cambanis, C. H. Houdré, H. L. Hurd, and J. Leskow, "Laws of large numbers for periodically and almost periodically correlated processes," *Stoch. Proc. Appl.*, **53**, pp. 37–54, 1994.

31. D. K. Chang and M. M. Rao, "Bimeasures and nonstationary processes" in *Real and Stochastic Analysis*, M. M. Rao, Ed., Wiley, Hoboken, NJ, 1987.

32. S.D. Chatterji, "Orthogonally scattered dilation of hilbert space valued functions ," *Lecture Notes in Mathematics*, No. 920, Springer Verlag, pp. 570–580, New York, 1982.

33. C. Corduneanu, *Almost Periodic Functions*, Chelsea Publishing Company, New York, 1989.

34. H. Cramér, *Methods of Mathematical Statistics*, Princeton University Press, Princeton, NJ, 1961.

35. H. Cramér, "On the theory of stationary random processes," *Math. Ann.*, **41**, pp. 215–230, 1940.

36. H. Cramér, "On some classes of nonstationary stochastic processes," *Proc. Fourth Berkeley Symp. Math. Stat, Prob.*, **2**, pp. 55–77, 1961.

37. A. V. Dandawate and G. B. Giannakis, "Statistical test for presence of cyclostationarity," *IEEE Trans. Signal Proc.*, **42**, pp. 2355–2369, 1994.

38. D. Dehay, "On a class of asymptotically stationary harmonizable processes," *J. Multivariate Anal.*, **22**, pp. 251–257, 1987.

39. D. Dehay, "Nonlinear analysis for almost periodically correlated strongly harmonizable processes," presented at *2nd World congress of the Bernoulli Society at Uppsala, Sweden,* August 13–18, 1990.

40. D. Dehay, "Processus bivariés presque périodiquement corrélés: analyse spectrale et estimation des densités spectrales croisées," in *Journées de Statistiques Strasbourg*, **XXIII**, pp. 187-189, 1991.

41. D. Dehay, "Estimation de paramètres fonctionnels spectraux de certains processus non-nécessairement stationnaires," *C. R. Acad. Sci. Paris*, **314(4)**, pp. 313–316, 1992.

42. D. Dehay and R. Moché, "Trace measures of a positive definite bimeasure," *J. Multivariate Anal.*, **40**, pp. 115–131, 1992.

43. D. Dehay, "Asymptotic behavior of estimators of cyclic functional parameters for some nonstationary processes," *Stat. and Decisions*, **13**, pp. 273–286, 1995.

44. D. Dehay, "Spectral analysis of the covariance of the almost periodically correlated processes," *Stoch. Proc. Appl.*, **50**, pp. 315–330, 1994.

45. R. L. Devaney, *An Introduction to Chaotic Dynamics*, Benjamin, San Francisco, 1986.

46. D. Dehay and H. L. Hurd, "Representation and estimation for periodically and almost periodically correlated random processes," in *Cyclostationarity in Communications and Signal Processing*, W. A. Gardner, Ed., IEEE Press, New York, 1993.

47. D. Dehay and H.L. Hurd, "Spectral estimation for strongly periodically correlated random fields defined on $\mathbf{R}^2$", *Math. Methods Stat.*, **11**, No. 2, pp. 135 – 151, 2002.

48. J. Diestel and J. J. Uhl, Jr., *Vector Measures*, Mathematical Surveys, No. 15, American Mathematical Society, Providence, RI, 1977.

49. J. L. Doob, *Stochastic Processes*, Wiley, Hoboken, NJ, 1953.

50. Y. P. Dragan, *Structure and Representation of Stochastic Signal Models* (in Russian), Naukova Dumka, Kiev, 1980.

51. Y. P. Dragan and I. N.Yavorskiy, *Rythmics of Sea Waves and Underwater Acoustic Signals* (in Russian), Naukova Dumka, Kiev, 1982.

52. Y. P. Dragan, V. A. Rozhkov, and I. N.Yavorskiy, *Methods of Probabilistic Analysis of Rhythms of Oceanological Processes* (in Russian), Gidrometeoizdat, Leningrad, 1987.

53. N. Dunford and J. T. Schwarz, *Linear Operators, Part I: General Theory*, Wiley-Interscience, Hoboken, NJ, 1958.

54. S. N. Elaydi, *An Introduction to Difference Equations*, 2nd ed., Academic Press, New York, 1999.

55. L. D. Enochson and N. R. Goodman, "Gaussian approximation to the distribution of sample coherence," Measurement Analysis Corporation, Technical Report AFFDL-TR-65-57, AD620987, June 1965.

56. C. J. Everett and H. J. Ryser, " The Gram matrix and Hadamard theorem," *Am. Math. Monthly*, **53**, No. 1 , pp. 21–23, 1946.

57. R. A. Fisher, "Tests of significance in harmonic analysis," *Proc. R. Soc. London Ser. A*, **125**, No. 796, pp. 54–59, 1929.

58. L. E. Franks, *Signal Theory*, Prentice-Hall, Englewood Cliffs, NJ, 1969.

59. L. E. Franks, "Polyperiodic linear filtering," in *Cyclostationarity in Communications and Signal Processing*, W. A. Gardner, Ed., IEEE Press, New York, 1994.

60. P. H. Franses, *Periodicity and Stochastic Trends in Economic Time Series*, Oxford University Press, New York, 1996.

61. R. Gangolli, "Wide sense stationary sequences of distributions on hilbert space and the factorization of operator-valued functions," *J. Math. Mech.*, **12**, pp. 893–910, 1963.

62. V. F. Gaposhkin, "Criteria for the strong law of large numbers for some classes of second order stationary processes and homogeneous random fields," *Theory Probab. Appl.*, **XXII**, No. 2, pp. 286–310, 1977.

63. W. A. Gardner, "Representation and estimation of cyclostationary processes," Ph.D. Dissertation, Department of Electrical and Computer Engineering, University of Massachusetts, August, 1972, reprinted as *Signal and Image Processing Lab Technical Report No. SIPL-82-1*, Department of Electrical and Computer Engineering, University of California at Davis, 1982.

64. W. A. Gardner and L. E. Franks, "Characterization of cyclostationary random signal processes," *IEEE Trans. Inf. Theory*, **IT–21**, pp. 4–14, 1975.

65. W. A. Gardner, "Stationarizable random processes," *IEEE Trans. Inf. Theory*, **IT–24**, pp. 8–22, 1978.

66. W. A. Gardner, *Introduction to Random Processes with Application to Signals and Systems*, Macmillan, New York, 1985.

67. W. A. Gardner, *Statistical Spectral Analysis: A Nonprobabilistic Theory*, Prentice Hall, Englewood Cliffs, NJ, 1987.

68. W. A. Gardner, "Signal interception: a unifying theoretical framework for feature detection," *IEEE Trans. Commun.*, **COM–36**, pp. 897–906, 1988.

69. W. A. Gardner, "Two alternative philosophies for estimation of the parameters of time-series," *IEEE Trans. Inf. Theory*, **37**, pp. 216–218, 1991.

70. W. A. Gardner, "Exploiting spectral redundancy in cyclostationary signals," *IEEE ASSP Mag.*, **8**, pp. 14–36, 1991.

71. W. A. Gardner and C. M. Spooner, "Signal interception: performance advantages of cyclic feature detectors," *IEEE Trans. Commun.*, **COM–40**, pp. 149–159, 1992.

72. W. A. Gardner and C. M. Spooner, "Detection and source location of weak cyclostationary signals: simplification of the maximum likelihood receiver, *IEEE Trans. on Commun.*, **COM–41**, pp. 905–916, 1993.

73. W. A. Gardner, "An introduction to cyclostationary signals," in *Cyclostationarity in Communications and Signal Processing*, W. A. Gardner, Ed., IEEE Press, New York, 1994.

74. W. A. Gardner, A. Napolitano, and L. Paura, "Cyclostationarity: half a century of research," *Signal Processing*, **86**, pp. 639–697, 2006.

75. M. J. Genossar, H. Lev-Ari and T. Kailath, "Consistent estimation of the cyclic autocorrelation," *IEEE Trans. Signal Proc.*, **42**, pp. 595–603, 1994.

76. E. G. Gladyshev, "On Multi-dimensional stationary random processes," *Theory Probab. Appl.*, **3**, pp. 425–428, 1958.

77. E. G. Gladyshev, "Periodically correlated random sequences," *Sov. Math.*, **2**, pp. 385–388, 1961.

78. E. G. Gladyshev, "Periodically and almost periodically correlated random processes with continuous time parameter," *Theory Probab. Appl.*, **8**, pp. 173–177, 1963.

79. G. Golub and C. Van Loan, *Matrix Computations*, Johns Hopkins Press, Baltimore, 1987.

80. N. R. Goodman, "On the joint estimation of the spectrum, co-spectra and quadrature spectrum of a two-dimensionsal stationary Gaussian process," Dissertation, Princeton University, 1957, Also Scientific Paper No. 10, Engineering Scientific Laboratory, New York University, AD134919, 1957.

81. N. R. Goodman, "Statistical analysis based on the multivariate complex Gaussian distribution," *Ann. Math. Stat.*, **34**, pp. 152–177, 1963.

82. N. R. Goodman, "Statistical tests for nonstationarity within the framework of harmonizable processes," Rocketdyne Research Report No. 65–28, AD619270, August 2, 1965.

83. L. Gu and L. Miranian, "Strong rank revealing Cholesky factorization," *Electron. Trans. Numer. Anal.*, **17**, pp. 76–92, 2004.

84. L. I. Gudzenko, "On periodically nonstationary processes," *Radiotekhnika i elektronika*, **4**, No. 6, pp. 1062–1064, 1959.

85. http://www.flmnh.ufl.edu/natsci/ornithology/sounds.htm.

86. P. R. Halmos, *Measure Theory*, Van Nostrand, Princeton, NJ, 1950.

87. P. R. Halmos, *Introduction to Hilbert Space*, Chelsea Publishing Company, New York, 1957.

88. E. J. Hannan, *Multiple Time Series*, Wiley, Hoboken, NJ, 1970.

89. H. Helson and G. Szegö, "A problem in prediction theory ," *Ann. Math. Pure Appl.*, **51**, pp. 107–138, 1960.

90. L. J. Herbst, "Almost periodic variances," *Ann. Math. Stat.*, **34**, pp. 1549–1557, 1963.

91. L. J. Herbst, "Periodogram analysis and variance fluctuations," *J. R. Stat. Soc. B*, **25**, pp. 442–450, 1963.

92. L. J. Herbst, "A test for variance heterogeneity in the residuals of a Gaussian moving average," *J. R. Stat. Soc. B*, **25**, pp. 451–454, 1963.

93. L. J. Herbst, "Spectral analysis in the presence of variance fluctuations," *J. R. Stat. Soc. B*, **26**, pp. 354–360, 1964.

94. L. J. Herbst, "Stationary amplitude fluctuations in a time series," *J. R. Stat. Soc. B*, **26**, pp. 361–364, 1964.

95. L. J. Herbst, "The statistical fourier analysis of variances," *J. R. Stat. Soc. B*, **27**, pp. 159–165, 1965.

96. L. J. Herbst, "Fourier methods in the study of variance fluctuations in time series analysis," *Technometrics*, **11**, pp. 103–113, 1969.

97. I. Honda, "On the spectral representation and related properties of periodically correlated stochastic processes," *Trans. IECE Japan*, **E65**, pp. 723–729, 1982.

98. I. Honda, "On the ergodicity of Gaussian periodically correlated stochastic processes," *Trans. IEICE Japan*, **E73**, pp. 1729–1737, 1990.

99. C. H. Houdré, "Harmonizability, V-boundedness, (2, p)-boundedness of stochastic processes," *Prob. Theory Relat. Fields*, **84**, pp. 39–54, 1987.

100. C. H. Houdré, "Linear Fourier and stochastic analysis," *Prob. Theory Relat. Fields*, **87**, pp. 167–188, 1990.

101. H. L. Hurd, "An investigation of periodically correlated stochastic processes," PhD dissertation, Duke University deptartment of Electrical Engineering, Nov., 1969.

102. H. L. Hurd, "Periodically correlated processes with discontinuous correlation functions," *Theory Probab. Appl.*, **19**, pp. 834–838, 1974.

103. H. L. Hurd, "Stationarizing properties of random shifts," *SIAM J. Appl. Math.*, **26**, pp. 203–211, 1974.

104. H. L. Hurd, "Representation of strongly harmonizable periodically correlated processes and their covariances," *J. Multivariate Anal.*, **29**, pp. 53–67, 1989.

105. H. L. Hurd, "Nonparametric time series analysis for periodically correlated processes," *IEEE Trans. Inf. Theory*, **IT–35**, pp. 350–359, 1989.

106. H. L. Hurd, "Correlation theory of almost periodically correlated processes," *J. Multivariate Anal.*, **37**, pp. 24–45, 1991.

107. H. L. Hurd and N. L. Gerr, "Graphical methods for determining the presence of periodic correlation in time series," *J. Time Series Anal.*, **12**, pp. 337–350, 1991.

108. H. L. Hurd and J. Leskow, "Estimation of the Fourier coefficient functions and their spectral densities for $\phi-$mixing almost periodically correlated processes," *Stat. Prob. Lett.*, **14**, pp. 299–306, 1992.

109. H. L. Hurd and J. Leskow, "Strongly consistent and asymptotically normal estimation of the covariance for almost periodically correlated processes," *Stat. Decisions*, **10**, pp. 201–225, 1992.

110. H. L. Hurd and V. Mandrekar, "Spectral theory of periodically and quasi-periodically stationary $S\alpha S$ sequences," Technical Report No. 349, Center for Stochastic Processes, Department of Statistics, UNC at Chapel Hill, Sept. 1991.

111. H. L. Hurd and G. Kallianpur, "Periodically correlated and periodically unitary processes and their relationship to $L_2[0, T]$-valued stationary sequences," in *Nonstationary Stochastic Processes and Their Appllication*, J. C. Hardin and A. G. Miamee, Eds., World Scientific Publishing, Singapore, 1992.

112. H. L. Hurd, "Almost periodically unitary stochastic processes," *Stoch. Proc. Appl.*, **43**, pp. 99–113, 1992.

113. H. L. Hurd and A. Russek, "Almost periodically correlated and almost periodically unitary processes in the sense of Stepanov," *Theory Probab. Appl.*, **41**, 1996.

114. H. L. Hurd and A. Russek, "Almost periodically correlated processes in LCA groups," Technical Report No. 369, Center for Stochastic Processes, Department of Statistics, UNC at Chapel Hill, 1992.

115. H. L. Hurd and C. H. Jones, "Dynamical systems with cyclostationary orbits," in *The Chaos Paradigm: Developments and Applications in Engineering and Science*, R. Katz, Ed., AIP Press, New York, 1994.

116. H. L. Hurd and T. Koski, "The Wold isomorphism for cyclostationary sequences," *Signal Processing*, **84**, No. 5, pp. 813–824, 2004.

117. H. L. Hurd and T. Koski, "Cyclostationary arrays: their unitary operators and representations," in *Stochastic Processes and Functional Analysis: A volume of recent advances in honor of M. M. Rao*, Lecture Notes in Pure and Applied

Mathematics No. 238 , Alan Krinik and Randall Swift, Eds., Marcel Dekker, New York, 2004.

118. H. L. Hurd, G. Kallianpur and J. Farshidi "Correlation and spectral theory for periodically correlated random fields indexed on $\mathbb{Z}^2$," *J. Multivariate Anal.*, **90**, No. 2, pp. 359–383, 2004.

119. H. L. Hurd, "Periodically correlated sequences of less than full rank," *J. Stat.Planning Inference*, **129**, pp. 279–303, 2005.

120. I. A. Ibragimov, "Some limit theorems for stationary processes," *Theory Probab. Appl.*, **12**, pp. 349–382, 1962.

121. Y. Isokawa, "An identification problem in almost and asymptotically almost periodically correlated processes," *J. Appl. Prob.*, **19**, pp. 53–67, 1982.

122. R. H. Jones and W. M. Brelsford, "Time series with periodic structure," *Biometrika*, **54**, pp.403–408, 1967.

123. K. L. Jordan, "Discrete representations of random signals, Technical Report No. 378, MIT Research Laboratory of Electronics, 1961.

124. G. Kallianpur and V. Mandrekar, "Spectral theory of stationary H-valued processes, *J. Multivariate Anal.*, **1**, pp. 1–16, 1971.

125. R. E. Kalman, "A new approach to linear filtering and prediction problems," *Trans. ASME J. Basic Eng.*, **83D**, pp. 35–45, 1960.

126. J. Kampé de Fériet, "Correlation and spectrum of asymptotically stationary random functions," *Math. Stud.*, **30**, pp. 55–67, 1962.

127. J. Kampé de Fériet, and F.N. Frenkiel, "Correlation and spectra for nonstationary random functions," *Math. Comp*, **16**, pp. 1–21(1962).

128. Y. Katznelson, *An Introduction to Harmonic Analysis*, Dover, New York, 1976.

129. A. Khintchine, "Korrelations theorie de stationaren stochastischen prozesse," *Math. Ann.*, **109**, pp. 604–615, 1934.

130. K. Kim, G. North and J. Huang, "EOFs of one-dimensional cyclostationary time series: computations, examples and stochastic modeling," *J. Atmos. Sci.*, **53**, pp. 1007–1017, 1996.

131. K. Kim and G. North, "EOFs of harmonizable cyclostationary processes," *J. Atmos. Sci.*, **54**, pp. 2417–2427, 1997.

132. K. Kim and Q. Wu, "A comparison study of EOF techniques: analysis of nonstationary data with periodic statistics," *J. Climate*, **12**, pp. 185-199, 1999.

133. A. Kolmogorov and Y. Rozanov, "On strong mixing conditions for stationary Gaussian processes," *Theory Probab. Appl.*, **5**, pp. 204–208, 1960.

134. A. Kolmogorov, "Stationary sequences in hilbert space," (in Russian), *Bull. Math. Univ. Moscow*, **2**, 1941. Translated in report CN/74/2, J. F. Barrett, trans., Department of Engineering, Cambridge University, pp. 1–24, 1974.

135. J. Leskow, "Asymptotically normality of the spectral density estimators for almost periodically correlated processes.," *Stoch. Proc. Appl.*, **52**, pp. 351-360, 1994.

136. B. M. Levitan and V. V. Zhikov, *Almost Periodic Functions and Differential Equations*, Cambridge University Press, London, 1982.

137. W. K. Li and Y. V. Hui, "An algorithm for the exact likelihood of periodic autoregressive-moving average (PARMA) models," *Commun. Stat. Simulation*, **17**, No. 4, pp. 1484–1494, 1988.

138. M. Loève, "Fonctions Aléatories du Second Order," in *P. Lévy's Processus Stochastiques et Mouvement Brownien*, pp. 228–252 Gauthier-Villars, Paris, 1948.

139. M. Loève, *Probability Theory*, Van Nostrand: New York, 1965.

140. R. Lund and I. V. Basawa, "Recursive prediction and likelihood evaluation for periodic ARMA models," *J. Time Series Anal.*, **21**, pp. 75–93, 2000.

141. H. Lütkepohl, *Introduction to Multiple Time Series Analysis*, 2nd ed., Springer-Verlag, New York, 1993.

142. A. Makagon and H. Salehi, "Structure of periodically distributed stochastic sequences", in *Stochastic Processes, A Festschrift in Honour of Gopinath Kallianpur*, Springer-Verlag, 245-251, 1993.

143. A. Makagon, A. G. Miamee and H. Salehi, "Continuous time periodically correlated processes: spectrum and prediction," *Stoch. Proc. Appl.*, **49**, pp. 277–295, 1994.

144. A. Makagon, and H. Salehi, "Spectral dilation of operator valued measures and its application to infinite dimensional harmonizable processes," *Studia Math.*, **85**, pp. 254–297, 1987.

145. A. Makagon, "Induced stationary process and structure of locally square integrable periodically correlated processes," *Studia Math.*, **136**, pp. 71–85, 1999.

146. A. Makagon and A. Weron, " Wold-Cramer concordance theorems for interpolation of q-variate stationary processes over locally compact Abelian groups ," *J. Multivariate Anal.*, **6**, pp. 123–137, 1976.

147. A. Makagon, A. G. Miamee and H. L. Hurd, "On AR(1) models with periodic and almost periodic coefficients," *Stoch. Proc. Appl.*, **100**, pp. 167–185, 2002.

148. K. V. Mardia, J. T. Kent, and J. M. Bibby, *Multivariate Analysis*, Academic Press, New York, 1979.

149. V.A. Markelov, "Axis crossings and relative time of existence of periodically nonstationary random processes," *Sov. Radiophys.*, **9** , pp. 440–443, 1966.

150. D. E. K. Martin, "Estimation of the minimal period of periodically correlated sequences," Ph.D. dissertation, Department of Mathematics, University of Maryland at College Park, 1990.

151. G. Maruyama, "The harmonic analysis of stationary stochastic processes," *Mem. Fac. Sci. Kyushu Univ. Ser. A*, **4**, 1949. Reprinted in *Gisiro Maruyama Selected Papers*, Kaigai Publications, Tokyo, 1988.

152. P. Masani, "The prediction theory of multivariate stochastic processes III," *Acta. Math.*, **104**, pp. 141–162, 1960.

153. P. Masani, "Recent trends in multivariate prediction theory," in *Multivariate Analysis V*, P.R. Krishnaiah, Ed., pp. 351–382, Academic Press, New York, 1966.

154. A. G. Miamee, "Spectral dilation of $L(\mathcal{B}, \mathcal{H})$-valued measures and its application to stationary dilation for Banach space valued processes," *Indiana Univ. Math. J.*, **38**, pp. 841–860, 1989.

155. A. G. Miamee and H. Salehi, "Harmonizability, V-boundedness and stationary dilation of stochastic processes ," *Indiana Univ. Math. J.*, **27**, pp. 37–50, 1978.

156. A. G. Miamee and H. Salehi, "On the bilateral prediction error matrix of a multivariate stationary stochastic process," *SIAM J. Appl. Math.*, **10**, pp. 247–253, 1979.

157. A. G. Miamee and H. Salehi, "On the prediction of periodically correlated stochastic processes," in *Multivariate Analysis.*, P.R. Krishnaiah, Ed., pp. 167–179, North-Holland, Amsterdam,1980.

158. A. G. Miamee and H. Salehi, "On an Expicit Representation of the Linear Predictor of a weakly Stationary Stochastic Sequence," *Bol. Soc. Mat. Mexicana* **28**, pp.81–93, 1983.

159. A. G. Miamee, "On determining the predictor of nonfull-rank multivariate stationary random processes," *SIAM J. Appl. Math.*, **18**, pp. 909–918, 1987.

160. A. G. Miamee, "Periodically correlated processes and their stationary dilations," *SIAM J. Appl. Math.*, **50**, pp. 1194–1199, 1990.

161. A. G. Miamee and M. Pourahmadi, "Best approximation in $L^p(d\mu)$ and prediction problems of Szegö, Kolmogorov, Yaglom and Nakazi," *J. London Math. Soc.*, **38**, pp. 133–145, 1988.

162. S. Mittnik, "Computation of theoretical autocovariance matrices of multivariate autoregressive moving average time series," *J. R. Stat. Soc. B* , **52**, pp. 151–155, 1990.

163. S. Mittnik, "Computing theoretical autocovariances of multivariate autoregressive moving average models by using a block Levinson method," *J. R. Statist. Soc.* B, **55**, pp. 435–440, 1993.

164. W. Mlak, "Dilation of Hilbert Space Operators (Genera Theory)", *Dissertationes Math.*, **CLIII**, pp. 1–61, 1978.

165. A. S. Monin, "Stationary and periodic time series in the general circulation of the atmosphere," in *Proceedings of Symposium on Time Series Analysis*, M. Rosenblatt, Ed., Wiley, Hoboken, NJ, 1963.

166. A. Napolitano and J. Leskow, "Quantile prediction for time series in the fraction-of-time probability framework," *Signal Proc.*, **82**, pp. 1727–1741, 2002.

167. A. R. Nematollahi and T. Subba Rao, "On The spectral density estimation of periodically correlated (cyclostationary) time series," *Sankhya*, **67**, Part 3, pp. 568–589, 2005.

168. H. Niemi, "On Stationary dilations and the linear prediction of certain stochastic processes ," *Soc. Sci. Fenn. Comment Phys. Math.*, **45**, pp. 111–130, 1975.

169. H. Niemi, "Stochastic processes as Fourier transforms of stochastic measures," *Ann. Acad. Sci. Fenn. Ser. A.I. Math.*, **591**, pp. 1–47, 1975.

170. H. Niemi, "On orthogonally scattered dilations of bounded vector measures," *Ann. Acad. Sci. Fenn. Ser. A.I. Math.*, **3**, pp. 43–52, 1977.

171. H. Niemi, "Diagonal Measure of a Positive Definite Bimeasure," in *Lecture Notes in Mathematics*, No. 945, pp. 237–246, Springer-Verlag, New York, 1982.

172. H. Niemi, "Grothendieck's inequality and minimal orthogonally scattered dilations," in *Lecture Notes in Mathematics*, No. 1080, pp. 175–187, 1984.

173. A. H. Nutall and G. Clifford Carter, "Approximation to the cumulative distribution function of the magnitude-squared coherence estimate," *IEEE Trans. ASSP*, **ASSP–29**, No. 4, pp. 932–936, 1981.

174. H. Ogura, "Spectral Representation of periodic nonstationary random processes," *IEEE Trans. Inf. Theory*, **IT–17**, pp. 143–149, 1971.

175. M. Pagano, "On Periodic and multiple autoregressions," *Ann. Stat.*, **6**, pp. 1310–1317, 1978.

176. A. Papoulis, *Probability, Random Variables and Stochastic Processes*, McGraw-Hill, New York, 1962.

177. E. Parzen, "On Consistent estimates of the spectrum of a stationary time series," *Ann. Math. Stat.*, **28**, pp. 24–43, 1957.

178. E. Parzen, "Spectral analysis of asymptotically stationary time series," *Bull. Int. Stat. Inst.*, **39**, No. 2, pp. 87–103, 1962.

179. E. Parzen, *Stochastic Processes*, Holden-Day, San Francisco, 1962.

180. E. Parzen, "On Spectral analysis with missing observations," *Sankhya, Ser. A*, **25**, pp. 383–392, 1963.

181. M. Pourahmadi and H. Salehi, "On subordination and linear transformation of harmonizable and periodically correlated processes," in *Probability Theory on Vector Spaces III*, pp. 195–213, Springer-Verlag, New York/Berlin, 1984.

182. M. Pourahmadi, "Taylor expansion of $\exp(\sum_{k=0}^{\infty} a_k z^k)$ and some applications," *Am. Math.Monthly*, **91**, pp. 303–307, 1984.

183. M. Pourahmadi, *Foundations of Time Series Analysis and Prediction Theory*, Wiley, Hoboken, NJ, 2001.

184. J. S. Prater and C. M. Loeffler, "Analysis and design of periodically time-varying IIR filters, with applications to transmultiplexing," *IEEE Trans. Signal Proc.*, **40**, pp. 2715–2725, 1992.

185. M. B. Priestley, "Evolutionary spectra and nonstationary process.", *J. R. Stat. Soc., Ser. B*, **27**, pp. 204–237, 1965.

186. B. Quinn, *The Estimation of Frequency*, Academic Press, New York, 2001.

187. H. Radjavi and P. Rosenthal , *Invariant Subspaces*, Springer-Verlag, New York/Berlin, 1973.

188. M. M. Rao "Harmonizable Processes: structure theory," *L'Enseign Math.* **28**, pp. 295–351, 1982.

189. M. M. Rao and K. Chang, "Bimeasure and nonstationary processes," in *Real and Stochastic Analysis*, M.M. Rao , Ed., pp.7–118, Wiley, Hoboken, NJ, 1986.

190. J. Ramanathan and O. Zeitouni, "On the wavelet transform of fractional Brownian motion," *IEEE Trans. Inf. Theory*, **IT–37**, pp. 1156–1158, 1991.

191. G. C. Reinsel, *Elements of Multivariate Time Series Analysis*, Springer-Verlag, New York, 1997.

192. F. Riesz and B. Sz.-Nagy, *Functional Analysis*, Fredrick Ungar, New York, 1965.

193. R. A. Roberts, W. A. Brown and H. H. Loomis, "A Review of digital spectral correlation analysis: theory and implementation," in *Cyclostationarity in Communications and Signal Processing*, W. A. Gardner, Ed., IEEE Press, New York, 1994.

194. G. H. Robertson, "Operating characteristics for a linear detector of CW signals in narrowband Gaussian noise," *Bell Syst. Tech. J.*, **46**, pp. 755–774, 1967.

195. H. L. Royden, *Real Analysis*, Macmillan, New York, 1968.

196. M. Rosenberg, "Quasi-isometric dilations of operator-valued measures and Grothendieck's inequality," *Pacific J. Math*, **103**, pp. 135–161, 1982.

197. M. Rosenblatt, "A central limit theorem and a strong mixing condition," *Proc. NAS*, **42**, pp.43–47, 1956.

198. M. Rosenblatt, *Stationary Sequences and Random Fields*, Birkhäuser, Boston, 1985.

199. Y. A. Rozanov, "Spectral theory of multi–dimensional stationary random processes with discrete time," *Usp. Mat. Nauk*, **13**, No. 2, pp. 93–142, 1958.

200. Yu.A. Rozanov, "Spectral analysis of abstract functions," *Theory Probab. Appl.*, **4**, pp.271–287, 1959.

201. Y. A. Rozanov, *Stationary Random Processes*, Holden Day, San Francisco, 1967.

202. W. Rudin, *Real and Complex Analysis*, McGraw-Hill, New York, 1987.

203. W. Rudin, *Fourier Analysis on Groups*, Wiley, Hoboken, NJ, 1990.

204. H. Sakai, "Circular lattice filtering using Pagano's method," *IEEE Trans. ASSP*, **30**, pp. 279 – 287, 1982.

205. H. Sakai, "On the spectral density matrix of a periodic ARMA process," *J. Time Series Anal.*, **12**, pp. 73 – 82, 1991.

206. Q. Shao and R. Lund, "Computation and characterization of autocorrelations and partial autocorrelations in periodic ARMA models," *J. Time Series Anal.*, **25**, No. 3, pp. 359–372, 2004.

207. L. Sharf, *Statistical Signal Processing*, Addison-Wesley, New York, 1990.

208. A. Shuster, "On lunar and solar periodicities of earthquakes," *Proc. R. Soc.*, **61**, pp. 455–465, 1897.

209. A. Shuster, "On the investigation of hidden periodicities with application to a supposed 26 day period of meteorological phenomena," *Terr. Magn.*, **3**, pp. 13–41, 1898.

210. M. H. Stone, "On one parameter unitary groups in hilbert space," *Annals of Math.*, **33**, pp. 643–648, 1932.

211. Taconite Inlet Project, http://eclogite.geo.umass.edu/climate/TILPHTML/TILPhome.html.

212. C. J. Tian, "A limiting property of sample autocovariances of perodically correlated processes with application to period determination," *J. Time Series Anal.*, **9**, pp. 411–417, 1988.

213. G. C. Tiao and M. R. Grupe, "Hidden periodic autoregressive moving average models in time series data," *Biometrika* **67**, pp. 365–373, 1980.

214. D. Tjöstheim and J. B. Thomas, "Some Properties and Examples of Random Processes that are Almost Wide Sense Stationary", *IEEE Trans.*, **IT–21**, pp. 257–262, 1975.

215. D. Tjöstheim, "On the analysis of a class of multivariate nonstationary stochastic processes," in *Prediction Theory and Harmonic Analysis*, V. Mandrekar and H. Salehi, Eds., pp. 403–416, North Holland, Amsterdam, 1983.

216. A. Trujillo-Ortiz and R. Hernandez-Walls, "Btest: Bartlett's test for homogeneity of variances," see URL http://www.mathworks.com/matlabcentral/fileexchange.

217. A. Vecchia, "Maximum likelihood estimation for periodic autoregressive moving average models," *Technometrics*, **27**, pp. 375–384, 1985.

218. A. Vecchia, "Periodic autoregressive-moving average (PARMA) modeling with applications to water resources," *Water Resour. Bull.*, **21**, No. 5, 1985.

219. A. Vecchia and R. Ballerini, "Testing for periodic autocorrelations in seasonal time series data," *Biometrika*, **78**, pp. 53–63, 1991.

220. S. Wang and M. Tang, "Exact confidence interval for magnitude-squared coherence estimates," *IEEE Signal Proc. Lett.*, **11**, No. 3, pp. 326–329, 2004.

221. A. D. Whalen, *Detection of Signals in Noise*, Academic Press, New York, 1971.

222. P. Westfall and S. Young, *Resampling-Based Multiple Testing*, Wiley, Hoboken, NJ, 1993.

223. N. Wiener, "Generalized harmonic analysis," *Acta Math.*, **55**, pp. 117–258, 1930.

224. N. Wiener and P. Masani, "The prediction theory of multivariate stochastic processes I," *Acta. Math.*, **98**, pp. 111–150, 1957.

225. N. Wiener and P. Masani, "The prediction theory of multivariate stochastic processes II," *Acta. Math.*, **99**, pp. 93–137, 1958.

226. H. O. A. Wold, "On prediction in stationary time series," *Ann. Math. Stat.*, **19**, pp. 558 – 567, 1948.

227. A. M. Yaglom, *Correlation Theory of Stationary and Related Random Functions*, Springer-Verlag, New York, 1987.

228. I. N. Yavorskiy, "The statistical analysis of periodically correlated random processes," (in Russian), *Radiotekhnika i elektronika*, **30**, No. 6, pp. 1096–1104, 1985.

229. V. N. Zasuhin, "On the theory of multi-dimensional stationary random processes," *Dokl. Akad. Nauk SSSR*, **116**, pp.435–437, 1941.

INDEX

WILEY SERIES IN PROBABILITY AND STATISTICS

ESTABLISHED BY WALTER A. SHEWHART AND SAMUEL S. WILKS

The ***Wiley Series in Probability and Statistics*** is well established and authoritative. It covers many topics of current research interest in both pure and applied statistics and probability theory. Written by leading statisticians and institutions, the titles span both state-of-the-art developments in the field and classical methods.

Reflecting the wide range of current research in statistics, the series encompasses applied, methodological and theoretical statistics, ranging from applications and new techniques made possible by advances in computerized practice to rigorous treatment of theoretical approaches.

This series provides essential and invaluable reading for all statisticians, whether in academia, industry, government, or research.

† ABRAHAM and LEDOLTER · Statistical Methods for Forecasting
AGRESTI · Analysis of Ordinal Categorical Data
AGRESTI · An Introduction to Categorical Data Analysis, *Second Edition*
AGRESTI · Categorical Data Analysis, *Second Edition*
ALTMAN, GILL, and McDONALD · Numerical Issues in Statistical Computing for the Social Scientist
AMARATUNGA and CABRERA · Exploration and Analysis of DNA Microarray and Protein Array Data
ANDĚL · Mathematics of Chance
ANDERSON · An Introduction to Multivariate Statistical Analysis, *Third Edition*
* ANDERSON · The Statistical Analysis of Time Series
ANDERSON, AUQUIER, HAUCK, OAKES, VANDAELE, and WEISBERG · Statistical Methods for Comparative Studies
ANDERSON and LOYNES · The Teaching of Practical Statistics
ARMITAGE and DAVID (editors) · Advances in Biometry
ARNOLD, BALAKRISHNAN, and NAGARAJA · Records
* ARTHANARI and DODGE · Mathematical Programming in Statistics
* BAILEY · The Elements of Stochastic Processes with Applications to the Natural Sciences
BALAKRISHNAN and KOUTRAS · Runs and Scans with Applications
BALAKRISHNAN and NG · Precedence-Type Tests and Applications
BARNETT · Comparative Statistical Inference, *Third Edition*
BARNETT · Environmental Statistics
BARNETT and LEWIS · Outliers in Statistical Data, *Third Edition*
BARTOSZYNSKI and NIEWIADOMSKA-BUGAJ · Probability and Statistical Inference
BASILEVSKY · Statistical Factor Analysis and Related Methods: Theory and Applications
BASU and RIGDON · Statistical Methods for the Reliability of Repairable Systems
BATES and WATTS · Nonlinear Regression Analysis and Its Applications

*Now available in a lower priced paperback edition in the Wiley Classics Library.
†Now available in a lower priced paperback edition in the Wiley–Interscience Paperback Series.

BECHHOFER, SANTNER, and GOLDSMAN · Design and Analysis of Experiments for Statistical Selection, Screening, and Multiple Comparisons
BELSLEY · Conditioning Diagnostics: Collinearity and Weak Data in Regression
† BELSLEY, KUH, and WELSCH · Regression Diagnostics: Identifying Influential Data and Sources of Collinearity
BENDAT and PIERSOL · Random Data: Analysis and Measurement Procedures, *Third Edition*
BERRY, CHALONER, and GEWEKE · Bayesian Analysis in Statistics and Econometrics: Essays in Honor of Arnold Zellner
BERNARDO and SMITH · Bayesian Theory
BHAT and MILLER · Elements of Applied Stochastic Processes, *Third Edition*
BHATTACHARYA and WAYMIRE · Stochastic Processes with Applications
BILLINGSLEY · Convergence of Probability Measures, *Second Edition*
BILLINGSLEY · Probability and Measure, *Third Edition*
BIRKES and DODGE · Alternative Methods of Regression
BISWAS, DATTA, FINE, and SEGAL · Statistical Advances in the Biomedical Sciences: Clinical Trials, Epidemiology, Survival Analysis, and Bioinformatics
BLISCHKE AND MURTHY (editors) · Case Studies in Reliability and Maintenance
BLISCHKE AND MURTHY · Reliability: Modeling, Prediction, and Optimization
BLOOMFIELD · Fourier Analysis of Time Series: An Introduction, *Second Edition*
BOLLEN · Structural Equations with Latent Variables
BOLLEN and CURRAN · Latent Curve Models: A Structural Equation Perspective
BOROVKOV · Ergodicity and Stability of Stochastic Processes
BOULEAU · Numerical Methods for Stochastic Processes
BOX · Bayesian Inference in Statistical Analysis
BOX · R. A. Fisher, the Life of a Scientist
BOX and DRAPER · Response Surfaces, Mixtures, and Ridge Analyses, *Second Edition*
* BOX and DRAPER · Evolutionary Operation: A Statistical Method for Process Improvement
BOX and FRIENDS · Improving Almost Anything, *Revised Edition*
BOX, HUNTER, and HUNTER · Statistics for Experimenters: Design, Innovation, and Discovery, *Second Editon*
BOX and LUCEÑO · Statistical Control by Monitoring and Feedback Adjustment
BRANDIMARTE · Numerical Methods in Finance: A MATLAB-Based Introduction
BROWN and HOLLANDER · Statistics: A Biomedical Introduction
BRUNNER, DOMHOF, and LANGER · Nonparametric Analysis of Longitudinal Data in Factorial Experiments
BUCKLEW · Large Deviation Techniques in Decision, Simulation, and Estimation
CAIROLI and DALANG · Sequential Stochastic Optimization
CASTILLO, HADI, BALAKRISHNAN, and SARABIA · Extreme Value and Related Models with Applications in Engineering and Science
CHAN · Time Series: Applications to Finance
CHARALAMBIDES · Combinatorial Methods in Discrete Distributions
CHATTERJEE and HADI · Regression Analysis by Example, *Fourth Edition*
CHATTERJEE and HADI · Sensitivity Analysis in Linear Regression
CHERNICK · Bootstrap Methods: A Guide for Practitioners and Researchers, *Second Edition*
CHERNICK and FRIIS · Introductory Biostatistics for the Health Sciences
CHILÈS and DELFINER · Geostatistics: Modeling Spatial Uncertainty
CHOW and LIU · Design and Analysis of Clinical Trials: Concepts and Methodologies, *Second Edition*
CLARKE and DISNEY · Probability and Random Processes: A First Course with Applications, *Second Edition*
* COCHRAN and COX · Experimental Designs, *Second Edition*

*Now available in a lower priced paperback edition in the Wiley Classics Library.
†Now available in a lower priced paperback edition in the Wiley–Interscience Paperback Series.

CONGDON · Applied Bayesian Modelling
CONGDON · Bayesian Models for Categorical Data
CONGDON · Bayesian Statistical Modelling
CONOVER · Practical Nonparametric Statistics, *Third Edition*
COOK · Regression Graphics
COOK and WEISBERG · Applied Regression Including Computing and Graphics
COOK and WEISBERG · An Introduction to Regression Graphics
CORNELL · Experiments with Mixtures, Designs, Models, and the Analysis of Mixture Data, *Third Edition*
COVER and THOMAS · Elements of Information Theory
COX · A Handbook of Introductory Statistical Methods
* COX · Planning of Experiments
CRESSIE · Statistics for Spatial Data, *Revised Edition*
CSÖRGŐ and HORVÁTH · Limit Theorems in Change Point Analysis
DANIEL · Applications of Statistics to Industrial Experimentation
DANIEL · Biostatistics: A Foundation for Analysis in the Health Sciences, *Eighth Edition*
* DANIEL · Fitting Equations to Data: Computer Analysis of Multifactor Data, *Second Edition*
DASU and JOHNSON · Exploratory Data Mining and Data Cleaning
DAVID and NAGARAJA · Order Statistics, *Third Edition*
* DEGROOT, FIENBERG, and KADANE · Statistics and the Law
DEL CASTILLO · Statistical Process Adjustment for Quality Control
DeMARIS · Regression with Social Data: Modeling Continuous and Limited Response Variables
DEMIDENKO · Mixed Models: Theory and Applications
DENISON, HOLMES, MALLICK and SMITH · Bayesian Methods for Nonlinear Classification and Regression
DETTE and STUDDEN · The Theory of Canonical Moments with Applications in Statistics, Probability, and Analysis
DEY and MUKERJEE · Fractional Factorial Plans
DILLON and GOLDSTEIN · Multivariate Analysis: Methods and Applications
DODGE · Alternative Methods of Regression
* DODGE and ROMIG · Sampling Inspection Tables, *Second Edition*
* DOOB · Stochastic Processes
DOWDY, WEARDEN, and CHILKO · Statistics for Research, *Third Edition*
DRAPER and SMITH · Applied Regression Analysis, *Third Edition*
DRYDEN and MARDIA · Statistical Shape Analysis
DUDEWICZ and MISHRA · Modern Mathematical Statistics
DUNN and CLARK · Basic Statistics: A Primer for the Biomedical Sciences, *Third Edition*
DUPUIS and ELLIS · A Weak Convergence Approach to the Theory of Large Deviations
EDLER and KITSOS · Recent Advances in Quantitative Methods in Cancer and Human Health Risk Assessment
* ELANDT-JOHNSON and JOHNSON · Survival Models and Data Analysis
ENDERS · Applied Econometric Time Series
† ETHIER and KURTZ · Markov Processes: Characterization and Convergence
EVANS, HASTINGS, and PEACOCK · Statistical Distributions, *Third Edition*
FELLER · An Introduction to Probability Theory and Its Applications, Volume I, *Third Edition,* Revised; Volume II, *Second Edition*
FISHER and VAN BELLE · Biostatistics: A Methodology for the Health Sciences
FITZMAURICE, LAIRD, and WARE · Applied Longitudinal Analysis
* FLEISS · The Design and Analysis of Clinical Experiments
FLEISS · Statistical Methods for Rates and Proportions, *Third Edition*

*Now available in a lower priced paperback edition in the Wiley Classics Library.
†Now available in a lower priced paperback edition in the Wiley–Interscience Paperback Series.

† FLEMING and HARRINGTON · Counting Processes and Survival Analysis
FULLER · Introduction to Statistical Time Series, *Second Edition*
† FULLER · Measurement Error Models
GALLANT · Nonlinear Statistical Models
GEISSER · Modes of Parametric Statistical Inference
GELMAN and MENG · Applied Bayesian Modeling and Causal Inference from Incomplete-Data Perspectives
GEWEKE · Contemporary Bayesian Econometrics and Statistics
GHOSH, MUKHOPADHYAY, and SEN · Sequential Estimation
GIESBRECHT and GUMPERTZ · Planning, Construction, and Statistical Analysis of Comparative Experiments
GIFI · Nonlinear Multivariate Analysis
GIVENS and HOETING · Computational Statistics
GLASSERMAN and YAO · Monotone Structure in Discrete-Event Systems
GNANADESIKAN · Methods for Statistical Data Analysis of Multivariate Observations, *Second Edition*
GOLDSTEIN and LEWIS · Assessment: Problems, Development, and Statistical Issues
GREENWOOD and NIKULIN · A Guide to Chi-Squared Testing
GROSS and HARRIS · Fundamentals of Queueing Theory, *Third Edition*
* HAHN and SHAPIRO · Statistical Models in Engineering
HAHN and MEEKER · Statistical Intervals: A Guide for Practitioners
HALD · A History of Probability and Statistics and their Applications Before 1750
HALD · A History of Mathematical Statistics from 1750 to 1930
† HAMPEL · Robust Statistics: The Approach Based on Influence Functions
HANNAN and DEISTLER · The Statistical Theory of Linear Systems
HEIBERGER · Computation for the Analysis of Designed Experiments
HEDAYAT and SINHA · Design and Inference in Finite Population Sampling
HEDEKER and GIBBONS · Longitudinal Data Analysis
HELLER · MACSYMA for Statisticians
HINKELMANN and KEMPTHORNE · Design and Analysis of Experiments, Volume 1: Introduction to Experimental Design, *Second Edition*
HINKELMANN and KEMPTHORNE · Design and Analysis of Experiments, Volume 2: Advanced Experimental Design
HOAGLIN, MOSTELLER, and TUKEY · Exploratory Approach to Analysis of Variance
* HOAGLIN, MOSTELLER, and TUKEY · Exploring Data Tables, Trends and Shapes
* HOAGLIN, MOSTELLER, and TUKEY · Understanding Robust and Exploratory Data Analysis
HOCHBERG and TAMHANE · Multiple Comparison Procedures
HOCKING · Methods and Applications of Linear Models: Regression and the Analysis of Variance, *Second Edition*
HOEL · Introduction to Mathematical Statistics, *Fifth Edition*
HOGG and KLUGMAN · Loss Distributions
HOLLANDER and WOLFE · Nonparametric Statistical Methods, *Second Edition*
HOSMER and LEMESHOW · Applied Logistic Regression, *Second Edition*
HOSMER and LEMESHOW · Applied Survival Analysis: Regression Modeling of Time to Event Data
† HUBER · Robust Statistics
HUBERTY · Applied Discriminant Analysis
HUBERTY and OLEJNIK · Applied MANOVA and Discriminant Analysis, *Second Edition*
HUNT and KENNEDY · Financial Derivatives in Theory and Practice, *Revised Edition*

*Now available in a lower priced paperback edition in the Wiley Classics Library.
†Now available in a lower priced paperback edition in the Wiley–Interscience Paperback Series.

HURD and MIAMEE · Periodically Correlated Random Sequences: Spectral Theory and Practice
HUSKOVA, BERAN, and DUPAC · Collected Works of Jaroslav Hajek—with Commentary
HUZURBAZAR · Flowgraph Models for Multistate Time-to-Event Data
IMAN and CONOVER · A Modern Approach to Statistics
† JACKSON · A User's Guide to Principle Components
JOHN · Statistical Methods in Engineering and Quality Assurance
JOHNSON · Multivariate Statistical Simulation
JOHNSON and BALAKRISHNAN · Advances in the Theory and Practice of Statistics: A Volume in Honor of Samuel Kotz
JOHNSON and BHATTACHARYYA · Statistics: Principles and Methods, *Fifth Edition*
JOHNSON and KOTZ · Distributions in Statistics
JOHNSON and KOTZ (editors) · Leading Personalities in Statistical Sciences: From the Seventeenth Century to the Present
JOHNSON, KOTZ, and BALAKRISHNAN · Continuous Univariate Distributions, Volume 1, *Second Edition*
JOHNSON, KOTZ, and BALAKRISHNAN · Continuous Univariate Distributions, Volume 2, *Second Edition*
JOHNSON, KOTZ, and BALAKRISHNAN · Discrete Multivariate Distributions
JOHNSON, KEMP, and KOTZ · Univariate Discrete Distributions, *Third Edition*
JUDGE, GRIFFITHS, HILL, LÜTKEPOHL, and LEE · The Theory and Practice of Econometrics, *Second Edition*
JUREČKOVÁ and SEN · Robust Statistical Procedures: Aymptotics and Interrelations
JUREK and MASON · Operator-Limit Distributions in Probability Theory
KADANE · Bayesian Methods and Ethics in a Clinical Trial Design
KADANE AND SCHUM · A Probabilistic Analysis of the Sacco and Vanzetti Evidence
KALBFLEISCH and PRENTICE · The Statistical Analysis of Failure Time Data, *Second Edition*
KARIYA and KURATA · Generalized Least Squares
KASS and VOS · Geometrical Foundations of Asymptotic Inference
† KAUFMAN and ROUSSEEUW · Finding Groups in Data: An Introduction to Cluster Analysis
KEDEM and FOKIANOS · Regression Models for Time Series Analysis
KENDALL, BARDEN, CARNE, and LE · Shape and Shape Theory
KHURI · Advanced Calculus with Applications in Statistics, *Second Edition*
KHURI, MATHEW, and SINHA · Statistical Tests for Mixed Linear Models
KLEIBER and KOTZ · Statistical Size Distributions in Economics and Actuarial Sciences
KLUGMAN, PANJER, and WILLMOT · Loss Models: From Data to Decisions, *Second Edition*
KLUGMAN, PANJER, and WILLMOT · Solutions Manual to Accompany Loss Models: From Data to Decisions, *Second Edition*
KOTZ, BALAKRISHNAN, and JOHNSON · Continuous Multivariate Distributions, Volume 1, *Second Edition*
KOVALENKO, KUZNETZOV, and PEGG · Mathematical Theory of Reliability of Time-Dependent Systems with Practical Applications
KOWALSKI and TU · Modern Applied U-Statistics
KVAM and VIDAKOVIC · Nonparametric Statistics with Applications to Science and Engineering
LACHIN · Biostatistical Methods: The Assessment of Relative Risks
LAD · Operational Subjective Statistical Methods: A Mathematical, Philosophical, and Historical Introduction
LAMPERTI · Probability: A Survey of the Mathematical Theory, *Second Edition*

*Now available in a lower priced paperback edition in the Wiley Classics Library.
†Now available in a lower priced paperback edition in the Wiley–Interscience Paperback Series.

LANGE, RYAN, BILLARD, BRILLINGER, CONQUEST, and GREENHOUSE · Case Studies in Biometry
LARSON · Introduction to Probability Theory and Statistical Inference, *Third Edition*
LAWLESS · Statistical Models and Methods for Lifetime Data, *Second Edition*
LAWSON · Statistical Methods in Spatial Epidemiology
LE · Applied Categorical Data Analysis
LE · Applied Survival Analysis
LEE and WANG · Statistical Methods for Survival Data Analysis, *Third Edition*
LePAGE and BILLARD · Exploring the Limits of Bootstrap
LEYLAND and GOLDSTEIN (editors) · Multilevel Modelling of Health Statistics
LIAO · Statistical Group Comparison
LINDVALL · Lectures on the Coupling Method
LIN · Introductory Stochastic Analysis for Finance and Insurance
LINHART and ZUCCHINI · Model Selection
LITTLE and RUBIN · Statistical Analysis with Missing Data, *Second Edition*
LLOYD · The Statistical Analysis of Categorical Data
LOWEN and TEICH · Fractal-Based Point Processes
MAGNUS and NEUDECKER · Matrix Differential Calculus with Applications in Statistics and Econometrics, *Revised Edition*
MALLER and ZHOU · Survival Analysis with Long Term Survivors
MALLOWS · Design, Data, and Analysis by Some Friends of Cuthbert Daniel
MANN, SCHAFER, and SINGPURWALLA · Methods for Statistical Analysis of Reliability and Life Data
MANTON, WOODBURY, and TOLLEY · Statistical Applications Using Fuzzy Sets
MARCHETTE · Random Graphs for Statistical Pattern Recognition
MARDIA and JUPP · Directional Statistics
MASON, GUNST, and HESS · Statistical Design and Analysis of Experiments with Applications to Engineering and Science, *Second Edition*
McCULLOCH and SEARLE · Generalized, Linear, and Mixed Models
McFADDEN · Management of Data in Clinical Trials, *Second Edition*
* McLACHLAN · Discriminant Analysis and Statistical Pattern Recognition
McLACHLAN, DO, and AMBROISE · Analyzing Microarray Gene Expression Data
McLACHLAN and KRISHNAN · The EM Algorithm and Extensions, *Second Edition*
McLACHLAN and PEEL · Finite Mixture Models
McNEIL · Epidemiological Research Methods
MEEKER and ESCOBAR · Statistical Methods for Reliability Data
MEERSCHAERT and SCHEFFLER · Limit Distributions for Sums of Independent Random Vectors: Heavy Tails in Theory and Practice
MICKEY, DUNN, and CLARK · Applied Statistics: Analysis of Variance and Regression, *Third Edition*
* MILLER · Survival Analysis, *Second Edition*
MONTGOMERY, PECK, and VINING · Introduction to Linear Regression Analysis, *Fourth Edition*
MORGENTHALER and TUKEY · Configural Polysampling: A Route to Practical Robustness
MUIRHEAD · Aspects of Multivariate Statistical Theory
MULLER and STOYAN · Comparison Methods for Stochastic Models and Risks
MURRAY · X-STAT 2.0 Statistical Experimentation, Design Data Analysis, and Nonlinear Optimization
MURTHY, XIE, and JIANG · Weibull Models
MYERS and MONTGOMERY · Response Surface Methodology: Process and Product Optimization Using Designed Experiments, *Second Edition*
MYERS, MONTGOMERY, and VINING · Generalized Linear Models. With Applications in Engineering and the Sciences

*Now available in a lower priced paperback edition in the Wiley Classics Library.
†Now available in a lower priced paperback edition in the Wiley–Interscience Paperback Series.

† NELSON · Accelerated Testing, Statistical Models, Test Plans, and Data Analyses
† NELSON · Applied Life Data Analysis
NEWMAN · Biostatistical Methods in Epidemiology
OCHI · Applied Probability and Stochastic Processes in Engineering and Physical Sciences
OKABE, BOOTS, SUGIHARA, and CHIU · Spatial Tesselations: Concepts and Applications of Voronoi Diagrams, *Second Edition*
OLIVER and SMITH · Influence Diagrams, Belief Nets and Decision Analysis
PALTA · Quantitative Methods in Population Health: Extensions of Ordinary Regressions
PANJER · Operational Risk: Modeling and Analytics
PANKRATZ · Forecasting with Dynamic Regression Models
PANKRATZ · Forecasting with Univariate Box-Jenkins Models: Concepts and Cases
* PARZEN · Modern Probability Theory and Its Applications
PEÑA, TIAO, and TSAY · A Course in Time Series Analysis
PIANTADOSI · Clinical Trials: A Methodologic Perspective
PORT · Theoretical Probability for Applications
POURAHMADI · Foundations of Time Series Analysis and Prediction Theory
POWELL · Approximate Dynamic Programming: Solving the Curses of Dimensionality
PRESS · Bayesian Statistics: Principles, Models, and Applications
PRESS · Subjective and Objective Bayesian Statistics, *Second Edition*
PRESS and TANUR · The Subjectivity of Scientists and the Bayesian Approach
PUKELSHEIM · Optimal Experimental Design
PURI, VILAPLANA, and WERTZ · New Perspectives in Theoretical and Applied Statistics
† PUTERMAN · Markov Decision Processes: Discrete Stochastic Dynamic Programming
QIU · Image Processing and Jump Regression Analysis
* RAO · Linear Statistical Inference and Its Applications, *Second Edition*
RAUSAND and HØYLAND · System Reliability Theory: Models, Statistical Methods, and Applications, *Second Edition*
RENCHER · Linear Models in Statistics
RENCHER · Methods of Multivariate Analysis, *Second Edition*
RENCHER · Multivariate Statistical Inference with Applications
* RIPLEY · Spatial Statistics
* RIPLEY · Stochastic Simulation
ROBINSON · Practical Strategies for Experimenting
ROHATGI and SALEH · An Introduction to Probability and Statistics, *Second Edition*
ROLSKI, SCHMIDLI, SCHMIDT, and TEUGELS · Stochastic Processes for Insurance and Finance
ROSENBERGER and LACHIN · Randomization in Clinical Trials: Theory and Practice
ROSS · Introduction to Probability and Statistics for Engineers and Scientists
ROSSI, ALLENBY, and McCULLOCH · Bayesian Statistics and Marketing
† ROUSSEEUW and LEROY · Robust Regression and Outlier Detection
* RUBIN · Multiple Imputation for Nonresponse in Surveys
RUBINSTEIN · Simulation and the Monte Carlo Method
RUBINSTEIN and MELAMED · Modern Simulation and Modeling
RYAN · Modern Engineering Statistics
RYAN · Modern Experimental Design
RYAN · Modern Regression Methods
RYAN · Statistical Methods for Quality Improvement, *Second Edition*
SALEH · Theory of Preliminary Test and Stein-Type Estimation with Applications
* SCHEFFE · The Analysis of Variance
SCHIMEK · Smoothing and Regression: Approaches, Computation, and Application
SCHOTT · Matrix Analysis for Statistics, *Second Edition*
SCHOUTENS · Levy Processes in Finance: Pricing Financial Derivatives

*Now available in a lower priced paperback edition in the Wiley Classics Library.
†Now available in a lower priced paperback edition in the Wiley–Interscience Paperback Series.

SCHUSS · Theory and Applications of Stochastic Differential Equations
SCOTT · Multivariate Density Estimation: Theory, Practice, and Visualization
† SEARLE · Linear Models for Unbalanced Data
† SEARLE · Matrix Algebra Useful for Statistics
† SEARLE, CASELLA, and McCULLOCH · Variance Components
SEARLE and WILLETT · Matrix Algebra for Applied Economics
SEBER · A Matrix Handbook For Statisticians
† SEBER · Multivariate Observations
SEBER and LEE · Linear Regression Analysis, *Second Edition*
† SEBER and WILD · Nonlinear Regression
SENNOTT · Stochastic Dynamic Programming and the Control of Queueing Systems
* SERFLING · Approximation Theorems of Mathematical Statistics
SHAFER and VOVK · Probability and Finance: It's Only a Game!
SILVAPULLE and SEN · Constrained Statistical Inference: Inequality, Order, and Shape Restrictions
SMALL and McLEISH · Hilbert Space Methods in Probability and Statistical Inference
SRIVASTAVA · Methods of Multivariate Statistics
STAPLETON · Linear Statistical Models
STAUDTE and SHEATHER · Robust Estimation and Testing
STOYAN, KENDALL, and MECKE · Stochastic Geometry and Its Applications, *Second Edition*
STOYAN and STOYAN · Fractals, Random Shapes and Point Fields: Methods of Geometrical Statistics
STREET and BURGESS · The Construction of Optimal Stated Choice Experiments: Theory and Methods
STYAN · The Collected Papers of T. W. Anderson: 1943–1985
SUTTON, ABRAMS, JONES, SHELDON, and SONG · Methods for Meta-Analysis in Medical Research
TAKEZAWA · Introduction to Nonparametric Regression
TANAKA · Time Series Analysis: Nonstationary and Noninvertible Distribution Theory
THOMPSON · Empirical Model Building
THOMPSON · Sampling, *Second Edition*
THOMPSON · Simulation: A Modeler's Approach
THOMPSON and SEBER · Adaptive Sampling
THOMPSON, WILLIAMS, and FINDLAY · Models for Investors in Real World Markets
TIAO, BISGAARD, HILL, PEÑA, and STIGLER (editors) · Box on Quality and Discovery: with Design, Control, and Robustness
TIERNEY · LISP-STAT: An Object-Oriented Environment for Statistical Computing and Dynamic Graphics
TSAY · Analysis of Financial Time Series, *Second Edition*
UPTON and FINGLETON · Spatial Data Analysis by Example, Volume II: Categorical and Directional Data
VAN BELLE · Statistical Rules of Thumb
VAN BELLE, FISHER, HEAGERTY, and LUMLEY · Biostatistics: A Methodology for the Health Sciences, *Second Edition*
VESTRUP · The Theory of Measures and Integration
VIDAKOVIC · Statistical Modeling by Wavelets
VINOD and REAGLE · Preparing for the Worst: Incorporating Downside Risk in Stock Market Investments
WALLER and GOTWAY · Applied Spatial Statistics for Public Health Data
WEERAHANDI · Generalized Inference in Repeated Measures: Exact Methods in MANOVA and Mixed Models
WEISBERG · Applied Linear Regression, *Third Edition*

*Now available in a lower priced paperback edition in the Wiley Classics Library.
†Now available in a lower priced paperback edition in the Wiley–Interscience Paperback Series.

WELSH · Aspects of Statistical Inference
WESTFALL and YOUNG · Resampling-Based Multiple Testing: Examples and Methods for p-Value Adjustment
WHITTAKER · Graphical Models in Applied Multivariate Statistics
WINKER · Optimization Heuristics in Economics: Applications of Threshold Accepting
WONNACOTT and WONNACOTT · Econometrics, *Second Edition*
WOODING · Planning Pharmaceutical Clinical Trials: Basic Statistical Principles
WOODWORTH · Biostatistics: A Bayesian Introduction
WOOLSON and CLARKE · Statistical Methods for the Analysis of Biomedical Data, *Second Edition*
WU and HAMADA · Experiments: Planning, Analysis, and Parameter Design Optimization
WU and ZHANG · Nonparametric Regression Methods for Longitudinal Data Analysis
YANG · The Construction Theory of Denumerable Markov Processes
YOUNG, VALERO-MORA, and FRIENDLY · Visual Statistics: Seeing Data with Dynamic Interactive Graphics
ZELTERMAN · Discrete Distributions—Applications in the Health Sciences
* ZELLNER · An Introduction to Bayesian Inference in Econometrics
ZHOU, OBUCHOWSKI, and McCLISH · Statistical Methods in Diagnostic Medicine

*Now available in a lower priced paperback edition in the Wiley Classics Library.
†Now available in a lower priced paperback edition in the Wiley–Interscience Paperback Series.